FROM LIVED EXPERIENCE
TO THE WRITTEN WORD

PAMELA H. SMITH

From Lived Experience to the Written Word

RECONSTRUCTING PRACTICAL
KNOWLEDGE IN THE EARLY
MODERN WORLD

THE UNIVERSITY OF
CHICAGO PRESS
Chicago & London

The University of Chicago Press, Chicago 60637
The University of Chicago Press, Ltd., London
© 2022 by The University of Chicago
All rights reserved. No part of this book may be used or reproduced
in any manner whatsoever without written permission, except in
the case of brief quotations in critical articles and reviews. For more
information, contact the University of Chicago Press, 1427 E. 60th St.,
Chicago, IL 60637.
Published 2022
Printed in China

31 30 29 28 27 26 25 24 23 22   2 3 4 5

ISBN-13: 978-0-226-80027-1 (cloth)
ISBN-13: 978-0-226-81824-5 (paper)
ISBN-13: 978-0-226-81823-8 (e-book)
DOI: https://doi.org/10.7208/chicago/9780226818238.0001

Published with the assistance of the Getty Grant Program

Library of Congress Cataloging-in-Publication Data

Names: Smith, Pamela H., 1957– author.
Title: From lived experience to the written word : reconstructing
    practical knowledge in the early modern world / Pamela H. Smith.
Description: Chicago : University of Chicago Press, 2022. | Includes
    bibliographical references and index.
Identifiers: LCCN 2021042946 | ISBN 9780226800271 (cloth) |
    ISBN 9780226818245 (paperback) | ISBN 9780226818238 (e-book)
Subjects: LCSH: Technical writing—History. | Artisans—History.
Classification: LCC T11.S575 2022 | DDC 808.06/66—dc23
LC record available at https://lccn.loc.gov/2021042946

Contents

Lived Experience and the Written Word

How-To Manuals

When I sat down to write this book a little over a decade ago, I began with what seemed a truism: we all take so-called how-to manuals for granted; they come with new appliances, computers, and Ikea furniture. Today, this is no longer the case. Computers have built-in "Help" features, and online communities solve our problems and answer our questions; we learn how to put together furniture and repair our appliances with YouTube videos. For some years after they stopped being printed, I missed those thick, seemingly comprehensive computer manuals, but now it seems remarkable to me that we once relied upon such printed instructions. The nature of software development has made the definitive printed manual a thing of the past, and YouTube has proven to many that it is easier to learn hand knowledge and techniques by watching an experienced practitioner. The founding in 2006 of the online *Journal of Visualized Experiments* (JoVE, www.jove.com), in which highly specialized laboratory and clinical procedures are narrated through video footage, indicates that these changes are not confined to hobbies and do-it-yourself (DIY) projects. In the natural sciences, such problems of capturing how-to knowledge in writing can contribute to the difficulties involved in the replication of experiments by other researchers.

At a moment when instructional how-to writing appears to be becoming less ubiquitous, *From Lived Experience to the Written Word* traces the early modern origins of such writing, locating the attempt to capture technique and skill in the transition from embodied practice and lived experience to the written word and textual description. In about 1400, European artists and craftspeople began writing down their techniques in texts often titled "books of art" or, in German, *Kunstbücher* or *Kunstbüchlein* (little books of art). Some of these texts and the recipes and instructions they contained were published with the advent of the printing press in Europe, while others remained in manuscript for centuries. Despite the difficulty of writing down techniques, let alone using the written accounts to learn a craft, many of these books became bestsellers for their printers. What was the appeal of these texts, and who read them? This book provides one answer to that question. It also considers technique and skill: why are they so hard to learn and teach? what kinds of knowledge are they? And why try to translate what are essentially bodily gestures into words in the first place?

Lived Experience and the Written Word

Writing things down seems second nature to a literate and text-centered society. Yet, the debate over the merit of the written word is as old as writing itself. Despite its expression almost thirteen hundred years ago, a sentiment such as that of the eighth-century Bedouin poet Dhu'l Rumma (c. 696–c. 735) resonates for us: "Write down my poems, because I favour the book over memory [. . .] the book does not forget and does not exchange any word for another."[1] Dhu'l Rumma's view of the advantages of writing seems self-evident to us, however, it was a position that had to be argued for. In a well-known passage from Plato's (427–347 BCE) dialogue *Phaedrus*, Socrates condemned the discovery of writing by the god Theuth:

If men learn [writing], it will implant forgetfulness in their souls; they will cease to exercise memory because they rely on that which is written, calling things to remembrance no longer from within themselves, but by means of external marks. What you have discovered is a recipe not for memory, but for reminder. And it is no true wisdom that you offer your disciples, but only its semblance, for by telling them of many things without teaching them you will make them seem to know much, while for the most part they know nothing, and as men filled, not with wisdom, but with the conceit of wisdom.[2]

Socrates sees painting and writing as similar because their products "stand before us as though they were alive, but if you question them, they maintain a most majestic silence [. . .] if you ask them anything [. . .] they go on telling you just the same thing forever." Moreover, "once a thing is put in writing, the composition, whatever it may be, drifts all over the place, getting into the hands not only of those who understand it, but equally of those who have no business with it." In this remarkable passage, which seems so strange to our text-centered culture, Socrates distinguishes between true embodied wisdom and a feeble externalized collection of data, a listlike reminder.

Many of the artisan authors discussed in *From Lived Experience to the Written Word* also expressed a distrust of writing. They declared—somewhat paradoxically, *in writing*— that writing was inadequate to their task. The goldsmith and sculptor Benvenuto Cellini (1500–1571) wrote in the 1550s: "How careful you have to be with this cannot be told in words alone—you'll have to learn that by experience."[3] More than a century after Cellini, after decades of published technical treatises had established the genre, the printer and mapmaker Joseph Moxon (1627–91) wrote in his history of trades, *Mechanick Exercises*, that "Craft of the Hand […] cannot be taught by Words, but is only gained by Practise and Exercise."[4]

Moxon's point was made over and over again by practitioners when they sat down to write out their techniques, and YouTube and JoVE have made clear again for us that it is often far more time-consuming to attempt to describe handwork in words than simply to demonstrate it. Most important, skilled and expert performance of techniques can never be taught in writing; they take time and much practice, and they necessitate communities of practitioners both to develop and define what constitutes skilled practice and to teach and transmit it.

In short, the relationship between writing and experience was and is fraught, and statements such as Moxon's help us recognize the significance of the act among craftspeople of putting pen to paper to record the experience of the workshop. Their texts proliferated from about 1400, and the success and popularity of these texts both reflected and fostered a new interest in practice that celebrated the potential and power of practical knowledge. It also informed one of the most significant transformations in the engagement of human beings with nature—the founding and growth of a "new philosophy" that proclaimed a hands-on approach to the study of nature. What began in unceasing trials of the craft workshop ended in the experimentation of the natural scientific laboratory.

Mind over Hand

This book traces a key development in the history of knowledge and epistemology, but the story of practice gaining a voice, or at least a written form, has social dimensions as well. Through their writing, skilled artisans gained intellectual and social authority. The people who formed an audience for texts of practice sometimes even tried their own hands at making things—turning on lathes or metal casting, among other types of handwork. Although the period from 1400 is marked by this burgeoning interest in practical knowledge, and by greater interaction among those individuals trained by texts in schools and universities, and those trained by hands-on experience under a master within an organized community of artisans, an intellectual and social hierarchy continued to put theory above practice, abstract thinking above bodily experience, and mind above hand. Denis Diderot's article "Art" in the *Encyclopédie ou Dictionnaire raisonné des sciences, des arts et des métiers* (1751–72) testifies to this ambivalent attitude to the arts and artisans by the late eighteenth century on the part of learned writers: on the one hand, artisans are selfish and ignorant, contributing only bodily labor; on the other, they hold valuable knowledge for the kingdom:

we invite the artists to take counsel with learned men and not to allow their discoveries to perish with them. The artists should know that to lock up a useful secret is to render oneself guilty of theft from society. It is just as despicable to prefer the interest of one individual to the common welfare in this case as in a hundred others where the artists themselves would not hesitate to decide for the common good. If they communicate their discoveries they will be freed of several preconceptions and especially of the illusion, which almost all of them hold, that their art has reached its ultimate perfection. Because they have so little learning they are often inclined to blame the nature of things for a defect that exists only in themselves. Obstacles seem insuperable to them whenever they do not know the means of overcoming them. Let them carry out experiments and let everyone make his contribution to these experiments: the artist should contribute his work, the academician his knowledge and advice, the rich man the cost of materials, labor, and time; soon our arts and our manufactures will be as superior as we could wish to those of other countries.[5]

Although this passage emerges out of a particular economic moment in French history, it expresses assumptions and prejudices about artisans that were common throughout the early modern period, even as fascination with the products and potential of art and handwork grew.

That these preconceptions are still alive can be seen in Walter J. Ong's argument in "Writing Is a Technology That Restructures Thought" that literacy increases objectivity, makes it possible for an individual to transcend their own time-bounded being, and provides an escape from the narrativity of oral cultures to attain the analytical thinking and philosophy of literate cultures.[6] In contrast to Plato in the *Phaedrus*, Ong does not think philosophy is possible in the absence of writing. What is often forgotten in such text- and writing-centered accounts is that practices do not need to be articulated verbally to be conceptualized and controlled by the practitioner.[7] Diverging from both these writers, I argue in *From Lived Experience to the Written Word* that artists and artisans in early modern Europe did not need writing to produce things and make knowledge, yet they nevertheless turned increasingly to writing to argue for a new place in the hierarchy of knowledge, to convey their "material imaginary," their epistemology, and a theory of skill.

Histories of Science and Art

As the labor of making came into view for elites as a powerful component of material production and natural knowledge, "new philosophers" proclaiming a new science in early modern Europe recognized that bodily engagement with natural materials was an essential part of gaining knowledge about nature. Many historians of science over the last thirty years or so have recounted the intersection of vernacular and scholarly cultures that resulted in a new union of hand and mind and over the following centuries built the remarkable system of producing knowledge of nature in what we term "modern science."[8] They also trace the beginnings in the late seventeenth century of the ways in which the "modern" half of that phrase gave rise to its opposite, "primitive," and the "science" half was defined against a new category of "pseudo-science." The new intellectual hierarchies consolidated by these new categories resulted in the historical vernacular knowledge systems of European artisans and non-European peoples being labeled "beliefs" and "old wives' tales," sometimes even "charlatanry," or just the stuff of bare and repetitive practice, uninformed by larger knowledge systems or by investigative practices.

Such hierarchies have decisively altered among scholars, if not among the general public, in the wake of forty years of scholarship in the sociology of knowledge and new histories of science and art. Historians of art and science increasingly consider how artists and artisans "think with their hands."[9] Their works, cited throughout this book, are showing how much "theorizing" went on in studios and workshops, the philosophical products of which could be material or textual. Although focused on practice and technology, this now substantial body of scholarship is just beginning to give deeper attention to the place of skill. Skill has been overlooked in the history of science and art, and also in today's experimental natural sciences. Embodied—and sometimes tacit—empirical methods have not been as rigorously examined and theorized as might be expected.[10] The short article format that has come to typify scientific writing in the present day gives little space to methods sections, one of the reasons that replicability of experiments is now under scrutiny. One of the thrusts of the present book is to make clear that a fuller study of bodily skill can be relevant for understanding the embodied and tacit dimensions of experimental natural science today. It also asks us, more generally, to pay attention to craft and skill in educational systems, and argues for the value of training of the attention by observation and imitation, and learning by apprenticeship.

Philosophers and anthropologists of technology have led the way in writing about skill, and I draw upon their work especially in chapter 5.[11] Recent forays across disciplinary boundaries—which do not happen often enough—among philosophers, historians, anthropologists, and theorists of embodied cognition have resulted in suggestive treatments of skill.[12] The francophone approach in combining the *histoire des techniques* (historical accounts of techniques and skill) and *histoire de la technologie* (the historical account of the conceptualization of skill and technique) in the long history of *la geste*, or gestural knowledge, has been especially productive.[13]

Nature and Art

Craft and technical writings in early modern Europe fed into a particular fascination with the continuum between the processes of nature and the ability of humans to shape and manipulate natural materials, called at the time *ars, arte, art*, and *Kunst*. The contin-

1.1. Willow trees two years after pollarding. In pollarding, all branches are cut from the trunk. Willows are pollarded every year or every few years in order to produce willow rods and switches of different sizes, useful for many purposes, including for stirring molten copper. Stirring metal with willow releases salicylic acid, which acts as a flux and aids the alloying of the metal (see Motture 2001, 30). CC BY-SA 3.0.

uum between nature and human skill also forms one theme of the book, which calls for reintegrating skilled craft knowledge into a deep history of human interaction with the environment.

All human societies interact in a variety of ways with their environment for survival, and, out of the experience gained through that interaction, certain skills and knowledge emerge.[14] Tool use and the development of skilled practice can perhaps ultimately be viewed as such an interface between humans and their environment, one that contributes to cultural evolutionary processes.[15] Many illustrations of such interaction can be drawn from agricultural practices, the most obvious being the "domestication" of plants and animals for human consumption and use. We can even find examples in which making involves exploiting processes of biological growth. In the human cultivation and management of tree growth, for instance, the physical matter of made things is *grown* by practices like pollarding every year for narrow poles (fig. 1.1), coppicing every five to twenty years, depending on the size needed, and fostering the decades-, sometimes century-long growth of individual tall trees for ship masts and house frames. In the early modern period, joiners and carpenters employed the properties of living wood in their practices, for example, in splitting wood along its growth rays, or in making a joint strong by inserting pins of dry wood into holes of green wood. As the green wood dried, it gripped the pin ever more tightly.[16] The still-living wood is part of the process of making. In this and other processes—the fostering of fermentation, for example—natural growth merges indistinguishably with the artifice of human manipulation.

This engagement with nature, and the resulting accumulation of skills and knowledge, has occurred throughout the long human past. The use of tools in working natural materials during the Paleolithic era may have been a factor in the emergence of language (which took on written form about seven thousand years ago).[17] The use of fire among early hominids allowed the manipulation of materials such as clay and native metals. The increasing ability to control fire itself led to high-temperature furnaces in which minerals could be smelted, and new materials could be made, such as bronze (an alloy of copper and tin) around 3500 BCE and glass around 1500 BCE—two artificial products made through engaging and experimenting with natural materials. Long before this, the domestication of plants about ten thousand years before the present, and of animals further in the past, and their subsequent cultivation and breeding for food, medicine, textile production, and other survival uses, was based on long-term systematic observation of patterns in nature. Celestial observations began to be recorded in grave and religious monuments, and eventually in calendrical form (often pictorial or megalithic constructions such as Stonehenge) in various periods in Africa, ancient Mesopotamia, Asia, Europe, and the Americas, and eventually in written form in Egypt and Babylon. These observations enabled the codification of a planting schedule divorced from immediate meteorological observations, and, by this means, numbers, arithmetic, and the calculation and prediction of celestial events emerged.

Such skills—what we often refer to as "crafts"—thus grew out of a collective human interaction with the material world, sometimes apparently emerging spontaneously in different places, and at other times traceable as they spread throughout various regions by means of migration and trade. Some see the modern scientific and technological capability of human beings as essentially different from these earlier developments, but there is no reason to regard the present development and accumulation of techniques and knowledge by means of the natural sciences as radically different from the very long-enduring human engagement with the environment. The major difference is one of scale. The system of knowledge production that we now call "science" involves a much more rapid accumulation and communication of knowledge on a global scale.

Codifying Collective Knowledge

Some might point to the recent history of science and the discoveries of individual scientists as evidence of new processes of knowledge production, but such a view grows out of a limited perspective on human history arising from historians' overwhelming reliance upon the written word. The existence of a written record for the exceedingly brief present of human history (only about 7000 of the approximately 200,000 years since *homo sapiens* emerged) gives historians the sense that they can pin scientific discoveries and inventions to individuals. Indeed, many discoveries of modern science have been revealed to have been the result of collective and collaborative processes,[18] much like those that produced bronze, writing, glass, and moveable type, to name only a few technological innovations from the deeper past. Insight into the process by which knowledge is ascribed to individual discovery is provided by the identification of the Gulf Stream in the eighteenth century,

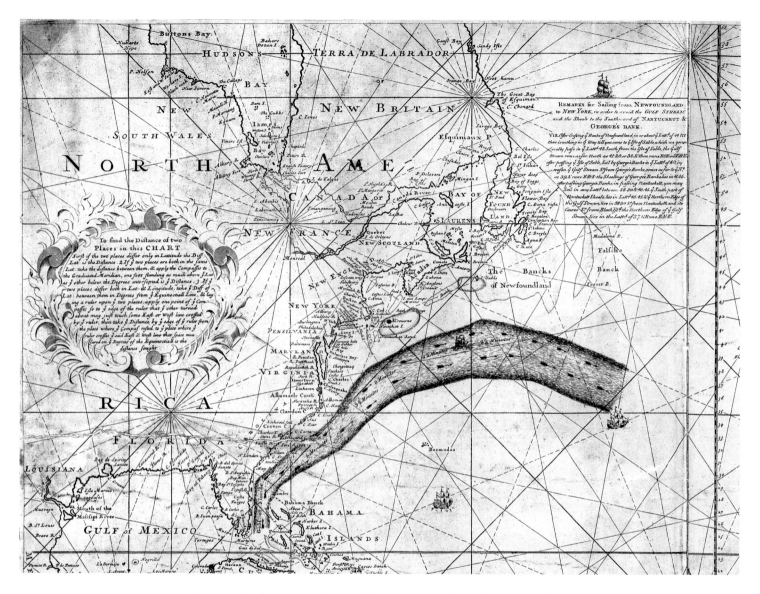

Within the map image:

To find the Diftance of two Places in this CHART

REMARKS for Sailing from NEWFOUNDLAND to NEW YORK, in order to avoid the GULF STREAM and the Shoals to the Southward of NANTUCKETT & GEORGES bank.

1.2. Benjamin Franklin and Timothy Folger, Chart of the Gulf Stream, 1769, detail. Library of Congress, Geography and Map Division, Washington, DC. Timothy Folger was Franklin's cousin and a Nantucket whale boat captain. He informed Franklin of the existence of the Gulf Stream and mapped it for him. CC0 1.0.

often credited to Benjamin Franklin (1706–90) who published the first printed map of it. At that time, however, the existence of the Gulf Stream was already common knowledge among whalers, and among mariners before them. Timothy Folger, Franklin's cousin and a Nantucket whale boat captain, informed Franklin of the existence of the Gulf Stream and mapped it for him (fig. 1.2).[19] No doubt, the transition of this knowledge from oral to written form is an extremely important change, but writing was not necessary to transmit this information from one generation or region to another. In this case, collective knowledge of the Gulf Stream was made and passed on in the community of whalers and sailors. Today, however, the map is known by Franklin's name.

Practical knowledge can move in and out of textual form, sometimes lost to the written record and then resurfacing in writing, like an iceberg, only its tip visible above a vast reservoir of collective knowing. Techniques could survive in practice for centuries, even millennia, only to be "discovered" by an administrator or scholar, who did no more than codify in written form a long-practiced piece of knowledge.[20] Such "discoveries" have usually been regarded as the work of inviduals, but another way to look at them is as developing within and emerging from relational social fields. Reliance by historians on written records and on identifiable individuals (whose papers or works are still extant) living in the recent human past has had the effect of misleading us about the collective and distributed nature by which all knowledge, and especially knowledge of our natural environment, is produced. Recent institutions, such as the Nobel Prize, have worked to reinforce a view today that innovations in the natural sciences are the inventions of a small number of individuals.

The evidence of collective perceptual and technical engagement with the environment can be traced in the historical and archaeological record through myriad forms of codification. Cuneiform tablets recording the risings and settings of the stars and planets, brought together in the library at Nineveh in the seventh century BCE, codified patterns related to pursuit of health, agriculture, and hunting. Lists and taxonomies can also encode such activities, as can stories, myths, and rituals. Polynesian peoples employed songs and mnemonic devices to codify and transmit from one generation to the next the knowledge needed to navigate thousands of miles over the open ocean.[21]

The processes by which natural materials are manipulated to produce substances and objects were codified and passed down as well: in written recipes, of course, but also in bodily practices and rituals, and sometimes in the objects themselves. The transmission of practical knowledge has also been facilitated or organized by the body politic, as in the cases of mining, controlled in ancient Egypt by the pharaoh-gods as early as the third millennium BCE, and of agriculture in China, where, throughout two millennia of imperial rule (221 BCE–1911 CE), agriculture formed a part of statecraft.[22] When transmission of skill is part of a hierarchy of ruling, it often results in the formation of culturally specific identities connected to skilled practice, such as shamans, medical doctors, and, today, scientists and engineers. These groups construct hierarchies of knowledge, sometimes asserting the kin-based, esoteric, or superior intellectual, nature of their skill. Such stratification in recent times has led to an assertion of the superiority of "modern" modes of understanding, including "modern science," over those, sometimes labeled "premodern" (or even "primitive"), that are based in handwork. These labels refer not so much to different activities as they function to claim the supremacy of mind over hand.[23]

Viewing human engagement in the natural world as *longue durée*, collective, and involving both material and social forces has important implications: first, it means that the various sites of human engagement with the natural world in the deep past, including agriculture and industry, can be placed with early modern workshops and modern scientific laboratories along a millennia-long continuum; second, it allows us to see that activities we call "science," "technique," and "technology" are not separate spheres of action, but components of a whole; [24] third, it implies that historians must think beyond written texts, for knowledge can be codified in many ways; and finally, and most significantly, if we conceive of skills as arising at the interface of the human senses and human body with the natural environment, we must conclude (with the "early moderns") that nature and culture cannot be understood to be separate realms.

Reconstructing Practical Knowledge

To unpack this complex interface between the human body and the natural environment, *From Lived Experience to the Written Word* turns to the historical moment when that interface was first articulated in writing by the practitioners themselves. As I noted, there is a paradoxical challenge in attempting to capture the experience of practice in the written word. This challenge was not only faced by the practitioners who attempted to articulate their techniques, but is also encountered by historians who wish to understand the skills and practices of early modern artisans. In a 2004 book, *The Body of the Artisan: Art and Experience in the Scientific Revolution*, I argued that early modern European artisans made claims to knowledge of nature in their naturalistic works of art, as well as in their written texts. In that book, I "read" the philosophical claims that sixteenth-century artisans like the Nuremberg goldsmith Wenzel Jamnitzer (1508–85) and the French potter Bernard Palissy (c. 1510–90) made in their objects (I did so in part through reading their texts), but I could not ground their claims in the actual processes and materials by which they produced objects. *The Body of the Artisan* highlighted how practice was valued and argued for by its practitioners, but I was not satisfied with my articulation of the content of artisans' bodily and material knowledge. I realized that I needed to get to the core of craft knowledge itself—that is, into knowledge of the behavior of natural materials and the skilled practices by which craftspeople manipulated these materials to produce objects. How could I effectively gain this knowledge? In seeking out collections of recipes that had some relationship to these processes, I sought to turn away from the *idea* of artisanal practice to that practice itself. As I read these books of practice, full of recipes, however, I quickly felt at sea: what should I pay attention to? how did I know what was important? where was "knowledge"?

I realized I must, as the sixteenth-century reformer known as Paracelsus (1493–1541) urged, "hasten to experience," for I could not assess recipe texts—practice recorded in writing—without knowing if they in fact reflected past practice, nor, in many cases, could I understand fully what the practices were without having a better understanding of the techniques of different crafts. I had the good fortune at this time of meeting Malcolm Baker (then head of the Medieval and Renaissance Galleries Project of the Victoria and Albert

1.3. The author, during a black-
smithing course, West Dean
College of Arts and Conservation,
West Sussex, UK, 2004.

Museum [V&A]), who suggested that I spend a year at the V&A Research Department. My subsequent year there (2003–4) began in my asking many questions of busy conservators and curators, and ended by my accepting their advice to undertake an apprenticeship of sorts myself, which I did through a remarkable course on historical techniques of painting led by Renate Woodhuysen, blacksmithing and silversmithing courses at West Dean College (fig. 1.3), and brief bronze-casting experiences with Andrew Lacey and Francesca Bewer. Naturally, these short immersion courses in historical techniques could not replicate any kind of true apprenticeship, and my skills in all these areas remain rudimentary.

It was during this time that I began studying in more detail a remarkable manuscript: Bibliothèque nationale de France, Ms. Fr. 640. This anonymous manuscript, apparently written after 1580 by a practitioner in the environs of Toulouse, includes detailed instructions for many techniques, as well as accounts of workshop practice, containing more than nine hundred "recipes" for objects of art and daily life. The largest group of recipes in the manuscript is devoted to metalworking, including casting; the second largest group deals with color-making practices of many kinds, including pigment application, dyeing, staining wood, coloring and painting metal, and making artificial gems. In addition, the manuscript contains much information on weapons production; tree grafting; land surveying; preservation of animals, plants, and foodstuffs; distillation; and much more. The resulting manuscript is a record of practices that gives unique insight into craft and artistic techniques, daily life in the sixteenth century, and material and intellectual understandings of the natural world. It is not, however, a straightforward set of instructions. Indeed, it sometimes seems a stream-of-consciousness collection: on the same page, the text includes a brief recipe for a medicine to counter eye diseases alongside instructions for the coloring of metals and wood.[25] It seems a random miscellany of recipes and processes as they occurred to the author. This epitomizes what has been seen as one of the frustrations of getting at artisanal knowledge, that such recipe collections and handbooks are "simply" more or less precise instructions to accomplish an action— "just recipes," with no overarching order or significance. At the same time, many written recipes leave out crucial information or are contradicted by other recipes in the same collection. It is thus difficult to take these historical recipe collections, as some do, at face value as utilitarian lists of ingredients and procedures. I sought instead to examine Ms. Fr. 640 and other recipe collections as testaments of practice, and as ways to enter into the "knowledge" of craft.

Concurrently with writing *From Lived Experience to the Written Word*, I established the Making and Knowing Project (www.makingandknowing.org), which spent six years producing a digital critical edition of this manuscript, now openly accessible at https:// edition640.makingandknowing.org. Working collaboratively with postdoctoral scholars and graduate students at Columbia University, as well as with practitioners and scholars from across the globe, we reconstructed the recipes in the manuscript. I describe this work of reconstruction most fully in chapters 9 and 10, but my experience with Ms. Fr. 640 informs this book throughout. In *The Body of the Artisan*, my focus was especially on well-known artists, such as Albrecht Dürer (1471–1528), Jamnitzer, and Palissy, who explicitly theorized their knowledge of nature, both in writing and in their objects. Even as I return to these figures, this book broadens its purview, and looks further afield to the large number of often anonymous practitioners, like the author-practitioner of Ms. Fr. 640, who wrote down recipes for craft and art processes, observed the practices of other workshops, and recorded their firsthand experiences. In doing so, my goal is to demonstrate that for a full understanding of the knowledge of handwork, it is crucial to engage *processes* of making, rather than solely texts and objects.

Plan of the Book

From Lived Experience to the Written Word thus seeks insight about the knowledge of handwork, and considers how the epistemic dimensions of early modern artisansal practice can be elucidated. Four sections treat, in turn, the material and intellectual world of artisans, the nature of their writings, the readers and collectors of those writings, and, finally, how these texts can be used by historians and scholars who wish to grasp the nature of hands-on knowledge-making more fully. Part 1, "Vernacular Theorizing in Craft," considers the claims that artisans, beginning around 1400 and continuing well into the sixteenth century, made about possessing a special form of knowledge. The act of *making* with natural materials could also form an act of *knowing* nature, that is, it was underpinned by a corpus of beliefs about nature and the behavior of natural materials, and these helped guide workshop practices. In this section and throughout the book, I refer to "philosophizing" and "theorizing," by which I mean making knowledge of a higher level of generalization—sometimes fully developed in a "knowledge system"—which involves conceptualizing natural processes and reasoning about why material transformations happen in particular ways. Such theorizing does not necessarily involve a conceptual system being imposed on natural particulars, but rather, can emerge out of the experience of practitioners working with materials. I call such a knowledge system a "material imaginary." In part 2, "Writing Down Experience," I survey the ways in which writing became important to artisans. Even as I trace the wide range of reasons why artisanal writings were both produced and read, I consider how we might articulate what kind of knowledge is actually contained in texts of practice and collections of recipes. In part 3, "Reading and Collecting," I investigate the readers and users of such texts, and consider the growing enthusiasm for them from 1400 through the eighteenth century. In part 4, "Making and Knowing," I lay out a path for present-day scholars (especially historians) and their students, used to working in the highly text-based modes of academia, to experience artisanal skill and its associated epistemic practices. I ground this approach in my own experiments in the Making and Knowing Project.

Guide to the Chapters

Chapter 1, "Is Handwork Knowledge?" begins by examining attitudes to handwork and surveying their formation from antiquity to early modern Europe, making clear the challenge that some craftspeople posed to social and intellectual hierarchies when they began to express themselves in writing. In becoming authors, these handworkers asserted their value within the historical record, sharing the moving plea made by a citizen of Cologne in the sixteenth century:

I hope that nobody will reproach me for writing so much about simple folk, sisters, brothers, friends, neighbours, citizens, peasants, journeymen, domestic, plain, and childish things and about myself. But who will do it but for me? In the Bible, in Roman history books and chronicles, in the Holy Scriptures, in herbals, in the seven liberal and other arts as well as in philosophy and poetry books one won't really find us. Thus, if my book and notes are preserved and kept up with the times, our descendants will know about us; otherwise we are as if we had never existed.[26]

In addition to making a statement about the worth of their existence, early modern artisans also made an intellectual and epistemic claim simply by writing down their knowledge. But what is this "knowledge" that their texts contain? To make clear both their challenge and some of the characteristics of their knowledge, this chapter focuses on the Venetian navigator Michael of Rhodes, then turns to the Paduan painter Cennino d'Andrea Cennini.

In chapters 2 and 3, I treat various forms of natural knowledge among artisans in the sixteenth century. Where my *Body of the Artisan* examined the epistemic claims made mainly in objects and images of nature by early modern painters and sculptors, these two chapters examine the philosophizing that a variety of artisan-practitioners codified in writing, in practices, and in objects. These chapters survey the lifeworld of the workshop and delineate what might be called a "philosophy"—or, as I explain there, a "material imaginary"—showing that artisanal practices and techniques were informed by a deeper systemic understanding of materials. Chapter 2, "The Metalworker's Philosophy," considers how metalworkers across Europe articulated their knowledge. Starting from their ingestion of butter before going to work, I show how their "philosophy" or "material imaginary" depended on the interaction between the human body and the terrestrial and celestial environment in their constant observation and investigation of natural materials.

In Chapter 3, "Thinking with Lizards," I turn to life-casting, the process through which artisans, including the author-practitioner of Ms. Fr. 640, made molds of living creatures and produced extremely lifelike metal objects. To produce these highly valued objects, practitioners needed an intimate knowledge of their mold materials, metals, and of the animals—especially reptiles such as snakes and lizards—that they were casting. This chapter describes the constant testing and trying of materials through which they obtained this knowledge, and the material imaginary that helped to guide their investigations. This material imaginary not only facilitated the production of objects, but also gave insight into the powers of nature, transformation, and generation.

After these three chapters on the philosophizing of artisans, I turn more fully to the nature of their texts. Chapter 4, "Artisan Authors," provides a short history of written accounts of technical processes, beginning before the turning point around 1400 when artisans became authors. This chapter's study of books of practice and recipe collections—the "little books of art," sometimes also called "books of secrets"—shows that the move from lived experience to the written word involved more than the simple recording of recipes: it could also convey attitudes and actions necessary in learning handwork, as well as the foundations of craft techniques and the systems of knowledge that underpinned them.

Chapter 5, "Writing *Kunst*," directly addresses the central problem that drives this book: how can embodied knowledge—what I refer to as *Kunst*—be captured in writing? Drawing on theoretical discussions among anthropologists, cognitive psychologists, and philosophers, I show how skill must be acquired alongside an experienced practitioner, performed methodically by means of observing, attending, and repeating through years of experience until the coordination of perception and action becomes habit. Interestingly, descriptions of skill by contemporary theorists often sound very similar to early modern artisans' articulations of skilled practice and how to achieve it. Among early modern artisans, such knowledge was often formulated as a set of particular instances and contingent instructions, a point elaborated in chapter 6, "Recipes for *Kunst*," which proposes that recipe collections embodied a model of the knowledge at the heart of craft practice. Recipe collections contain traces of the workshop and the type of knowledge produced there: they emphasize particular cases and discrete instances, but at the same time indicate the modes by which such discrete instances can be employed to produce a more generalized, higher order form of knowledge that is materialized by a tangible product or object (the *proof* of this type of knowledge).

Chapters 7 and 8 treat the larger social and intellectual field in which books of practice were received. Judging by the large numbers of editions and the frequent compilation and recompilation of these texts, printers apparently viewed practical knowledge as best selling material. Why? Who bought these books? Who read them? There is only scattered evidence for the audience of these books, but it is clear that they must be understood within a larger context of interest in practice and in the productive potential of human art and industry—an interest that extended far beyond the social spheres of artisans themselves. Chapter 7, "Who Read and Used Little Books of Art?" considers what we can know about artisans as owners and readers of books, and shows that scholars, patrons, and political elites formed an important segment of their readership. That these texts were not exclusively of interest to artisans themselves is revealed, for example, by the provenance of Ms. Fr. 640 itself, which entered the king's library (now the national library of France) not from a workshop, but from the collection of Philippe de Béthune, Count of Selles and Charost (1561–1649). A military and political adviser to Henri III, Henri IV, and Louis XIII, Béthune considered the development of manufacturing and crafts an important element of political economy.[27] Béthune is only one example of the audience for such texts, but many others discussed in chapter 7 reveal the wide range of uses to which little books of art could be put and the ways in which they could be appreciated.

Chapter 8, "*Kunst* as Power: Making and Collecting," focuses in particular on the librarian Samuel Quiccheberg (1529–67), who helped organize the *Kunstkammer* (chamber of art) of both the Fugger family of long-distance merchants and the Wittelsbach Dukes of Bavaria. These collections brought together natural objects and the products created by the artisanal transformation of nature; in doing so, they revealed the productive power of nature and the capacity of craftspeople to harness that power. Analyzing one of the artworks of a collection—a chamber fountain created by goldsmith Wenzel Jamnitzer alongside the collection of Quiccheberg—this chapter argues that elites supported, promoted, appropriated, and strove to gain control of the realm of the human hand represented by the collections. The view that the manipulation of nature by art (and, eventually, by science as well) could yield productive knowledge came to be fundamental to the exercise of statecraft, a view that supports the large-scale public funding of the natural sciences today.

In the sixteenth century, historical research was seen as empirical in ways similar to the investigation of nature. Today, however, historical study has largely been cordoned off from the natural sciences. I argue in chapters 9 and 10 that a full understanding of the relationship between making and knowing demands from the historian an approach that incorporates processes of experimentation. In chapter 9, "Reconstructing Practical Knowledge: Hastening to Experience," I explain how the skills and epistemologies of early modern artisans cannot be studied solely by reading their writings, but instead must be sought in trying out their recipes. The chapter briefly surveys instances of historical reconstruction among conservators, archaeologists, and others, before discussing my own efforts to reconstruct the methods and materials of Ms. Fr. 640. These attempts have been deeply collaborative, developed alongside craftspeople with knowledge of historical techniques, as well as students and fellow scholars. Describing this experimental work, the chapter explores the potential pitfalls and rewards of reconstruction as one tool for producing knowledge about the past.

Chapter 10, "A Lexicon for Mind-Body Knowing," uses the knowledge that the Making and Knowing Project gained from its reconstructions of the recipes in Ms. Fr. 640 to address one of the knottiest problems that faces scholars and researchers from a wide range of disciplines: how can we begin to overcome the stubborn division between mind/thought and body/action that continues to inform our conceptual frameworks? I believe that we need to come up with new frameworks and vocabularies in which to express the amalgam of "hand" and "mind" that constitutes cognition and knowledge. The chapter unpacks the mental and material processes of Ms. Fr. 640's author-practitioner to try out some language by which to describe craft knowledge, including "knowledge that works," "materialized theory," a set of "fundamental structuring categories," and a "framework for action."

This book considers lived experience and the written word, vernacular and craft knowledge, the issues of skill and technique, and the kinds of sources and methodologies by which we can gain knowledge of such practices. But it is also an unexpected story about the routes by which skill and knowledge travel. My epilogue considers the flows of matter, knowledge, commodities, and people and their skills across geographic and epistemic distance.

As a whole, this book chronicles a pivotal moment in European history during which handworkers and practitioners of the mechanical arts asserted the power of their knowledge. From about 1400, practitioners expressed their self-consciousness and articulated their knowledge, and in doing so, they made novel claims about the true sources of knowing, about proof and certainty, which they set alongside long-held assumptions that certainty could only be obtained through abstract mental work. And, as artisans crossed the divide of the written word, their knowledge formed a new type of intellectual culture that combined practical knowledge and written texts, and their techniques and processes became part of the toolbox by which modern forms of knowledge about nature, or "science," would come to be made. Artisanal experience and skill thus underpin the complex developments that led to the making of the modern natural sciences. It is hard to overestimate the importance of this new form of knowledge and knowledge-making, and this book is about the crucial part that artisan authors had in its construction.

Return to Practice

We appear now to be in the midst of a moment when practice and craft are being celebrated again, from academia to self-help.[28] Popular books on craft bring to light a variety of personal benefits of craft work, including habits of self-governing instilled by bodily practice. Economic historians emphasize the efficacy of embodied knowledge and early modern guilds for transmitting knowledge and fostering innovation.[29] Economists engage in experimentation with real world human practices, law theorists prioritize practice and norms over principles of law, and cognitive philosophers and neuroscientists study habits and skills. And this is not even to touch on recent phenomena such as Wikipedia and open-source software as examples of "hive mind" and collective knowledge production.

This focus on various aspects of craft and practice is not new to the last twenty years—this book provides an overview of the repeated return of such an interest up through the early eighteenth century. Such movements obviously did not end then, either—in modern memory, for example, the Arts and Crafts movement emerged in the wake of industrialization, and, in the twentieth century, phenomenologists made philosophical arguments for thought and knowledge emerging through the processes of bodily dwelling in the material world, paying special attention to making practices.[30] At such moments, what we might call the politics of craft and practice, often having a strong component of reform—reform of individual behavior, of the public sphere, or of knowledge hierarchies—comes to the fore. As we shall see, writing about embodied practice can be a language in which to make arguments about hierarchies of knowledge. Writing about craft is to take on a form of knowledge rooted in particular human capacities that is intractably difficult to articulate in words and texts. This brings us back to our repeated question—why try? Although their texts appear to give more or less transparent instructions, the artisan authors surveyed here sought both to upend a social and intellectual hierarchy and to articulate a particular way of knowing that, long before the research on embodied cognition, laid out the workings and power of this form of knowledge. One of my aims in this book is to highlight their view that intelligence is not held by the mind alone, but, instead, emerges from the work of the hand. This book returns to a key historical moment to reveal this alternative conception of human intelligence and knowledge. Like these artisan authors, *From Lived Experience to the Written Word* attempts to convey how the work of *making* is related to *knowing*, and the ways in which making is an epistemic activity and an active form of knowledge in its own right. By recounting this crucial development around 1400 in the long history of writing about craft, especially at this present moment of increased attention to craft and making, I could be seen as participating in the politics of craft that I chronicle. I am convinced, however, that the modest works of practice I examine in this book contain rich historical evidence for an account of human creativity—one in which art, or *Kunst*, understood as embodied practical knowledge, plays a central role.

PART 1

Vernacular Theorizing in Craft

CHAPTER 1

Is Handwork Knowledge?

The Work of the Hand

The book of Ecclesiastes contains a beautifully observed account of the activities and hardships of craftspeople, but it is framed by an argument that such people who are "occupied in their labors" can never attain wisdom:

The wisdom of a learned man cometh by opportunity of leisure: and he that hath little business shall become wise.

How can he get wisdom that holdeth the plough, and that glorieth in the goad, that driveth oxen, and is occupied in their labours, and whose talk is of bullocks?[1]

After describing how the carpenter "laboureth night and day," how the smith "fighteth with the heat of the furnace," and how the potter "fashioneth the clay with his arm, and boweth down his strength before his feet," the passage notes that each of them "trusts to their hands" and "is wise in his work," and "their desire is in the work of their craft." Without them, "a city shall not be inhabited," for "they will maintain the state of the world." While crafts sustain life and uphold the state of the world, however, their practitioners "shall not be sought for in public counsel, nor sit high in the congregation: they shall not sit on the judges' seat, nor understand the sentence of judgment: they cannot declare justice and judgment; and they shall not be found where parables are spoken." This vision reflects an enduring prejudice against handwork that was common throughout ancient Eurasia. In ancient China, the Confucian Mencius (372–289 BCE) stated that "[s]ome labor with their hearts and minds; some labor with their strength. Those who labor with their hearts and minds govern others. Those who labor with their strength are governed by others."[2]

Sophia and Sophists

Attitudes to handwork in Europe were shaped by such views, and also by the survival of Greek writings about the relationship between what we call the "mind" and the "hand." These ideas, found in the writings of Plato (427–347 BCE), Xenophon (430–354 BCE), and Aristotle (384–22 BCE), emerged within the particular social and political structure of Athens. Insight into their views can be gained by examining their dismissal of many of their contemporaries and teachers of rhetoric as "mere sophists." We still use the terms "sophist" and "sophistry" to demean a person or an argument. For Plato, the sophists' unsavory qualities included a wandering life not tied to a single polis or to the collectivity of Greek city-states. Sophists engaged in practice rather than theory, focusing on the "technique" of "rhetoric" rather than the "knowledge" of "philosophy," and thereby, according to Plato, teaching an appearance of wisdom rather than "real" wisdom. Sophists taught "how to do" rather than "how to think," and, worst of all, they were paid for this teaching—for Xenophon this was akin to prostitution.[3] Although many ancient philosophers called each other by the name of sophist, Plato was the first to give the sophists a persona, and he did so, not because sophists actually existed as an identifiable school, but because they were a convenient foil against which to establish a genealogy of his own "true philosophy." Plato made the sophists the counterfeit background to the emergence of true wisdom in the teachings of Socrates.

As historians of philosophy have shown, Plato's project was not uncontested in the ancient world, but because Plato wrote (and his writings so remarkably survived), his genealogy has become our dominant narrative. Historians of ancient philosophy, most recently Håkan Tell, have pieced together a clearer picture of this ancient *sophia* tradition from scattered and meager documents that have come down to us.[4] Ancient *sophia* seems to have been oriented toward the *practice* of wisdom and expertise of a practical kind—the performance of religious ritual, architecture, city planning, water catchment, medicine, poetry—oriented to community life in the polis and an active life in politics. In fragmentary ancient texts, this pre-Socratic *sophia* was referred to as a kind of cunning wisdom, a *metis*, celebrated in the fables of Aesop (620–564 BCE), sometimes portrayed as wise counsel, akin to prudence, but most often oriented toward doing and practice. In an expression of this view, Anaxagoras (500–428 BCE) claimed that "[m]an is the wisest of all living creatures because he has hands."[5]

As we shall see, after nearly two millennia, around 1400 CE this submerged tradition surfaced with an explosion of practical and technical writing. The years around 1400 represent a pivotal "practical moment" at which, with remarkable suddenness, a flood of written accounts of practice—that is, handbooks, or books of art—began to be penned by European craftspeople and practitioners. These individuals, who previously had lived out

their lives without recording their experiences and knowledge, creating and producing in relative obscurity, suddenly began to write. Artisans' lived experience moved from orality and tacit knowledge to the written word.

This practical moment began with expressions of self-consciousness among craftspeople, expressed not in words but in self-portraits. In 1392 Peter Parler (1333–99) made a self-portrait in stone on the Prague Cathedral where he served as master mason (fig. 1.1). Just before 1400, Cola Petruccioli (1360–1401) painted a remarkable self-portrait in the Dominican church in Perugia, dipping his paintbrush into a paint pot of vermilion (fig. 1.2).[6] Many artists and artisans followed suit in the decades after, including Adam Krafft (c. 1460?–1509), who carved himself in 1485, life-size and dressed in his mason's working garb, holding his stoneworking tools, at the base of his astonishing sacrament house which reaches into the vaults of the St. Lorenz Church in Nuremberg (fig. 1.3). Others displayed this self-consciousness by marking their creations with their name. The Lyonese cannon founder, Jean Barbet (fl. 1475–d. 1514), who commemorated his casting of a monumental sculpture on the inside of an angel's left wing: "le xxviii jour de mars / lan mil cccc lx + xv jehan barbet dit de lion fist cest angelot" (on the 28th day of March in the year 1460 + 15, Jean Barbet, called of Lyon, made this angel) (fig. 1.4).[7] Better known is the

1.1. Peter Parler, self-portrait, 1392, St. Veit's Cathedral, Prague. Sandstone. Parler served as master mason on the cathedral. His smock is adorned with the mason's square. CC0 1.0.

1.2. Cola Petruccioli (1360–1401), self-portrait, Salone del Vestito, Basilica of San Domenico, Beato Benedetto XI Museum, Perugia. Petruccioli dips his brush into a paint pot of vermilion. CC0 1.0.

1.3. Adam Krafft, c. 1493–96, self-portrait at base
of monumental ciborium he sculpted, St. Lorenz,
Nuremberg. Bridgeman Images.

1.4. Jean Barbet (fl. 1475–d. 1514), angel, 1475.
Bronze, H 113 cm. Frick Collection, New York.
Barbet, a cannon founder, inscribed his signature or
perhaps a votive petition under the wing. Purchased
by The Frick Collection, 1943. 1943.2.82. © The
Frick Collection.

work of Jan van Eyck (1395–1441) who left no texts, but declared in writing in his Arnolfini portrait that "Jan van Eyck was here." Others, such as Lorenzo Ghiberti (1378?–1455) and Antonio Averlino (c. 1400–1469), called Filarete, portrayed themselves in their works of art and also wrote texts. Ghiberti's self-portrait on the doors of the Florence Baptistery is well known, as is his manuscript work, *I Commentarii*. Filarete, trained as a goldsmith, portrayed himself around 1455, with his apprentices, on the back of a bronze door in St. Peter's in Rome, and in the early 1460s, followed this up with an architectural treatise dedicated to Francesco Sforza (1401–66).[8]

Handwork from *Sophia* to 1400

To understand this sudden explosion of practical texts, this chapter briefly surveys how a distinction between knowing and doing was conceptualized in ancient Greece, classical Rome, and medieval Europe. It then turns to how craftsmen came to develop a new mode of writing in which they both expressed and theorized their own practices and manipulations of nature. Focusing in particular on the remarkable manuscript produced by the Venetian oarsman and navigator, Michael of Rhodes (c. 1380–1445), I argue that Michael converted his embodied experience into textual form in order to make it conform to the expectations of his aristocratic patrons about what knowledge should look like, and that the seeming jumble of discrete pieces of information contained in his book reveals Michael's effort to model and codify the wellsprings of his knowledge. A comparison of Michael's work to the well-known "book of the art" (*Libro dell'arte*) of his immediate contemporary, the painter Cennino d'Andrea Cennini (active c. 1390–1437), shows how they expressed in words, theorized, and sanctified their labor within a religious worldview.

EPISTĒMĒ AND *TECHNĒ*

The ancient practical wisdom tradition of the sophists was overshadowed, if not completely suppressed, by the success of Plato and Xenophon's founding figure of Socrates and his philosophical tradition that emphasized abstract, universalizing, and disembodied *sophia* over practice. Plato's student Aristotle opened his *Metaphysics* with the observation that "[a]ll men naturally desire knowledge," before going on to develop a fuller picture of what constituted knowledge. Human knowledge began in sensations, he wrote, especially sight, which humans then processed through memory and experience into knowledge. Not all human knowledge was equal for Aristotle, however:

we consider that the master craftsmen in every profession are more estimable and know more and are wiser than the artisans, because they know the reasons of the things which are done; but we think that the artisans, like certain inanimate objects, do things, but without knowing what they are doing (as, for instance, fire burns); only, whereas inanimate objects perform all their actions in virtue of a certain natural quality, artisans perform theirs through habit. Thus the master craftsmen are superior in wisdom, not because they can do things, but because they possess a theory and know the causes.[9]

We see here how Plato's work to define true wisdom created a hierarchy of knowledge and knowledge-making that, even in Aristotle's more sensation-based epistemology, placed practical pursuits lower on the epistemic and social hierarchy than true philosophy. Aristotle differentiated between common technicians (*cheirotechnoi*, a term emphasizing their manuality) and master technicians (*architektones*), who "know the causes and do things according to the *logos*." Thus, while Aristotle viewed technicians as capable of varying degrees of rationality, his hierarchy, as one scholar has written, was "topped by speculative forms of knowledge, which prevail over productive, technical ones,"[10] or, as Aristotle said, true knowledge, or "wisdom," "is concerned with the primary causes and principles [. . .] and the speculative sciences [are held] to be more learned than the productive."[11]

In book 6 of the *Nicomachean Ethics*, Aristotle made clear that the pursuit of knowledge was also a seeking after virtue, and that different paths of pursuing knowledge resulted in knowledge and virtue of different kinds. Aristotle divided knowledge into three types: *epistēmē, pragmon*, and *technē*. *Epistēmē* (called *scientia* or *theoria* in Roman writings) was demonstrable knowledge that, at its most certain, could be compared to geometrical deduction from known axioms. Less certain knowledge was gained by the collection of instances, *exempla*, or experiences. This *pragmon* (*praxis* in Latin) was relevant to the life of action and politics, in which the collection of experiences, especially drawn from historical accounts, provided the basis for the citizen or ruler to take prudent and virtuous action in the active life of the polis. This form of experiential knowledge could never be as certain as that known by deduction or by syllogism, but it was sufficient for human action. The final type of knowledge was *technē* (*ars* in Latin), by which people produced valuable goods, but which possessed the least certainty because its practitioners had to deal with the uncertainties and idiosyncrasies of changeable matter.[12]

In a telling comment in the *Politics*, Aristotle made clear how much the contingency of materials and bodily involvement was bound up for him with moral deformation. The uncertainty and instability of chance and manual labor compromised the certainty and virtue of an endeavor:

The most scientific [i.e., certain] of these industries are those which involve the smallest element of chance, the most mechanic [a word with overtones of "cramped in body" and "vulgar in taste"] those in which the operatives undergo the greatest amount of bodily degradation, the most servile those in which the most uses are made of the body, and the most ignoble those in which there is the least requirement of virtue as an accessory. But while we have even now given a general description of these various branches [of wealth-getting], yet a detailed and particular account of them, though useful for the practice of the industries, would be illiberal [i.e., not worthy of a free man] as a subject of prolonged study.[13]

As is well known, Aristotle's theory of knowledge was formulated in a society in which slaves performed most bodily labor and the work of productive craft. Aristotle believed slaves possessed reason, but lacked its "deliberative part,"[14] and thus were different in nature from free men. Prejudice against bodily labor and the slaveholding system supported each other in his fundamentally hierarchical theory of knowledge. Based on this theory of knowledge, "doing" and "making" gradually came to be regarded as separate from "knowing."[15] Doing was regarded as knowledge of "how to" rather than causal knowledge that answered "why," and it was seen as goal-oriented "know-how," involving specific and particular practices, while "knowing" came to be associated with generalizable, abstract knowledge, expressed in propositions, general theories, and proofs that dealt with causes. Plato and Aristotle regarded "knowing" as a more powerful type of understanding, despite "doing" and productive knowledge being everywhere apparent in the society and public life of fourth-century Athens. Both referred to *technē* when discussing the practical knowledge needed in daily life, and Plato even stated that the philosopher-ruler of the *Republic* needed *technē*.[16] The problem of *technē* for them, as Serafina Cuomo has made clear, was that it carried the potential for both good and ill: the former, because it was the source of valuable and useful goods important in the *oeconomia* of the household and city state; the latter, because the labor of the craftsperson was a source of physical deformity and, as we have seen, in Greek culture this translated into moral deficiency. Moreover, the artifice of the craftsperson could shade into craftiness and potentially deception. The goods made by the artisan engendered "unnatural" commercial speculation that could bring about destabilizing social mobility. The social mobility that *technē* could and did engender in both Greek and Roman antiquity, in Cuomo's account, "spelt trouble" for elites.[17] For Plato, Aristotle, and other—although not all—thinkers in antiquity, *technē* was threatening to elites as a source of moral danger, unnatural commercial speculation, and social mobility, and, as Cuomo has elegantly made clear, their claims about *technē* must be understood within this power dynamic.[18] As in other periods treated in this book, statements about handwork were often statements about power and social hierarchy.

During Roman antiquity, craftspeople formed large collective organizations, called *collegia*, and, despite (or because of) the collective power represented by these groups, the disdain on the part of philosophers and members of the elite continued to be strongly expressed. Cicero (106–43 BCE) remarked that "[a]ll mechanics are engaged in vulgar trades, for no workshop can have anything liberal about it," and Seneca the Younger (4 BCE–65 CE) said that the arts, "which are common and low […] belong to workmen and are mere hand-work; they are concerned with equipping life; there is in them no pretense to beauty or honour," and "those alone are really liberal—or rather, to give them a truer name, 'free'—whose concern is virtue." Plutarch (46–120 CE) believed no one would long to be even a great sculptor or painter, "[f]or it does not of necessity follow that, if the work delights you with its grace, the one who wrought it is worthy of your esteem."[19]

TECHNĒ AND NATURAL PHILOSOPHY

A new and sustained engagement with *technē* appears in medieval Arabic writers who studied and elaborated on Aristotle's works. A taxonomy of craft in their texts was already developed by the ninth century, when Baghdadi Qusṭā ibn Lūqā elaborated four classes of crafts,[20] and gave a place to the mixed mathematics of mechanics and optics in natural philosophy.[21] In the *Enumeration of the Sciences*, Abû Nasr al-Fârâbî (c. 870–c. 943) listed a "science of *ingeniis* [machines or devices]," which considered the principles of things *as things*, rather than as theories, and taught modes of invention and manipulation of natural materials by means of the arts.[22] In *The Great Book of Music*, al-Fârâbî even asserts that embodied knowledge can overturn theory, remarking that, although music derives its principles from mathematics, the bodily sense of hearing can rightly contradict a mathematical principle (for example, in a semitone not precisely constituting half of a full tone).[23] We know too little about the training and culture of intellectuals writing in Arabic at this time, but their wandering lifestyle and their mix of practical and theoretical pursuits suggest a less hierarchical view of the relationship between *technē* and *epistēmē*.

When Islamic texts arrived in the backwater of Western Europe, Hugh of St. Victor (c. 1096–1141) relied on them for his enormously influential taxonomy of knowledge, *Didascalicon*, which included the mechanical arts within the category of the sciences. At the same time, Hugh subordinated all human science to divine science, and gave what one scholar has called an "essentially metaphysical view of the *artes mechanicae*."[24] In Hugh's view the mechanical arts were degraded ("counterfeit," as Hugh expressed it), but, at the same time, "human art […] can still evoke wonder."[25] It was not until thirteenth-century scholar-clerics, including Albertus Magnus (c. 1200–1280) and Roger Bacon (c. 1220–c. 1292), that actual craft manipulation of nature and natural materials came to be of deeper interest and importance in Latinate taxonomies of knowledge. Arabic treatises on perspective and mathematics, and new instruments arriving in Europe, such as astrolabes and magnets, impelled Bacon to rethink the place of the mechanical arts in natural philosophy.[26]

Scholars' taxonomies do not, however, tell the whole story. During the late Middle Ages, handwork was increasingly valorized as religious practice, in part as an imitation of Christ the carpenter. Monastic institutions enjoined monks to pray and to labor (*ora et labora*).[27] Fascinating and ingenious objects and new techniques entered Europe via the Iberian peninsula. Urban life concentrated artisans and made visible their vital place in society, where their works provoked wonder and their productivity was praised. In contrast to Aristotle's ideal polis, in some cities at this time, craftsmen could rise to hold power as citizens.[28]

By the early fourteenth century, then, writers of natural philosophy gave greater place to arts that engaged with natural materials and processes. Handwork had come to possess a higher status as monastic ideals and powerful urban guilds reinforced its place in salvation and social order. Nevertheless, the hierarchy of knowledge, from body and the sphere of the mechanical arts at the bottom, up to mind and the spirit of the liberal arts at the top, nevertheless endured. It was carried forward as the ancient corpus of texts became the core of European universities, where students were trained in the texts of philosophy and the liberal arts—the arts of the *homo liber*, the free man, as Seneca had phrased it. The practitioners of the mechanical arts, in contrast, were trained by apprenticeship, absorbing the gestural knowledge of their masters and learning by experience. Even today, despite the union of practice and theory in the natural sciences, the public perception of a hierarchy still endures, and many people view engineering and the "applied" sciences as lower in status than the "theoretical" sciences.

The Practitioner as Writer: Michael of Rhodes and His Book

When a practitioner set pen to paper, he challenged this tradition—not always explicitly, but rather by presenting himself as an *auctor*, an author and authority. A vivid example of such a challenge is a coat of arms a Venetian galley oarsman gave to himself in the 1400s (fig. 1.5). It depicts a world turned upside down: a giant mouse holds a bloodied cat in its paws, blood dripping from the cat's already stiffening sides. This striking image is flanked by oversized turnips—the rustic food of peasants—in place of the usual noble fleur-de-lis. This appears a work of some self-consciousness and no little irony on the part of its vernacular creator, Michalli da Rhuodo, or Michael of Rhodes. Michael joined the Venetian navy as a lowly galley oarsman in 1401, taking part in more than forty voyages in as many years across the Mediterranean and to Flanders, and rising to a high position. He documented his voyages in a remarkable manuscript, begun in the 1430s, which contains an unusual level of detail about a seafaring life.[29] It reveals his interest in mathematical and algebraic calculation for commerce, calendar computation, and navigation, as well as his deep piety, his concern with the movements and influences of the heavenly bodies (fig. 1.6), and his knowledge of technical details of ships, for which he provided many striking illustrations, filled with precise measurements (fig. 1.7).

All these components of his book convey a vivid but enigmatic glimpse into the concerns and ambitions of an informally educated handworker who was part of neither the intellectual nor social elite. In Michael's book, a working man engaged in "doing" all his life suddenly comes into focus through his foray into the written word, and he looks much as

1.5. Coat of arms of Michael of Rhodes, from *The Book of Michael of Rhodes: A Fifteenth-Century Maritime Manuscript* (Long, McGee, and Stahl 2009, vol. 1, fol. 147b). Image reprinted courtesy of MIT Press.

1.6. (*facing, left*) Zodiac man, indicating the influences exercised by the planets over various parts of the body, from *The Book of Michael of Rhodes* (Long, McGee, and Stahl 2009, vol. 1, fol. 103b). Image reprinted courtesy of MIT Press.

1.7. (*facing, right*) Drawing showing the bow of a galley of Flanders, with measurements, from *The Book of Michael of Rhodes* (Long, McGee, and Stahl 2009, vol. 1, fol. 138b). Image reprinted courtesy of MIT Press.

we might have hoped he would: while embedded in the deep piety and astrological world-view of his religious and cultural milieu, he seems to be self-conscious and mathematically knowledgeable, and, to judge by his remarkable coat of arms, apparently dreaming of over-turning the social order. It thus appears that Michael's book will be a treasure trove for illuminating the reassertion of the practical *sophia* tradition lost in antiquity. This view rapidly dissolves on reading Michael's book: it is a jumble, with no clear authorial voice or intention and with much contradiction. The math turns out to be mostly problem solving compiled from textbooks used by merchants in their so-called abacus schools; the schematic descriptions of sailing routes are so inaccurate that anyone actually following them

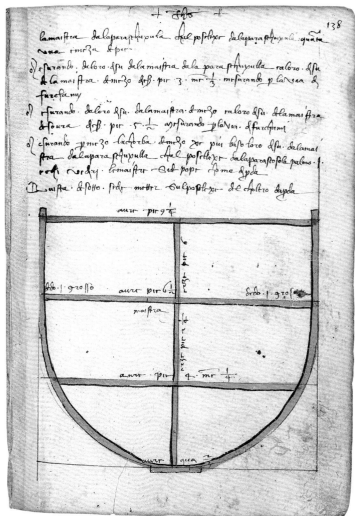

would have been shipwrecked; the images are not all original; the recipes often seem dry lists; magical incantations appear in the text with no rhyme or reason; and the descriptions of ships do not appear to form useful parameters or instructions for building vessels.[30] Many such texts of practice—both manuscript and printed—possess this character, being compiled from other texts with little conceivable order: barebones recipes; magic tricks; undigested, out-of-date, often inaccurate information that could not possibly be made use of by readers. What is such "useless" information doing in a practical how-to manual? So many of the characteristics we assign to knowledge we believe to be powerful—such as originality and innovation, precision, applicability beyond the particular circumstances of its formation, capable of leading to general rules, useful, and instructive—seem to be absent from many of these books. Perhaps Michael's book really gives little insight into this working person's life. How can we gain an understanding of craft and skilled practice as "knowledge" from something so flawed? Was Michael simply attempting to imitate an intellectual elite tradition of authorship without really being able to embody it?

To answer these questions, let us look more closely at Michael's book: What did writing *do* for him? Michael was low-born, educated informally, worked with his hands, and learned by experience. He came from a culture in which writing down techniques, processes, and personal information was unusual. What did writing mean to such a person? As James Amelang, in his book on artisanal autobiography, *The Flight of Icarus*, lays out, as writers in a nonwriting culture, vernacular authors were, almost by definition, "outsiders." As he notes, "Many of these outsiders were desperately trying to convince themselves that they would become insiders and sought to use writing in general […] as a means to this end. Writing was often (although not always) a transformative strategy, enabling authors to cross social as well as cultural boundaries."[31] Reflecting on their lives and recording their thoughts, ambitions, and practices clearly could be a challenging experience for manual workers—a fact that appears most clearly in later sources.[32] For example, an eighteenth-century Swiss manufacturer of saltpeter and cloth, Ulrich Bräker (1735–98), when he joined a local learned society, lamented, "It was no good: I was like the raven who wanted to fly with the ducks […] My neighbors […] and acquaintances, briefly, my own kind, looked askance at me whenever we met. Here I heard a sneering hiss; there I caught a contemptuous smile […] My wife tore into me and couldn't be calmed down for weeks; she even became disgusted and revolted at the sight of any book."[33] Just as seems to be evident in Michael's self-made coat of arms, Bräker was aware of crossing a deep social divide in translating his lived experience into the written word. By the eighteenth century, when Bräker wrote, writing had come to be an invention and exploration of self, and books served as mirrors or prisms in which individuals could observe and represent their own lives.[34] Michael's book, in contrast, gives a different kind of insight into what a book could do for an author.

Michael earned his livelihood by being signed on every year in some capacity to the Venetian navy or the trading fleet. From 1401 to 1422 he worked his way up from oarsman to *armiraio*, the highest position he could attain as a nonnoble. In order to be hired at the rank of nonnoble officer, Michael had to compete with dozens of other mariners. Such officer positions not only raised one to a position of command, but also paid better and gave more space to the individual to ship his own commodities for private sale. Due to changes in the rules of the competitions in 1418, the applicants for nonnoble officer positions were judged by Venetian patricians who had negligible maritime experience themselves. As Alan Stahl's careful reconstruction of Michael's service in the Venetian fleets has shown, the competitions from 1433–36 (the years during which Michael substantially completed his book) were particularly fraught by conflict between the noble patrons of the ships of the fleet and the nonnoble officers, as well as by various charges of influence and corruption. In these difficult years when Michael composed his text, a book might assist in this competition. There was no getting around the social and intellectual status of a book,[35] especially one that had the "look" of expertise, including a list of his voyages, portolans (inaccuracy not at issue because they were not actually used for navigating), extensive figures, and many impressive images of ships. Michael's book from this vantage point was a "proof" of his expertise. Today, the "look of proof" might seem a less powerful proof than actually navigating a ship safely across the Mediterranean, or being able to garner a profit by being able to calculate—to take an actual example from Michael's book—the correct price of a load of silk worth 11 $\frac{3}{11}$ soldi per pound being exchanged for cloth worth 60 soldi in cash and 66 soldi by barter, in which one-third of the price had to be paid in cash.[36] But in his world, a book and authorship conformed to the textual world and expectations of the competition's judges, many of whom had never been to sea. Michael's book may in fact have done the trick, for he was elected as *armiraio* of the Flanders galley in 1436.[37]

Shipboard Knowledge

On the strength of its images and its precise measurements for ships, Michael's text until recently was viewed as an instructional manual for shipbuilders, but closer study has shown that the audience for Michael's detailed pictures may have been patrician youths whose merchant families sent them to sea not so much to become acquainted with ships, as to receive an overview of the trade, including "completing ships for the sea," that is, fitting out the ships with all necessary components, including stores and equipment. Indeed, many well-illustrated, apparently instructional texts may have been employed by practical men explaining the tools and expertise of their trade to a social group above them, such as administrators or patrons.[38] Pictures are effective in organizing technical knowledge into an abbreviated form because the processes described are tedious and confusing to follow in writing, especially for those seeking an overview. Michael's book thus indicates that one aim of a practical text could be for use with patrons and officials who needed to survey a field of activity. The illustrations in a book, combined with oral elaboration, made complex procedures clearer, and probably helped the listeners see the activities as "knowledge." Yet, for actually doing and producing things, embodied nontextual knowledge was still best, as demonstrated when the Venetian Council of Forty specified in 1403 that on every ship construction over a certain size, an elderly carpenter be hired "for the benefit and improvement of the craft of ship carpenters," so that he could "give to the others the means to learn this craft."[39] Administrators well into the eighteenth century would continue to rely not upon books, but on the knowledge embodied in living craftspeople whenever they sought the actual production of things.[40]

Thinking through Writing

A large portion of Michael's book is taken up with calculations, most of them eminently practical problems familiar from merchants' textbooks for calculating partnership shares, bartering, or freightage, as well as the rule of three, a very ancient commercial technique based on proportions that allowed calculation of an unknown variable. This technique was already employed before the common era in Egypt and Mesopotamia, but first written down in Sanskrit in the seventh century by the Indian scholar Brahmagupta (598–668 CE). While Michael's copying out of these problems in his book may appear today as counterfeit authorship, the historian Raffaella Franci has revealed that these long mathematical sections are a kind of experimentation with the problems. Michael solves each of them in three ways: first by the rule of three, then by the rule of double false position (another ancient technique first written down in India for solving problems with unknown variables), and, finally, by means of algebra (what Michael called the "rules of the unknown").[41] Writing makes possible the ordering of information in a variety of ways and allows for comparison, thus providing a form of experimentation. Putting something into writing in some cases might have the effect of formalizing a dynamic oral tradition, but it could also allow for experimentation in a way that one scholar has characterized as Michael "thinking out loud."[42] Indeed, the recipe-like nature of many practical books—listing various methods and combinations as "another way," followed by yet "another way"—positively encourages the experimentation with materials and comparison of techniques (more on this in chapter 6, below).[43]

Examples of how Michael's mathematical experimentation allowed him to solve practical problems appear throughout his book. In 1431, after a battle against the Genoese, a violent storm arose, and the Venetian fleet was separated. In 1437 and 1439, Michael writes that he voyaged in one of the four ships of a convoy escorting the Byzantine emperor and his party of seven hundreds from Constantinople to Venice and back, and, as violent storms separated the ships of the convoy, all on board feared shipwreck.[44] In these and similar cases, how could they determine where they were? And how could they reach the other ships of the fleet? Michael used a technique to mentally calculate how far a vessel had been blown off course, using a table of numbers (*raxon del marteloio*) and a computational practice called dead reckoning by English mariners and "mental navigation [*navichar a mente*]" by Michael.[45] By these means, internalized through his endless practice in computation, Michael could master the contingencies of sea travel. He also devoted much computational practice to calculating the lunar and solar years and the date of Easter, and this clearly represented to him a means to control "the flow of time, as one might master the flow of money, or the flow of the ocean currents, to secure human safety and advantage."[46] Mastery of mathematics, like mastery of other techniques, was attained by constant practice, by doing all the problems in three different ways.

The mathematics in Michael's book—as in most other apparently instructional mathematics texts—was devoted only to problems and examples, rather than theorems or proofs, and this might be taken as a sure sign that we are dealing with the "how to" of practical knowledge rather than the "why" of science. But we can see in Michael's case that such problems were a way to gain practice in *thinking through* the forces of nature and the value of materials until the practice was internalized. This practiced "thinking and working through" allowed a higher order, intuitive response to tides and winds, or to the fluctuations in commodity prices. Such carefully computed problems, done three ways, and then copied into his book, are a demonstration of how to learn to improvise. Much of Michael's book, then, is a demonstration of the training of intuition, an extremely powerful combination of practice and thought that we might consider an embodied equivalent to the functioning of generalization in logical induction. Michael was thus not only presenting the content of his "doing" in his book, he was also modeling his cognitive journey to expertise. For Michael, intuition was rooted in his practical experience on board ships: "These are the points of the stars, by which are made the fortunes of the sea and the wind and the rain, and in the same way of great calm and great heat, and in addition you must always watch out not for the appropriate things but for the contrary things that can arise, so that if you want to navigate, always be prepared for every contrary thing."[47] Hours of practice made Michael able to respond to the unknown—"every contrary thing." This ability to respond to changeable and uncertain particulars, and to solve unknown variables, could be construed as a form of knowledge analogous to Greek conceptions of certain, demonstrable knowledge that is the result of generalizing from particulars. In the late sixteenth century, scholars would turn to algebra as, to use their phrase, an "art of thinking about thinking."[48] In this way, Michael, too, was thinking about cognition.

Experience and Judgment

In a text compiling construction rules, a master mason repeatedly emphasized that his written rules were not to be relied upon exclusively, for a mason must be able to exercise judgment based on his accumulated experience: "Give to this writing careful attention, just as I have written it for you. However, it is not written in such a way that you should follow it in all things. For [in] whatever seems to you that it can be better, then it is better, according to your own good thinking."[49] Because practical texts were often written as lists of rules and specific problems, they have been seen as prescriptive for particular cases rather than as descriptive of general methods. If we see these books of practice as intended to replicate the ways in which general methods were taught in apprenticeship, however, they gain new significance. Take, for example, the master mason Matthäus Roriczer (1440–93) (fig. 1.8). In his 1486 *Büchlein von der Fialen Gerechtigkeit* (Little Book of the Correctness of Pinnacles), he appears to be simply trying to provide a design for a well-formed pinnacle

1.8. Hans Holbein the Elder
(c. 1460–c. 1524), portrait of
Matthäus Roriczer, master mason
of the Regensburg Cathedral,
1490–95. Silver point on gray
prepared paper, 12.4 × 9.4 cm.
Kupferstichkabinett, KdZ 5008,
Staatliche Museen, Berlin. © bpk
Bildagentur / Art Resource NY.
Photo: Jörg P. Anders.

at the peak of a spire. In actuality, this piece of technical writing and its illustrations work through an exercise in deriving the elevation of a pinnacle from a ground plan, the most important "secret" of masons' lodges throughout Europe.[50] Maarten Prak has argued that the modular forms of templates and the prefabricated components of large-scale buildings in the Middle Ages and early modern period had a higher order, generalizing function, as would be found in general theorems, but with modularity substituted for theory.[51] Templates like Roriczer's, common to building construction sites and artists' model books back into the distant past, then, were not "designs" in our commonly understood sense today, but could be a means to represent the process of coming to expertise. These processes of an apprenticeship, in which examples were worked through, skills were modeled, and techniques and knowledge were passed on, resulted in an ability to generalize more broadly on the basis of practice and experience in a variety of circumstances.

But, of course, the written text could not substitute for actual practice. Vannoccio Biringuccio (1480–1539), metalworker and author of the practical manual of metalworking, *De la pirotechnia*, made clear that physical demonstration was more important than detailed written accounts in developing this ability (what he called "judgment"): "because the light of judgment cannot come without practice, which is the preceptress of the arts, I shall pass through this [process of melting and alloying metals] briefly with the idea of one day being able to supplement this further by demonstrating it to you."[52]

Authorship and Collective Knowledge

In 1444, Michael of Rhodes wrote a second book, an abbreviation and compilation of his 1434 manuscript with some additional material. At his death in 1445, both books passed into the possession of another mariner, who apparently carried the books with him on his voyages. This man carefully erased Michael's name on the second book, substituting his own.[53] Should we view this man as a plagiarist and intellectual thief? In reality, he was not plagiarizing. Rather, Michael's book had in fact become the book of its inheritor. It was never a book *by* Michael of Rhodes in our sense. It was, as Michael calls it, the Book *of* Michael of Rhodes.[54] It contained a collectively gained shared resource of maritime knowledge, shaped to Michael's own needs and laid out according to Michael's best efforts at demonstrating his techniques and his journey to expert knowledge. As has been noted, a book was not an optimal means of transmitting embodied knowledge. That was done better by the collective working conditions of the workshop and the course of "doing" and "making" in apprenticeship. Much practical writing expresses this collective creation of knowledge. Vannoccio Biringuccio, for example, viewed his book as a way to "see" the methods of others; it could be helpful in ways similar to traveling to other workshops and to watching other masters (as occurred in the journeyman phase of apprenticeship, known in German as *Wanderschaft*): "seeing the methods that others use is a wide gateway toward proceeding confidently by other paths in order to arrive at desired ends."[55] And he reinforced at the same time the inadequacy of learning by the book: "even though to advise you better might require that I should have told you more of these things than I have, you will learn many things by yourself while working and practicing it, such as how to make a choice of clays, stone, moulds, furnaces, seasons, weather, and the like, and it would take too long if I should wish to tell you all of them."[56]

The concepts of individual invention and creation thus imperfectly suit these varied texts of practice, for they are the voice of a collectivity. The same holds true for other early modern practical texts that we now think of as "literature," such as the plays of Shakespeare (1564–1616), whose authorship has become so important to Anglophonic self-understanding and identity: Shakespeare was a master compiler who hardly invented a plot. The practical activity of his plays is better understood in terms of a plurality of producers. As David Kastan has written:

Authorship is important to us, heirs of a romantic conception of writing as individual and originary, and if it was indeed important to some of Shakespeare's contemporaries, it was not particularly important to Shakespeare himself or to the publishers who first brought his plays to the reading public. They did not see their task as the preservation of the work of the nation's greatest writer as they set forth his plays; they were seeking only some small profit with limited financial vulnerability, as with their six-penny pamphlets they turned Shakespeare into a "man of print" and made his plays available to desiring readers.[57]

Indeed, "authorship" in Europe included much compilation and incorporation for many centuries. Roger Chartier and other historians of the book have shown how the idea of the author as the central component in the production of a text emerged fully in the eighteenth century out of a system that viewed the author as one of several cogs in a mechanism that included the dedicatee (often patron), the copyist, the printer (sometimes coordinating compilation), and the seller of the work.[58] It was only in the late fourteenth and early fifteenth centuries that the French terms *escrire* (to write) and *escripvain* (writer) haltingly came to mean not only the act of copying but also the composition of texts.[59] Even after 1400, then, authoring a text could be understood within several different frameworks, of which compilation continued to be a central one.[60]

The overlapping of categories of authorship can be seen in the history of the *Problemata* of Aristotle, a compilation of questions on natural subjects. This text called upon a composite and collective authority, despite its nominal attribution to the *auctor* Aristotle, right up until the last editions of the eighteenth century. At that point, the idea of the individual authorship of Aristotle comes to predominate. As Ann Blair observes,

whereas the *"Omnes homines"* [as the *Problemata* was also known] relied on a conception of the author as a collective of timeless, faceless (although not always nameless) authorities, the eighteenth-century preface of the *Problems of Aristotle* claims a direct, authenticated authorship for a text (and a set of works) that certainly could not meet the criteria for authenticity in learned circles. The extra prefatory boasting has become useful, it would seem, to counter rival projects and to sell a traditional text in an environment in which access to collective wisdom had become less attractive to readers than the promise of having before them what the most famous Aristotle had genuinely written himself.[61]

As these and other book historians point out, authors (as *auctores*, or authorities) are made by readers, not necessarily by writers.[62]

Recipe texts are another example of collective knowledge, but in the seventeenth century, when institutions, such as the Royal Society and the Académie royale des sciences, attempted to write down "histories of trades," in order to document craft practices, they often attributed recipes or processes to individual authors, which sometimes entailed crediting a rather generic individual, such as a "director of an indigo manufactory for many years," or a "merchant potter of Chartres," or just a "master carpenter."[63] Because identifiable authorship had by the late seventeenth century come more fully to imply authority, recipes needed an individual author.

Sanctified Labor

Michael's contemporary, the Paduan painter Cennino d'Andrea Cennini, also wrote a "book of the art," *Il Libro dell'arte*, that records painters' techniques current in his time, setting out a complete course of an apprenticeship, from picking up the chicken bones under the table and charring them for charcoal through pigment-making, panel painting, gilding, frescoing, and casting. In doing so, as he tells us, he seeks to establish that the art of painting requires both *"scienza* [science]" and *"operazione del mano* [work of the hand]." Cennino nowhere defines *scienza* very clearly, but for him it has two important characteristics: first, it is a superior form of activity and it lends this superior status to whoever articulates it; and second, it involves the work of imagination and intellect.

Cennino and Michael thus both put pen to paper within a hierarchy of knowledge that placed the written word and *scientia* higher than practice and embodied knowledge. Their movement from practicing to writing, like those of other artisan authors, was partly about articulating their identity, their skills, their journey to expertise, and their particular way of knowing. Like the practitioners illustrated above who included self-portraits in the body of their artworks, these individuals began to take the measure of themselves and sought to make their practical knowledge known and valued, and to articulate it in a form recognizable as "knowledge" within their culture.[64] Yet, just as Michael's book can be read as an epistemic tool of experimentation and a text that both models and demonstrates how to learn to improvise and intuit, so Cennino's book gives further insight into a crucial aspect of this form of knowledge.

Cennino's lived experience comes across vividly in the opening lines of his book, in which he declares his work one of religious piety and common good. Cennino begins and ends his book with a prayer to "God All-Highest, Our Lady, Saint John, Saint Luke, the Evangelist and painter, Saint Eustace, Saint Francis, and Saint Anthony of Padua," asking that the students of his book "study well and […] retain it well, so that by their labors they may live in peace and keep their families in this world, through grace, and at the end, on high, through glory, *per infinita secula seculorum.*"[65] This twin aim of religious piety—indeed redemption—and artisanal livelihood carries through the entire text. Earning an honest livelihood through labor was both a socially and spiritually sanctioned activity. Cennino's piety pervades the manuscript, praising those who become painters out of enthusiasm and desire to praise God, stressing the costume of "Enthusiasm, Reverence, Obedience, and Constancy" that the painter must don when he begins his apprenticeship.[66]

Cennino advocates the use of fine gold and good colors because they will make the painter's reputation, but he should not be concerned if he is not paid well, for "God and Our Lady will reward you for it, body and soul."[67] On beginning every panel, a painter should invoke "the name of the Most Holy Trinity […] and that of the Glorious Virgin Mary."[68] And when the painter begins the extremely delicate task of drawing with a sharp-pointed needle on gilded glass, being careful not to make a single mistake, he must begin "with the name of God."[69]

Michael's book also begins with an invocation "[i]n the name of God and of the Blessed Virgin Mary and of the Evangelist Saint Mark, our protector and governor," and the prayer to St. Sebastian immediately following this seems to be just as Michael recited it: "Saint Sebastian, your faith is great, intercede for me Michael, a miserable sinner to Lord Jesus Christ, and may I deserve to be freed from plague, epidemic and illness by your prayers. Pray for me St. Sebastian as you deem worthy to carry out the promises of Christ." Michael invoked God and the saints at important moments, such as when he "received the steelyard by special grant from our Signoria on January 28 of 144[5]," and the numerous prayers in his text foresaw every conceivable precarity: "St. Sebastian, and for fever, and for fear of serpents, and for a woman who can't give birth, and when you catch no fish, and to staunch a nosebleed, and when you are bitten by a venomous snake. And so you will not confess under torture," and "for the protection from drowning and for protection from bodily harm in battle." For extra protection, he carried a large St. Christopher image on his travels (fig. 1.9). In common with numerous other artisanal actions and texts, Michael began each page with a cross, the sign of Jesus.[70]

Cennino's book progresses from the prayer to an abbreviated recounting of Genesis, moving immediately to the story of the Fall. As he says, "[i]nasmuch as you have disobeyed the command which God gave you; by your struggles and exertions you shall carry on your lives."[71] The Fall of humankind imposed a necessity of laboring for salvation, thus Cennino is expressing more than formulaic piety: Cennino's work is itself an act of religious devotion and redemption through labor. We see this in an altarpiece from the same period in Wurzach, on which Hans Multscher (1400–1467) signed the work: "Intercede with God for Hans Multscher von Reichenofen, citizen of Ulm, who made this work in the year numbered 1437."[72] Numerous other works of art reveal that the maker viewed bodily labor as an act of devotion and a work of individual salvation.[73] Christ's incarnation as a humble craftsman and his bodily sacrifice provided an explicit model and valorization of manual labor for some of these craftspeople, particularly in the fifteenth and early sixteenth centuries, leading up to and through the early Protestant Reformation.

This spiritual dimension of his craft is also expressed in straightforward technical instructions. For example, Cennino describes the care with which the laying-in of the flesh tones of living individuals should be done, by making gradations of flesh color with vermilion and lead white, while the flesh of the dead, as detailed in his chapter on "How to paint a

1.9. St. Christopher carrying the Christ child, from *The Book of Michael of Rhodes* (Long, McGee, and Stahl 2009, vol. 1, fol. 202a). Image reprinted courtesy of MIT Press.

dead man," should contain no vermilion because "a dead person has no colour."[74] Cennino called the flesh color "*incarnazioni*," and described its use as akin to the incarnation of life in a body.[75] This giving life to (or "incarnating") an image for Cennino constituted an everyday artisanal technique by which the abstract principle and profound miracle of incarnating life in a body could be represented. Cennino thus proclaimed, through this nuts-and-bolts recipe, the transformative power of art and the artisan. His simultaneous material and spiritual understanding of the production of pigments, expressed in this recipe, illustrates one kind of "theory" that underlay artisanal practices—a lived and practiced "theory," rather than a written and abstracted one.[76]

Cennino's and Michael's piety was a commonplace component of their culture, self-evident at the time, but it is also possible to understand it in epistemic terms, as one piece of an underlying set of principles that gave meaning to and ordered the world and its materials; it was knowledge that explained "why," yet it was contained in practices, written down as compilations of recipes and techniques. We shall return to Cennino's (and other artisans') "theory" in the next two chapters, for it was not only artisanal piety that lay behind his apparently straightforward recipes, but also a system of vernacular knowledge. Cennino's recipes are thus deceptively simple: they provide a set of instructions to be followed by an aspiring painter, and, at the same time, they reveal an explanatory framework on the basis of which this practice was carried out.

This brief survey of the books of Michael of Rhodes and Cennino indicates that their effort to move from shipboard and workshop experience to the written word was more than a bid for higher intellectual and social status in authorship. In writing books, they chose a format that was recognizable as "knowing" to convey the foundations of their "doing" and "making." Their texts, which have been taken to be how-to instructions and recipes, in fact reveal a more general system of knowledge that underpinned their practice, and modeled the attitudes and actions that formed the path to their "knowing." As Michael enumerated those actions: pay attention, watch out, "be prepared for every contrary thing," and practice over and over again.

CHAPTER 2

The Metalworker's Philosophy

Butter

When the heirs of the Augsburg merchant, Christoph Fugger, commissioned a gilded bronze altar in the Dominican church of St. Magdalena in 1581 (fig. 2.1), they could have had no idea that it would take three long and troubled years to complete. An account book records the material side of this process—the artisans paid, the raw materials bought, and a number of surprising entries for the purchase of butter.[1] This chapter investigates what metalworkers did with all those pounds of butter. Many foodstuffs were part of making processes, like the soft bread used to erase drawings, the crushed garlic employed in mordant gilding, or the fishbones and fig juice used as glue. Provisions for workers' nourishment, such as beer, were also recorded in many contracts. Were these metalworkers eating the butter, or were they using it in making the sculpture? The contract notes that the butter was purchased "for the gilders against the evil smoke."[2] What might this signify?

In this chapter I survey many instances of butter in workshop accounts and technical recipes, with the aim of understanding how a connection between butter and noxious fumes could have emerged out of the metalworking shop. I delineate a "material imaginary," in order to reconstruct the worldview in which materials generated meaning through their use. I seek to understand how artisanal practices and techniques were informed by a deeper systemic understanding of materials. This chapter shows that, in early modern workshops, "making" with natural materials could also be about investigating and "knowing" the fundamental properties and behavior of natural materials. Well into the nineteenth century, artisans remained the preeminent experts on the processes of nature and the manipulation of natural materials; however, even sympathetic historians and theorists of craft have downplayed the view that such making can be knowing (or "theorizing" or "philosophizing"), or that it can be regarded as a knowledge system. The foremost editor of metalworking treatises in the twentieth century and a great admirer of craft expertise, Cyril Stanley Smith, wrote about early modern metalworkers: "the artisans were the true scientists of this period," but "they lacked the flash of genius to produce a consistent

2.1. Hubert Gerhard (sculptor), Carlo di Cesari del Palagio (b. 1540, caster), and Johann Müller (gilder), *Relief of the Resurrection*, central panel of an altarpiece, Augsburg, 1581–84. Gilt bronze, H 97.4 cm, W 63 cm, weight 103 kg. Victoria and Albert Museum, A.20–1964. Bequeathed by Mr George Weldon. © Victoria and Albert Museum, London.

theoretical framework."[3] Theorist of craft Peter Dormer says, "Craft and theory are oil and water."[4] In contrast, I show in this chapter that workshop practices were underpinned by a system of knowledge that comprised a flexible and accommodating—yet coherent—body of materialized principles about the behavior of natural materials, and I argue that, by virtue of their productive aims, artisans pursued this knowledge in an empirical and systematic way.

Historians and anthropologists have discussed practical knowledge as a form of "indigenous knowledge," "material intelligence," and "savoir prolétaire."[5] In previous work, I called it a "vernacular science of matter," but I came to see this term as potentially misleading because, although it made clear that craft could be a form of knowledge, it could also create the anachronistic impression that practical knowledge was a proto-science in the modern mold: a form of knowledge that broadly holds to skepticism, seeks to falsify, aims for clear definitions and expression *in words* (or numbers), and largely seeks to reduce the subject at hand to its simplest components in order to perform analysis. As we shall see, these are generally not the characteristics of vernacular philosophizing. I use "material imaginary" here to mean a system of knowledge that provides flexible parameters within which the exploration of material properties and behavior is undertaken, thus both informing and giving meaning to practices. This chapter focuses in particular on the material imaginary of metalworkers, which wove together metals, planets, health, and body—a particular instance of a widely shared view of the world in early modern Europe. The workshop use of butter provides a path into that world.

Evil Smoke

The purchase of butter "for the gilders against the evil smoke" in the Fugger account book is not the only place we find recommendations about butter and metalworking. In numerous vernacular texts of practice from the twelfth to the seventeenth century, we find evidence of metalworkers trying to protect themselves from the harmful nature of the "smoke," or vapor, of metals. The effects of the fumes inhaled by mercury amalgam gilders, who mixed mercury and gold, were well known, and the sixteenth-century goldsmith and sculptor Benvenuto Cellini observed that men involved in amalgam gilding lived but a very few years.[6] Already in the twelfth century, the metalworker Theophilus—possibly the goldsmith Roger of Helmarshausen (fl. early twelfth century)—cautioned, "Be very careful that you do not mill or apply gilding when you are hungry, because the fumes of mercury are very dangerous to an empty stomach and give rise to various sicknesses against which you must use zedoary [a medicinal rhizome] and bayberry, pepper and garlic and wine."[7] In the thirteenth century, cleric Albertus Magnus studied the practices of miners and smelters in order to flesh out Aristotle's brief account of minerals, noting that the odors and vapors of metals "are very injurious to the chest. Evidence of this is that when the miners go into the mines they cover their mouths and noses with two or three layers of cloth so that their breathing may not be too much injured by the vapour—for this is where the greatest damage is done."[8]

To protect against this harmful smoke, a variety of texts recommended the use of butter. The late sixteenth-century anonymous French compilation of craft practices, Ms. Fr. 640, includes an "antidote against the smoke of metals" that advises the practitioner: "In the morning, take a piece of thin toast with butter, neither antimony nor any other vapors will harm you. Or put half a pig's bladder in front of your face."[9] Another entry advises, "And before working at it, take in the morning good buttered toast, and hold the said butter, or zedoary, or gold coins, in your mouth, and cover your face with a cloth from the eyes down."[10] In a detailed seventeenth-century set of instructions for making the pigment vermilion (in which sulfur and mercury are heated together over a hot fire), the worker is advised to "eat a thick piece of bread and butter" before "roasting the cinnabar," to produce the red powder.[11] A medical manual from about the same period recommended a poultice made from butter to be used against viper bites and other "fiery poisonings," warning that butter should be used more as a medicine than a food, for in excess it could exacerbate phlegmatic illnesses.[12] The medical and religious reformer Paracelsus (1493–1541) recommended cream or butter to heal the diseases of miners, and set out instructions for protecting and healing those "workers who are set on fire by such *spiritus*" of the metals. His remedy involved combining herbs ground while green in a one-to-one ratio with cream, and boiling the mixture for one hour in the *balneum maris* (a type of double boiler). The mixture was then to be consumed in a sober condition. Only by following such a procedure, he asserted, could "the mineral diseases" be overcome.[13]

All these examples point to the use of butter as a widely used remedy against the "evil smoke" of minerals and metals. What was the "philosophy," or material imaginary, that underpinned these practices? We get a clue to the answer when Biringuccio notes that mercury miners often end up weakened and paralyzed because the nature or essence of mercury is "cold."[14] What did Biringuccio mean?

A Hierarchy of Metals

Biringuccio and other authors of metalworking manuals viewed the characteristics of each ore as determined by the particular influence rained down upon the ore body from its corresponding planet. Metalworking texts linked gold to the sun, silver to the moon, tin to Jupiter, copper to Venus, iron to Mars, lead to Saturn, and quicksilver to Mercury. The seven planets and seven metals were also sometimes related to the seven days of the week.[15] The hierarchy of the metals mimicked that of the planets: the noblest metal, gold, sat at the apex, its power imbued by the action of the sun, making it the purest and least corruptible of metals.[16] The sun and gold were also both frequently identified with Christ.

Metals were formed by the mixing of the relevant planet's emanations with exhalations

2.2. Prospecting for ore, with a divining rod, a forked hazel branch. From Georgius Agricola, *De re metallica: libri XII; quibus officia, instrumenta, machinae, ac omnia denique ad metallicam spectantia […] describuntur et per effigies […] ob oculos ponuntur* (Basil: Froben, 1561), book II, p. 28. Agricola was skeptical of the efficacy of the divining rod, stating that the prospector would do better to observe natural signs, such as plant growth. Bayerische Staatsbibliothek, Munich, Res/BHS II B 5#Beibd.1. http://mdz-nbn-resolving.de/urn:nbn:de:bvb:12-bsb10199044–2. CC NC 1.0.

of elemental sulfur and mercury rising out of the ground.[17] These exhalations revealed themselves in the fires, vapors, and miasmas (called *Witterung* in German) that billowed out from the earth around ore deposits.[18] Such gusts rising out of the earth could account for the action of the divining or dowsing rod, which some viewed as dipping and trembling when passed in the proximity of ores (fig. 2.2).[19]

Since the Middle Ages, scholars and alchemists had seen sulfur and mercury not only as physically occurring metals, but also as "principles" that were contained in all metals, giving them their distinctive qualities. The principle of sulfur was regarded as hot, fiery, and dry, which was connected to its characteristic form as a solid powder as well as to its combustibility. The principle of mercury resembled the liquidity of natural quicksilver, and it gave metals a quality characterized as wet, cold, and slimy. Because all substances recognized as metals at that time had liquid ("mercurial") and solid ("sulfuric") states, they were seen to incorporate both these principles, although their different melting points and combustibility showed that they possessed these qualities in varying degrees. This was explained in terms of the different degrees of perfection of the sulfur or mercury they contained. For example, the first practical text on mining noted that "gold [. . .] is made from the very finest sulphur—so thoroughly purified and refined in the earth through the influence of heaven, especially the Sun, that no fattiness is retained in it that might be consumed or burnt by fire, nor any volatile, watery moisture that might be vaporized by fire—and from the most persistent [fixed] quicksilver, so perfectly refined that the pure sulphur is not impeded in its influence on it and can thus penetrate and color it from the outside to its very core."[20] As for silver ores, they "are made through the influence of the Moon from clear quicksilver and expurgated strong sulphur, the Moon representing the power of the maker, and quicksilver and sulphur [forming the] matter."[21] Lead and quicksilver (i.e., elemental mercury) also contained mercury and sulfur, but the mercury (principle) in lead was unrefined, watery, heavy, and dirty, and the sulfur was boiling and burning. Quicksilver was composed of muddy, watery moisture mixed with the very finest sulfurous earth.[22]

The hierarchy of the planets and metals was thus based in part on material characteristics of the metals. The gradations in the material hierarchy also underlay and reflected a social order in which artisans who worked gold stood at the apex, while ironsmiths were near the bottom, and workers in lead even lower.[23] Artisans who worked and trafficked in noble metals were generally regarded as ennobled by their contact with these materials. Theophilus divided his ideal workshop into two spaces by "a wall rising to the top," one area for working copper, tin, and lead, sealed off from the other, for gold and silver.[24] This was a necessary material boundary as well as a social one, for the goldsmith had to guard against the noble metals becoming mingled in his shop with the ignoble ones, as the work of separating them was laborious. The Florentine shoemaker-turned-scholar Benedetto Varchi (1503–65) believed that the artisan who worked with noble metals by definition must be more refined than other metalworkers. Similarly, in the 1590s, the head of the goldsmiths' association in Augsburg told city council members that goldsmiths did not deal with just any material, but rather that theirs was a "craft that as one might say, did not have to do with *hobl spännen* [wood shavings, i.e., something of little value], but with gold and silver."[25] The metalworker Biringuccio, clearly admiring ironworkers, cleaves to the familiar hierarchy, although perhaps he also means to subvert it:

if it [working iron] were not an activity so laborious and without any delicacy, I would say that it is one greatly to be lauded, for when I consider that the masters make their works without moulds or pattern, letting only the eye and good judgment suffice for it, and that they make it exact and of good shape by hammering alone, it seems to me a great thing [. . .] [I]n my opinion, if it were not for the nobility of the material, I would say that the smith working in iron should justly take precedence over the goldsmith because of the great benefit that he brings.[26]

Influence of the Heavens

This natural and social hierarchy of metals had its pinnacle in the deity, who created all things to contain a particular gift or power.[27] This sounds a commonplace, but we have yet to fully appreciate the pervasiveness of the view that the heavens influenced the earth: it was reinforced by religion, scholarship, and every conceivable practice. Michael of Rhodes, introduced in chapter 1, oriented his life and practice according to his view that the heavens influence earthly affairs. This was both a natural and a religious view: Michael included in his book an entire section on the influences of the planets and the signs of the zodiac, laying out practices that should be engaged at different times. For example, under the heading "*A veder raxion dei pianetti* [See the reckoning of the planets]," he writes of how "the sun rules the first hour of Sunday, Venus the second, Mercury the third, the moon the 4th, Saturn the 5th, Jupiter the 6th, Mars the 7th," and so on up to the twenty-fourth hour, and the "planets are represented by the seven days of the week. Sunday is represented by the Sun, Monday is represented by the Moon, Tuesday is represented by Mars, Wednesday by Mercury, Thursday by Jupiter, Friday by Venus, Saturday is represented by Saturn." "And if you wish to undertake a journey," Michael continues, "begin it on the day that you choose according to its planet and continue according to the heavens, that is, Saturn, Jupiter, Mars, the Sun, Venus, Mercury, and the Moon."[28] Like his contemporaries, Michael saw sacred history interpenetrating daily life: "Here you shall see odious and perilous days to begin anything, so watch out for them, and this came out of the very mouth of St. Jerome and you should observe it: The first Monday of April, because on this day Cain killed his brother Abel, and this was the first blood shed in the world."[29] The movements of the celestial bodies directed the seasons and the cycles of life and death, and their influences permeated the human body. Human life thus had to be oriented by observing the celestial bodies and the periods they marked out.[30]

From mid-sixteenth-century handbooks of mining and metalworking, we know that on establishing a mine, the shareholders and miners first had the site baptized by a priest "in the name of God and a fortunate outcome."[31] Similarly, when potters loaded a kiln, they offered prayers "to God with the whole heart, ever thanking Him for all that He gives us," in addition, keeping "an eye however to the state of the moon [...] [for] if the firing happens to take place at the waning of the moon, the fire lacks brightness in the same manner as the moon its splendour."[32] Bronze casters invoked the name of God when they lit the fire to melt the metal. This cosmic connection was also manifested in the practice of miners praying in the chapel before going down the shafts—a practice that continued into the twentieth century[33] (figs. 2.3 and 2.4)—as well as marking the entrances to mine shafts with a cross (fig. 2.5). Such practices all express a mental world in which the divine heavenly influence (as a 1505 *Bergbuch*, or mining book, puts it, "heaven in its course, emanations, and influence")[34] is felt intimately on earth. The course of the planets through

2.3. *Ain Schicht* (The Start of the
Shift), from *Speculum metallorum*
(1575), fol. 117r. Miners, identifiable by their distinctive hoods,
emerge from a chapel before
entering two shafts, pictured in
the lower half of the illustration.
Stadtarchiv Calw.

2.4. Prayer room in the David
Richt shaft in the Freiberg coal
fields, 1911. Stadt- und Bergbau-
museum, Freiberg, Inv.-Nr. D542.
Photo: Karl Reymann. © Stadt-
und Bergbaumuseum Freiberg.

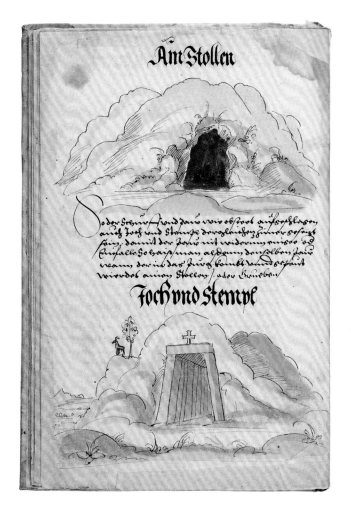

2.5. *Ain Stollen* (an adit, or shaft entrance) and *Joch und Stempl* (a reinforced shaft), from *Schwazer Bergbuch*, Bochumer Entwurfs-exemplar (1554), p. 16. From the 1554 draft of the *Schwazer Bergbuch*, this illustration shows the shaft opening before and after being shored up. Reinforcing the mine shaft included making a cross over the entrance. Montan-historisches Dokumentations-zentrum beim Deutschen Bergbau-Museum, Bochum, 040003313001. © Deutsches Bergbau-Museum.

the zodiac, the waxing and waning of the moon, and the cycle of the seasons determined all growing things, including metal ores.

Artisans' practices incorporated the connection between heaven and earth, and their objects sometimes embodied this link between life on earth and celestial events. The connection of divine, natural, and human order appears in *Handsteine*, works created out of impressive ore specimens, called *Stufen* in German. These specimens were collected by miners and shaped into works of art for noble patrons—especially the Habsburgs—beginning in the sixteenth century.[35] Naturally occurring mineral and metal formations, often including pieces of native silver, extracted especially from Tyrolean and Saxon mines, were worked into objects that combined the artifice of nature with the art of the human hand (fig. 2.6). Dendritic silver formations, resembling plant growth, formed striking material evidence for the conception of metals as growing and ripening over time in the mines (figs. 2.7 and 2.8).

2.6. Caspar Ulich (fl. 1555–76), *Handstein*, St. Joachimsthal, third quarter sixteenth century. Argentite, silver, gilding, minerals, enamel, and glass, 27.4 cm × 8.5 cm. Kunstkammer, KK 4149, Kunsthistorisches Museum, Vienna. The piece of silver ore is shaped into a Crucifixion scene set over a mine site, with shaft entrances and miners at work. The inscription reads "DAS BLVT IHESV CHRISTI REINIGET VNS VON ALLEN SVNDEN [The blood of Jesus Christ purifies us of all sins]." © KHM-Museumsverband.

2.7. Native silver, Freiberg District, Erzgebirge, Saxony, Germany, in Houston Museum of Natural Science. Native silver formations were prized by collectors and *Handstein* makers for their branching shapes, giving striking visual evidence for the conception of metals growing and ripening over time in the mines. CC BY-SA 2.5.

2.8. Caspar Ulich (fl. 1555–76), *Handstein*, St. Joachimsthal, third quarter sixteenth century. Gilded silver, minerals, enamel, silver ore, 22.1 × 10 cm. Kunstkammer, KK 4143, Kunsthistorisches Museum, Vienna. Dendritic silver is employed at the base of this scene of Jesus in the Garden of Gethsemane to mimic plant growth. © KHM-Museumsverband.

Biringuccio described mineral deposits as being like veins of blood or branches of a tree—they grow skyward, then in place of leaves and blossoms, they burst forth in colors at the surface, caused by the mineral vapors and fumes exiting the vein[36] (figs. 2.9 and 2.10). The natural formations found in the veins were regarded as nature's art and "gifts of God,"[37] which often did not need to be worked further than the state in which the powers emanating from the heavens had caused them to be formed. In some cases, the specimens were reshaped to make clearer the parallel between the divine order, in which God was creator of ores, and the human realm, in which miners and smelters were the creators of metals (fig. 2.11).

This understanding draws analogies between human and divine, and earthly and celestial phenomena, as the pastor of St. Joachimsthal, Johannes Mathesius (1504–65), wrote:

The most beautiful Handstein that in all my days I saw was a piece of silver ore [*glaß ertz*] worth several Marks in which the resurrection of the son of God, along with his grave and the guards, had been artfully carved. The ore [*gewechse*, implying a growing thing] was such that the body of the Lord came forth in white silver, the guards and grave were black like lead.[38]

Mathesius here explains how the earth produced a growing body ("*gewechse*"), shaped by human hands, that incarnated, or materialized, in noble white silver the nonmaterial divinity of the Lord and the black dross of the material world.

A *Bergkreye* (mining song) printed in 1530 used the analogy of a *Handstein* to formulate the connection between heaven and earth, and divine and human. Through its

2.9. Sulfur outcropping, Sierra Nevada, California, illustrating Biringuccio's description of the blossoming of minerals at the earth's surface. Author photo.

2.10. Minerals deposited by sulfur exhalations, Solfatara, Italy, illustrating Biringuccio's description of the colorful growth and bloom of minerals as their exhalations burst forth from the earth. Author photo.

2.11. (*facing, left*) *Handstein*, St. Joachimsthal or Tirol? c. 1550. Silver, gilding, wood, mineral specimens, including proustite, argentite, marcasite, lautite, malachite, quartz, and fluorite. H 27.1 cm. Kunstkammer, 4167, Kunsthistorisches Museum, Vienna. The *Handstein*, which is documented as having belonged to the *Kunstkammer* in Schloss Ambras of Archduke Ferdinand II (1529–95), an avid collector of such works, portrays huts covering mine shafts and machinery. Like the image in the *Speculum metallorum* (fig. 2.12), it expresses the analogy that the blood of Christ redeems and purifies the world of sin, just as miners and smelters redeem the noble metals from the dross of the ore body. © KHM-Museumsverband.

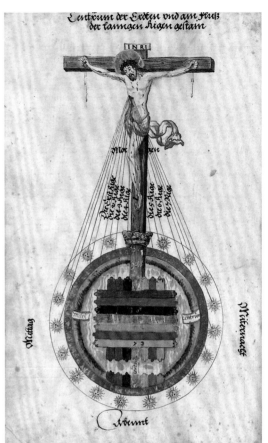

2.12. *Speculum metallorum* (1575), fol. 56v. The blood of Jesus on the Cross is shown falling on the planets, which in turn influence the formation of the ore bodies in the earth. Stadtarchiv Calw.

naturally growing proofs of divine celestial events and effects, the earth itself could make men pious:

Then I turn my heart and eyes alone to God
Who shows me *Handsteine*
native white silver ore
From whence my help will come
from God the lord
Who to make us pious, created heaven and earth.[39]

The view that nature displayed the power of God in material things is also found in the illustrations of a 1575 manuscript *Bergbuch* (mining book), the *Speculum metallorum* (figs. 2.12 and 2.13). This remarkable image represents the blood of Christ working through the influence of the seven planets on the globe of the world to form the seven metals at its center, which are represented by horizontal and vertical bands of different colors.

The influence of the heavenly bodies on earthly life was also articulated in the works of Paracelsus. Traveling throughout Europe and observing the diseases of miners, smelters, and metalworkers, he wrote vividly of the celestial influence on the bodies of humans and metal ores:

We are as loose and soft as a chick in its shell; for all the rays of the planets and of the limbus, which is the seed, enter into us and produce an essential action, for heaven and earth are the uterus and both are one thing. And man is the least thing and yet everything.[40]

Paracelsus also described the influence of the planets on the metals in a remarkable passage on the cycle of the seasons and the sequence of growth:

The seeds of the metals and the minerals have been sowed in the earth. They have their fall and their harvest in order to sprout sooner or later according to the arrangement of the godly order. In the same manner, roses then cherries, pears, nuts, grapes etc. so long until the year has passed and it was a year. Although one comes later than the other, thus here too, now gold blooms in a region, now silver, here iron, there lead etc., that is past, that is present and the other future, that one's spring is past, that one's May, the other's hay month etc. for those who were there at the beginning of the world, they have reached to gold and silver of the spring with the violets, their successors have taken the gold and silver with the clover and the crowfoot etc.

Metals grow according to the period of the year, just as plants do, but they grow over a much longer period, "for many thousands of years."[41]

2.13. *Speculum metallorum*
(1575), fol. 16r. A representation of
the metals in their ores, with Jesus
on the Cross identified with gold.
Stadtarchiv Calw.

Materials we now think of as inanimate were not regarded as such in this period. Mathesius suggested that mines and mining towns "had their time." He reported it as a matter of common experience that shafts grew smaller over time, as the minerals and earths matured within them. When miners struck bismuth, they believed they "came too early," before the ore was ripe. With this human disturbance, they had ruined their chances to find the precious ore.[42] Biringuccio speculated that antimony might be "a material that is about to reach metallic perfection, but is hindered from doing so by being mined too soon."[43]

These evocative written and visual expressions of the formation of metals and humans in the matrix of the material world were echoed by an English goldsmith two generations later. Discussing how stones are shaped by both climate and heavenly bodies, the goldsmith notes that "pretious stones are […] grown for the most parte in the east Countries, because the temperature of the Clymatt is there fitt to generate them […] and the influence of the heavens by continewance of tyme makes it a most pure stone." These stones "have goodly vertues and properties whereby we may with pleasure observe their consent and agreement with the heavens."[44]

Temper

In this cosmic relationship, the warmth and rays of the sun gave life to all things, just as they vivified gold. Like other growing things, minerals and metals, too, had to be purged and tempered to cure them and bring them into a healthy state. Minerals and humans alike received their "temper," or constituent mixture, from the movements of the heavens (fig. 2.14). This term, "temper"—a crucial one in early modern Europe—meant "to bring

2.14. Ulrich Rülein von Calw (attrib.), *Ein wolgeordnet und nutzlich Büchlein wie man berg-werck suchen […] soll* (Augsburg: Erhart Ratdolt, 1505). Orientation of the veins of ore, with the sun shining on them. École des Mines de Paris, 8° rés. 26, n.p. Licence ouverte Etalab.

into balance," often by mixing in other ingredients, and it was a common term in technical recipes. Health in humans depended on a balance (a "temperament," different for every individual) of the four humors: phlegm, black bile, yellow bile, and blood. Phlegm was cold and wet, black bile was cold and dry, yellow bile hot and dry, and blood, the noblest humor, was hot and moist. Each individual's unique combination of the four humors— their "complexion," or constitution—could be tempered by diet and other "nonnaturals," such as exercise, sleep, purging, and so on. As one vernacular cookbook put it: "Everything that grows from the earth and all humans take on the characteristics of the four elements, that is, hot, cold, dry and wet. He who wishes to live long will play one against the other, as the cold with hot and hot with cold, dry with wet and wet with dry."[45] Metals partook in this system, and their balance of qualities could be rectified by tempering just as human temperament was rectified by diet. We still refer to iron as being "tempered" when making steel from it. Paints, too, had to be tempered to achieve the right balance of pigment and media. Thus, the inclusion in recipe books of instructions for medicines alongside those for alloying, or mixing, of metals may reflect an underlying connection, for health in humans and in metals was based upon the same set of processes. The apparent jumble of recipes in collections may not always form a random assemblage.

The Metalworker's Butter

We return finally to butter. This view of cosmic order, in which the actions of the heavens were intimately involved in the life of human beings, provides a framework for the practice of eating butter to counteract the "evil smoke." Such smoke was conceived to be fumes given off by the metals as they grew in the earth and were heated by metalsmiths. The exhalations of the minerals rising up from the ground or billowing out from the smelting furnace (fig. 2.15) contained the soul or essence of the metal, and they had a power and subtlety that allowed them to penetrate very deeply into the bodies of miners and smelters.[46] As Paracelsus put it in his treatise on miners' diseases, the soul of the metals enters into the metalsmith's body like the "odor of a rose."[47] Hence the injunctions found in metalworking texts to prevent smoke from entering the body, to bind up the mouth and nose (fig. 2.16), and to work in the open.

A *Kunstbuch* (book of art), printed first in 1533 but probably older, and directed to the "common" gold- and silversmith, laid out the natures of metals and the diseases they cause. Like Biringuccio, this book specified that mercury, along with lead and silver, in which the principle of mercury predominates, is cold and damp in its complexion and effects. When these metal vapors are inhaled, they imbue the body with their essential coldness, causing harm to the chest, crippling the limbs, and destroying the internal organs. To counteract these effects, a person exposed to the vapors must consume foods with a warm essence. This *Kunstbuch* suggests that amalgam gilders, who work with a mercury mixture, keep a piece of musk with them, eating of it often, for the warmth of the musk counteracts the cold mercury vapor. Musk also helped to strengthen the noble organs of the heart and brain.[48] This book also recommends the more commonly available herb *"Elene Campana,"* called *Alantwurtzel* in German and scabwort in English, against the vapors; according to herbals, this herb was "hot and moist," and tempered the essential coldness of metals. Scabwort also eased metalworkers' coughs;[49] indeed it was still being used in the nineteenth century as a remedy for tuberculosis, despite a different paradigm of health at that time.

Driving out the vapors once they entered the body was achieved by sweating, bathing, and rubbing, and by eating foods to increase the body's heat. The herb wormwood (*Artemisia absinthium*), hot and dry in the second degree, could counteract the cold metal vapors, particularly if consumed after steeping in a light, dry wine (the astringent wine also working against the wet slimy nature of mercury and lead).[50] Garlic, classified as hot and dry,[51] also acted against cold metallic vapors. Employed for all kinds of poisoning,

2.15. Georgius Agricola, *De re metallica: libri XII* (1556), book 9, p. 341, iron smelter. Rare Book and Manuscript Library, Butler Library, Columbia University, New York.

2.16. Georgius Agricola, *De re metallica: libri XII* (1556), book 9, p. 341,
detail, showing the iron smelter's
face covering. Rare Book and
Manuscript Library, Butler Library,
Columbia University, New York.

crushed garlic was also bound to the soles of the feet to pull poison away from the heart
and neutralize it.[52] Another remedy called for drinking a mixture of Armenian bol (a type
of fine clay used especially in gilding), wine, and rosewater to temper the hot, sharp vapors
drawn into the body when working with saltpeter, sal ammoniac, or verdigris.[53]

An *Artzneybuch* (remedy book) connected the lung diseases of miners, smelters, and
alchemists with those suffered by people who live in deep valleys, because they are all
penetrated by cold, damp airs arising from the ores and metals. These people must not
go barefooted or bareheaded, and they should live in open, airy houses.[54] Their diet must
include hot, dry foods, including spices, raisins, figs, dates, and pine nuts, which were even
more effective if eaten with sugar sprinkled over them, as sugar was warm and moist. Even
more helpful was a roll moistened with subtle white wine; not a heavy, coarse roll but one
that tended toward sweetness, as the *Artzneybuch* specified.[55] Bread, as a food tempered by
leavening and baking, could bring the humors into balance, while the "subtle" white wine

heated and dried, counteracting the essence of the metals. The book also recommends the ingestion of lungs from animals, especially foxes, for their healing properties in lung diseases.[56] Indeed a mixture of fox lungs with various warming substances, including *Süßholtz, veyel wurtzel, alantwurtzel,* cinnamon, the blood of muscat grapes, fennel seeds, and ground sugar, was recommended to be taken every morning and evening by miners, smelters, and valley-dwellers.[57] A medical manual providing directions specifically to miners advises that before they go into the shafts, miners should eat a piece of bread with butter spread on it, and a little ground *"Alant wurtzel* or *Elena Campana* strewn over it. Always keep *Alantwurtzel* with you and chew it, and when you go out of the shafts, spit it out. It is good to hold in the mouth when you are in poisonous air and when the valley is covered in thick fog."[58]

Butter was particularly effective for metalworkers because it was classified as hot and wet—indeed, its heating qualities were such that it could even increase fever. The addition of hot, dry herbs, such as scabwort, could further amplify its heat. Honey, which was hot and dry, and sugar, hot and wet, were sometimes recommended to reinforce the effects of butter.[59] This classification of materials as wet, cold, dry, and hot rested on a long written tradition that had been slowly codified and systematized in Greek, Roman, and Arabic medical texts. All materials were made up of components—whether the Aristotelian elements earth, water, air, and fire, or alchemical sulfur and mercury—which were in a certain balance that could be subjected to human manipulation. Balance was the key to the metalworker's consumption of butter, and it formed the framework for understanding and intervening in natural materials, bodies, health, and much else in the material world.[60] But butter had an added advantage for counteracting the effects of working with metals: because the subtle metal vapors were drawn more rapidly into an empty stomach than into a full one,[61] butter had the effect of floating to the top of the stomach and sealing it off, like oil in a boiling pot.[62] This corking effect might be detrimental in a healthy person, but was just what the miner and smelter needed: a stomach filled up with the perfect temper of bread,[63] heated, and plugged by butter and a little scabwort.

Eating Butter as Practice and Theory

Butter and full-fat milk have continued to be used as a prophylactic against heavy metal poisoning today. Sculptors active today report that they were taught to drink milk before a brass pour to protect them from the zinc, which vaporizes very rapidly. Even more effective was to drink a bottle of milk, then throw the bottle into the molten zinc-containing metal; the bottle would explode with a startling bang and form a layer of molten glass over the metal that allegedly sealed in the fumes.[64] Such practices raise the question of whether butter has an effect observable by modern scientific methods. This question was investigated after World War II in order to establish whether milk subsidies should be paid to workers in lead industries. In the 1950s, research was undertaken on white rats, which both drink milk and show human-like symptoms of lead poisoning. It was found that milk did not protect from lead poisoning and appeared possibly to hasten poisoning.[65] An earlier study found that white mice fed on a diet that included 5 percent butter fat were more resistant to mercury poisoning than mice fed on a diet not including butter fat.[66] In mercuric chloride poisoning, standard treatment includes gastric lavage with milk.[67] Eating calcium-rich foods is advised by the Environmental Protection Agency to protect children from lead poisoning.[68] Perhaps this advice shows the usefulness of the early modern practice—or it may show the durability of the logics of early modern health practices even in the face of (admittedly inconclusive) scientific studies.

Conclusion

In early modern Europe, the metalworker's eating of butter allows us a glimpse of the pervasiveness and power of what we can call "the health worldview" as an organizing framework for thought and practice. People oriented themselves in relation to the natural world through the workings of their own bodies, and this understanding of nature was reinforced and given meaning by the celestial and social hierarchy lived out in everyday practices. This health worldview was not the sole preserve of physicians, but was, as we have seen, shared by craftspeople at this time.[69]

CHAPTER 3

Thinking with Lizards

The Workshop and Its Tools

Most handworkers, like the majority of people in early modern Europe, lived lives unrecorded by history. Guild and administrative documents give insight into their social world, and depictions of workshops can sometimes be useful,[1] but these do not give access to their intellectual outlook. The depictions of a goldsmith's workshop by the Parisian goldsmith Etienne Delaune (c. 1518–83/95) illustrate several important features that make artisanal knowledge systems difficult to access (figs. 3.1 and 3.2). Both workshops are dominated by the tools neatly arranged in every possible space. Tools codify techniques and knowledge of the trade, but require expert handling to display and give proof to that knowledge. The forge and fire form a core piece of the workshop, and are another tool that requires long experience and expertise. The diversity of ages, and presumably skill levels, among the men also stands out in these depictions. Both engravings include a graybeard in spectacles, three adults of various ages, and a boy learning the techniques of the trade, including wire drawing and pumping the bellows. Artisanal knowledge is largely embodied and often tacit, and apprentices learned their craft, not by reading texts—sometimes not by language at all—but rather by working alongside experienced masters, as the boy is doing in these engravings, learning by observation and repetitive bodily experience.

The men are absorbed in different techniques, of raising a silver vessel on a stake, chasing a charger, and performing various techniques at the fire. Such techniques, too, are repositories of knowledge for experts, as are the objects made by the workshop, sitting high on the shelf above the open window in figure 3.2. Objects produced by certain techniques, such as the metal vessels here, form repositories and records of these techniques. Objects are the tangible and enduring result of a culture—the residue of an enormous number of exchanges among individuals as well as of their belief systems, organized practices, networks, and accumulated knowledge. Objects can inscribe the memory of previous generations' innovations and cognitions. Their making requires significant expertise, which itself is the result of a "culture" that has multiple layers—of socialization within a craft, a network of workshops, and a pattern of consumption and production.[2] The prominent

3.1. (*facing, top*) Etienne Delaune, *A Goldsmith's Workshop*, c. 1576. Engraving, 80 × 120 mm. British Museum, London, PD 1951,1120.5. The walls of this goldsmith's workshop are lined with the tools of the craft: pliers, files, drills, gravers, and hammers. The boy turning the winch on the left is drawing wire. The worktable is placed perpendicular to the large window, in order to provide maximum natural light to the craftsmen. On the right, a young man holds a pair of tongs in a small forge, with a bellows and an anvil by his side. Each workman at the bench sits with a leather apron tucked into his belt and attached to the bench to catch filings of precious metal. © The Trustees of The British Museum. Photo: Erich Lessing.

3.2. (*facing, bottom*) Etienne Delaune, *A Goldsmith's Workshop*, c. 1576. Engraving, 80 × 120 mm. British Museum, London, PD 1863,1114.778. The man serving a client through the window is possibly a self-portrait by Delaune, who worked as a goldsmith in Paris in 1546. © The Trustees of The British Museum.

window in both engravings indicates not just the need for light in the detailed work, but also these multiple layers of social life in which the workshop and craft knowledge more generally are embedded. The objects of craft—which are objects of knowledge as well as the proofs of that knowledge—are demonstrated (or proven) within a community of makers and consumers. Robert Blair St. George called artifacts "the complex representations of latent values, sensual enactments of deeper cognitive structures breaking above the surface of felt reality."[3] To investigate these "cognitive structures" of craft knowledge, then, objects must become primary sources.

In chapter 1, we examined rare how-to texts of practitioners that on their face appear to instruct in specific techniques, and saw how they can also provide a window into conceptions of what practical knowledge is. Chapter 2 considered the practice of consuming butter as a way to enter into artisanal knowledge systems—knowledge systems that were also expressed in objects such as *Handsteine*. These knowledge systems provide the logics underlying the puzzling practice of eating butter before working metals. In this chapter, I return again to uses of butter to close out this first part of the book and to think about what the production of objects in relation to technical recipe texts can tell us about practical knowledge. With a special focus on the life-casting instructions in the anonymous French recipe text, Ms. Fr. 640, I show that its recipes make evident the constant testing and trying of materials in the workshop. The recipes also reveal networks of meaning within a rich material imaginary, which extends from butter to life-casting, to lizards, mercury, vermilion, gold, and blood. This first part aims, then, as does the book as a whole, not just to reconstruct (and *reenact*; see chapters 9 and 10) artisans' practices and experiential knowledge, but also to explore various tools by which to gain access to material and mental worlds.

Butter and Life-Casting

3.3. (*facing, top*) Wenzel Jamnitzer (attrib.), writing box, Nuremberg, c. 1560–70. Cast silver. 6.0 × 22.7 × 10.2 cm. KK 1155–64, Kunsthistorisches Museum, Vienna. The writing box is ornamented with small creatures and delicate plants, most cast from life. © KHM-Museumsverband.

3.4. (*facing, bottom*) Writing box, KK 1155, Kunsthistorisches Museum, Vienna, detail of lid. © KHM-Museumsverband.

Along with its use in tempering the bodies of metalworkers to protect them from the effects of metal fumes, as discussed in the last chapter, butter was also integral to the sculptural technique known today as life-casting (figs. 3.3 and 3.4). Ms. Fr. 640 contains many instructions for casting from life plants (figs. 3.5–3.7), insects (figs. 3.8 and 3.9), reptiles (figs. 3.10 and 3.11), and amphibians (fig. 3.12). According to this text, butter was essential in casting the most delicate creatures. Butterfly and grasshopper wings and some flower petals were to be coated with butter to give them substance for casting because "butter is amiable and handleable."[4] Thickening up such delicate components was necessary because they were likely to be crushed by the molding sand or insufficiently filled with molten metal. The delicate cicada's wings on a writing box show evidence of this technique (fig. 3.13). This procedure sounds a straightforward, if delicate operation, yet it opens up an interconnecting web of resonances that formed the early modern artisan's understanding of nature, a web that can give us further insight into the material imaginary of the metalworker.[5]

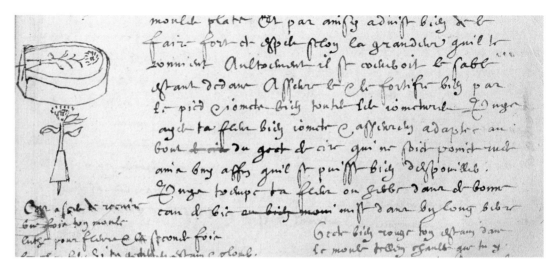

3.5. (*facing, left*) Drawing showing a method to mold plants and flowers for life-casting. Ms. Fr. 640, fol. 117r, Bibliothèque nationale de France, Paris. From top: a wax gate is shaped around the stem of the plant, then the plant is laid down horizontally into an oval mold, with vents (or risers) shaped from wax attached to the gate to channel air out of the mold as it is filled with metal. The mold is then filled with a plaster mixture, and, finally, fired to burn out the plant and melt out the wax. Metal is then poured into the void space where the plant has been burned out. Source: gallica.bnf.fr.

3.6. (*facing, top right*) Molding a flower for life-casting. Ms. Fr. 640, fol. 145v, Bibliothèque nationale de France, Paris. Source: gallica.bnf.fr.

3.7. (*facing, bottom right*) Life-cast marsh marigold, c. 1540–50. Cast silver, 5.5 cm. HG11140, Germanisches Nationalmuseum, Nuremberg.

3.8. Molding a fly for life-casting. Ms. Fr. 640, fol. 165v, Bibliothèque nationale de France, Paris. Two channels branch off the gate to feed the fly's wings with metal. Source: gallica.bnf.fr.

3.9. Life-cast fly, detail of writing box, KK 1155, Kunsthistorisches Museum, Vienna. Photo: Pamela H. Smith and Tonny Beentjes.

3.10. Molding a lizard for life-casting. Ms. Fr. 640, fol. 124v, Bibliothèque nationale de France, Paris. The sketch has the gate for filling the mold with metal entering a narrow section of the tail at the bottom of the drawing, with channels and vents shown as scored conduits. The outline of the mold is indicated by the dashed line. The infrastructural elements would have been shaped with wax and melted out of the mold at the same time the lizard was removed or burned out in order to prepare the mold for casting. Source: gallica.bnf.fr.

3.11. Life-cast lizards, c. 1540–50. Cast silver, 7 × 3.8 × 2.9 cm. HG11135–36, Germanisches Nationalmuseum, Nuremberg. The remarkable detail was achieved by molding and casting according to techniques detailed in Ms. Fr. 640, fol. 124v (see fig. 3.10). Evidence for this is visible on the farthest left curve of the large lizard's tail, where the slightly raised oval area indicates that the gate through which the mold was filled with metal had to be filed off. The casting seam along the lower right edge of the tail shows where the mold halves met.

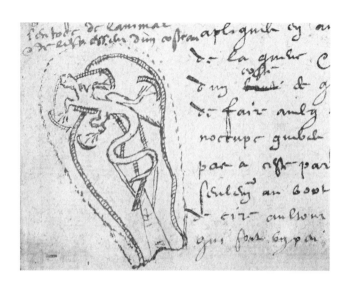

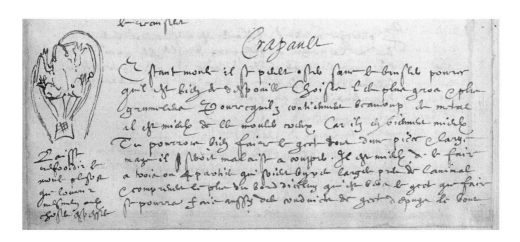

3.12. Molding a frog for life-casting. Ms. Fr. 640, fol. 143r, Bibliothèque nationale de France, Paris. This information-rich drawing shows four channels branching off the gate, two entering the back of the frog, and two entering the underside of the frog's feet, where balls of wax at the toes create void space when melted out to ensure air space when the metal enters the small spaces of the toes. The channels then run to the front feet, and the mold is vented through risers reaching from the frog's mouth to the other end of the mold. Source: gallica.bnf.fr.

3.13. Cicada, detail of writing box, KK 1155, Kunsthistorisches Museum, Vienna. The insect's wings show thickening, as advised in Ms. Fr. 640, fol. 142v, especially evident at the top of the wings, where traces of filing remain. Photo: Pamela H. Smith and Tonny Beentjes.

Casting from Life and the Investigation of Nature

To begin, let us look more closely at the process of casting from life in which butter was employed. Casting from life is accomplished by molding plants or a recently killed animal in plaster, burning out or removing the actual plant or animal (the pattern), then pouring molten metal into the void left by the animal or plant. It is sometimes referred to as lost-pattern (or even lost-lizard) casting, a form of direct casting in which the mold must be destroyed to remove the cast sculpture.[6] This technique resulted in a precise replica of the cast creature, creating stunningly lifelike objects. While this might seem simply a mechanical means of reproduction, we shall see, on the evidence of this sixteenth-century compilation of recipes, that it is in fact a means of investigating and understanding nature.

The manuscript provides instructions for catching, keeping, feeding, killing, and finally molding and casting plants, and also insects, birds, reptiles, and other animals. It is filled with observations on the behavior of reptiles—what might be called vernacular natural history—as well as with numerous accompanying accounts of experiments on the behavior of different materials for making molds and of metals for casting. We get a sense of the practitioner's close observation of—and even experimentation upon—the snake from which he will form a mold:

When you take them for molding, if it is possible, do not remove their teeth if you want to keep them. For, having had their teeth removed, they get sore gums & mouth, & can no longer eat. You can keep them in a barrel full of bran or even better of earth, in a cool place, or in a glass vessel. And give them any live frog or other little live animal, for they do not eat anything dead.[7]

At rare moments, his natural historical observation takes on a tone of moral allegory, as he notes that snakes attack their prey not "with a direct attack but with sinuous turns & from the side, as do Satan & his disciples. It has a small head but a very long body. The entryway to sin seems small & inconsequential, but the consequences of it are very great." And then he continues his phenomenological observation,

It abstains from eating seven or eight days, once it has devoured some frog. It can swallow three or four of them, one after the other, and what it devoured is neither corrupted nor consumed in one go in its stomach, but some part little by little, that is to say bones and everything. And the remains are found as fresh as when it had devoured them, such that sometimes, when one presses & torments it, it renders up what it has engulfed, parts of which are found totally consumed & others as fresh as if it were alive. It can keep a frog engulfed for two to three hours & renders it up completely alive.[8]

The value of knowing an animal's dietary habits also appears in the practitioner's observations of nightingales. "For trapping them," he notes, "one needs to observe their nature,

the food that they like the best, & the season of their pleasure." A careful strategy is required to capture them:

Approach, therefore, making as if searching the ground for something. And taking some
worms which come from old meal or from beneath kneading troughs or mills, which the nightingale is fond of, put some on your hat attached with a pin or otherwise, in order that it wiggles.
And at five or six paces from the hedge where it sings, make a hole in the ground & put in some
worms and your device of little crossed sticks. It will be anxious for you to leave so it can go see
what you have done, and seeing worms, it will enter.[9]

He then notes in the margin that it is easier to catch them in the "coolness of the evening
and the morning, near fountains & shaded places." When caught, the bird must "have a
cage made like a barn […] with green cloth, because it fears the cold." Attention must be
paid to its diet: "for making it accustomed to eating when first it is put in the cage, one
needs to give it ants with soil at the bottom of the cage […] and give it chopped sheep's
heart & immediately some eggs & mealworms."[10] Some pages along, he provides further
information on what types of foods the nightingale enjoys "in order to fill its belly & keep it
from diminishing & growing leaner," including "mutton heart or other delicate flesh," and
advises to give it "live mealworms, for it is very fond of them."[11]

Alongside such natural historical observations on the behavior of snakes and nightingales are numerous experiments on materials for molding and casting, the behavior of
plaster, sands, clays, firing techniques, and production of metal alloys. Most of the manuscript's more than three hundred recipes for making molds and casting metals testify
to the constant experimentation necessary to produce the ideal mold material: it must
be fine enough to take the imprint of the animal's or plant's delicate surface texture; light
enough not to flatten the animal during molding; durable enough to withstand the burn-
out, the heating of the mold before casting, and the pouring of the red-hot metal. Finally,
the material must be friable enough to crumble easily when the mold is broken to reveal
the finished cast object. Such qualities can be discovered only by repeated experiment
with natural materials. The author-practitioner of Ms. Fr. 640 experiments ceaselessly,
noting, for example, in attempting to identify effective mold-making ingredients, "I
have tried four kinds of sand for lead & tin: chalk, pestled glass, tripoli & burnt linen, all
4 excellent."[12] And on another day, he "tried the bone of oxen feet, thoroughly burned &
pulverized & ground on porphyry, until it is not felt between your fingers […] because on
its own it is very arid & lean, it wants to be well wetted & moistened with wine boiled with
elm root." His experiments with "iron dross, well burned bone of oxen feet, felt [cloth] also
well burned over a closed fire" revealed them to be excellent mold materials as well.[13] One
entry records several experiments with local Toulouse sand under the notation "*Sables
experimentes* [Experimented sands]."[14] In some places, he simply notes in the margin,

"Try." Based on his experience he advises to "try to mix it with a lean sand, such as pumice, scales [blacksmiths' hammerscale] & similar things."[15] And, again in the margin: "Try to mix ceruse [lead white pigment] or minium [red lead] with other sands."[16] Finally, "grain [...] makes a tawny powder, very delicate & very soft, which, once mixed, could mold very neatly," and ends by underlining "Try wheat flour burned over a closed fire."[17] We can see, then, how this practitioner sought out the behavior of natural materials through close observation, and through testing of different materials in order to identify and enhance their properties, aiming to make them function in ways that would produce objects or effects.

His repeated testing is echoed in other artisans' manuals, like that of Biringuccio, who advises constant trial: "It is necessary to find the true method by doing it again and again, [...] to have a superabundance of tests [...] not only by using ordinary things but also by varying the quantities, adding now half the quantity of the ore and now an equal portion, now twice and now three times."[18] Ms. Fr. 640 demonstrates that casting from life (and craft knowledge more generally) was not just productive but also investigative. This investigation extended seamlessly from natural historical observation and experiment to exploration of the properties and virtues of stones, woods, metals, and all manner of other materials, and it included, as well, inquiry into the transformation of matter by means of fire, acid, grinding, crushing, and other forms of disaggregation.

Bernard Palissy, a French Huguenot potter (c. 1510–90), also undertook myriad trials of firing and glazing to produce extraordinary glazed ceramics (fig. 3.14).[19] His works of art were more than a naturalistic representation of these animals; they were self-consciously an investigation into the natural processes that took place in the earth by which various kinds of earths, minerals, and precious materials came into being.[20] He viewed his attempts to imitate these natural processes in his workshop as not only creating remarkable objects, but also producing knowledge about the processes by which rocks, minerals, ores, and fossils were formed.

Palissy was far more self-conscious than the anonymous author-practitioner of Ms. Fr. 640 about the meaning of his creative activities: he described the making of these ceramics an "art of the earth," by which he investigated natural processes taking place in the earth, and, through this art, recapitulated natural generation.[21] Palissy explicitly claimed to come to know the workings of nature—its generative powers and forces of transformation—through the bodily experience of imitating it in experiments with clays, sands, and salts. He constructed a philosophy of nature on the basis of these experiments and the objects they produced, which he set forth in his two books: *True Recipe by Which All the Men of France Would be Able to Multiply and Augment Their Treasures* (1563) and *Admirable Discourses on the Nature of Waters and Fountains, Either Natural or Artificial, on Metals, Salts and Salines, on Rocks, Earths, Fire and Enamels* (1580). Palissy claimed to prove this philosophy by displaying natural objects in a "cabinet" in Paris, accompanied by explanatory labels to make clear the place of each in his philosophy.[22]

The investigation of nature carried out by Palissy and the author-practitioner of Ms. Fr. 640, in collecting, keeping, molding, and casting these animals, is explicitly referenced in Biringuccio's *Pirotechnia*, where he spoke of the "oft-repeated experience" by which the artisan would gain understanding of the innate, hidden powers of things.[23] Biringuccio wrote that God gave each natural thing a particular gift or power.[24] If we humans cannot see this divine power, he says, the fault lies in our "defective vision, in our little knowledge, and in our lack of careful thought concerning the necessity of seeking hidden things." He continues, "[T]hose things that have such inner powers, like herbs, fruits, roots, animals, precious stones, metals, or other stones, can be understood only through oft-repeated experience." Such "oft-repeated experience" enabled the extraction of the virtues implanted by God in nature. It was in such repeated experiencing of nature that artisans sought to

3.14. Bernard Palissy, platter, decorated with "*rustiques figulines*," second half sixteenth century. Lead-glazed earthenware, 74.6 × 45 × 16.9 cm. Musée des beaux-arts de Lyon. © Lyon MBA. Photo: Alain Basset.

identify and employ nature's powers in transforming materials for human use. Extracting these hidden forces and qualities could produce powerful effects resulting in both material objects and natural knowledge.

Generative Creatures

Palissy incorporated life-molded ceramic animals into his famed grottoes, in which he covered walls in generative rocks (fig. 3.15), natural and ceramic plants (fig. 3.16), and creatures—including birds, snakes, lizards (fig. 3.17), crayfish, frogs, and insects—that in nature were to be found around natural water sources.

Grottoes were popular in elite gardens in late sixteenth-century France, and Ms. Fr. 640, too, contains instructions for making and decorating grottoes.[25] Lizards, snakes, and the other creatures that Palissy created from clay and the author-practitioner of Ms. Fr. 640 made from metal could have spiritual or allegorical meaning, as we saw from the comments about snakes, quoted above,[26] but they also were distinctive within the natural world. Within the early modern cosmos, these creatures were seen as inhabiting more than one elemental zone: lizards, snakes, toads, and crabs lived both on land and in water, and insects and birds inhabited both air and land, and often water, too. Moreover, many of the creatures employed in life-casting appeared seemingly spontaneously from rotting materials and putrid mud. Lizards, snakes, frogs, toads, turtles, and all kinds of insects were often found in warm, wet, and muddy conditions, and swarmed around putrefying matter.

3.15. Bernard Palissy (workshop), brick decorated with jasper rocaille, third quarter sixteenth century. Glazed clay. From the grotto in the Tuileries gardens. OA2494, Musée national de la Renaissance, Château d'Écouen, France. © RMN-Grand Palais / Art Resource NY. Photo: Jean-Gilles Berizzi.

3.16. Bernard Palissy (workshop), fragment with a bouquet of five sage leaves. From the excavations at the Tuileries, Palais du Louvre. Glazed clay, 8 × 8 cm. EP506, Musée national de la Renaissance, Château d'Écouen, France. © RMN-Grand Palais / Art Resource NY. Photo: René-Gabriel Ojéda.

3.17. Bernard Palissy (workshop), architectural fragment with a green lizard, 1556–90. Life-molded, lead-glazed earthenware, 16 × 11 × 3 cm. From the grotto in the Tuileries gardens. EP871a, Musée national de la Renaissance, Château d'Écouen, France. © RMN-Grand Palais / Art Resource NY. Photo: Stéphane Maréchalle.

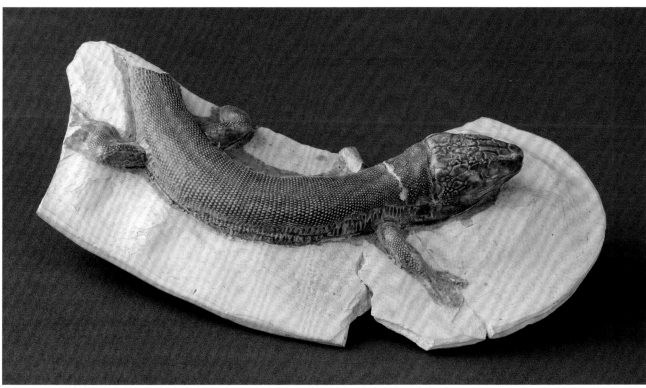

Their presence in these conditions gave support to the view that they were in fact generated in such wet, fertile sites, teeming with life, and that their generation involved processes of decay through which new life was generated.[27]

Many of these reptiles and insects were also known for their brief lifespan and the processes of transformation through which they passed in their short lives. Caterpillars and worms metamorphosed through a period of "sleep" or "death" into moths and butterflies; their metamorphosis from pupae was taken as evidence of the possibility of resurrection.[28] Lizards regenerated their tails when detached, snakes shed their skins, adult frogs and turtles emerged from the ground after freezing winters, and living crustaceans were reported to be found alive in solid stone well into the eighteenth century.[29] These animals were sometimes viewed as impure and associated with putrefaction, yet they were also connected to processes of transformation and generation, such as birth. The association of lizards with birth, for example, appears to have been very ancient,[30] and endured at least until the twentieth century, when Jewish silversmiths in Morocco adorned birth amulets with naturalistic lizards and salamanders.[31] The types of animals used for life-casting thus exemplified life processes and could give insight into processes of putrefaction, transformation, generation, and regeneration.

Lizards

Lizards possessed special powers and significance in early modern Europe, as, incidentally, they still do for scientists today. Scientists study lizards today for their ability to regenerate their tails and for the ability of some species to reproduce asexually.[32] Lizards in the early modern world were connected to metal processes. Among the many recipes for lighting a house contained in a medieval book of secrets, one calls for cutting off the tail of a lizard and collecting the liquid that bleeds from it, "for it is like Quicksilver," and when it is put on a wick in a new lamp, "the house shall seem bright and white, or gilded with silver."[33] A 1531 collection of pigment-making and metalworking recipes, *Rechter Gebrauch der Alchimei*, contains several recipes for creating noble metals through a process of catching, feeding, and burning lizards (*Mollen*; in modern German, *Molken*, salamanders or lizards). These processes involve human blood, goat's milk, and brass. In one recipe, the lizards are placed into the milk in a covered vessel, which is buried in damp earth. The recipe contains a mix of seemingly straightforward instructions—making sure that air holes are cut in the lid of the vessel, for instance, otherwise the lizards will die—and symbolic elements, such as the use of human blood and the injunction to let the vessel stand until the seventh day in the afternoon. By the seventh day, the lizards will have eaten the brass from hunger, says the recipe, and their strong poison will have compelled the brass to "transform itself to gold." The vessel is then to be heated at a low enough temperature to burn the lizards to ash without melting the metal. The metal is to be hung in the smoke of sal ammoniac (ammonium chloride), which finally will produce a "good *calx solis*," a calcined or powdered gold.[34]

The next recipe in the collection calls for mercury and lizards, which, if burned slowly, will cause the lizards to eat the mercury to produce "good silver." It specifies nine *Mollen* mixed with a pound of mercury and milk to be heated in manure for four weeks.[35] Manure is often used for a long, slow heating because it contains thermophilic bacteria capable of producing an even temperature of 50–70°C by their biological activity. These bacteria form a feedback loop that maintains a constant temperature—if the temperature rises too high, the activity of the bacteria is diminished, and the temperature falls.

At least twenty *Mollen*, mixed in a glass with a half-pound of mercury and old broken bricks, "broken up as fine as flour," constitute the ingredients of the next recipe. The *Mollen* will eat the mercury-soaked brick flour, with more being added until they consume no more. After twenty days, they are to be placed in a pot, covered with lime and burned to ashes. If the powder is then placed in an assay crucible with lead ("*auff einen test*"), it will result in a "*materia infallibile*," and pass the assayer's test as fine gold.[36]

These sixteenth-century lizard recipes resemble a recipe in the twelfth-century *De diversis artibus* by Theophilus. This recipe for "Spanish gold" involved concocting "red copper, basilisk powder, human blood, and vinegar." To produce the basilisk powder, two twelve- to fifteen-year-old cocks were put into a cage, walled with stones all around. These cocks were to be well fed until they copulated and laid eggs, at which point toads would replace the cocks to hatch the eggs, being fed bread throughout their confinement. Male chickens eventually emerged from the eggs, but after seven days they grew serpent tails. They were to be prevented from burrowing into the floor of their cage by the stones, and, to further reduce the possibility of escape, they were to be put into brass vessels "of great size, perforated all over and with narrow mouths." These were closed up with copper lids and buried in the ground. The serpent-chickens, or basilisks, fed on the fine soil that fell through the perforations for six months, at the end of which time the vessels were to be uncovered and a fire lit under them to completely burn up the basilisks. Their ashes were finely ground and added to a third part of the dried and ground blood of a red-headed man, which was then tempered with sharp vinegar. Red copper was to be repeatedly smeared with this composition, heated until red-hot, then quenched in the same mixture until the composition ate through the copper. According to the recipe, it thereby "acquire[d] the weight and color of gold" and was "suitable for all kinds of work."[37]

This recipe has been much debated by historians, being viewed variously as a garbled set of instructions for the making of brass or for the chemical process of cementation in which gold is purified, or as an alchemical recipe in which the fantastical substances are encoded "cover names" known only to alchemical adepts.[38] While these debates are as fascinating as the recipe itself, my purpose in bringing it up here is the association of lizards—and here, by extension, basilisks—with processes of putrefaction and generation (death and rebirth) that produced gold.

A recipe that promised to yield a gold pigment by mixing mercury with a fresh hen's egg and then putting it back under the hen for three weeks drew on perceived similarities between mineral and animal generative processes.[39] Biringuccio, too, saw mercury as generative: he noted that the prospector for mercury should be on the lookout for verdant mountainsides, since all places where mercury is engendered "have abundant water and trees, and the grasses are very green, because it has a moist coolness in it and does not give off dry vapors as sulphur [does]."[40]

An association between the regenerative powers of lizards and mercury was made as late as the eighteenth century by José Flores, a doctor at the Universidad de San Carlos in Guatemala City, who theorized that the medicinal specific of lizard meat, used with success by indigenous communities in otherwise hopeless maladies, worked according to the same principles as mixtures of antimony and mercury. He noted that the dosage of one lizard per day had no side effects, whereas the dosages of the mercury mixture had to be precisely calibrated to avoid harm.[41]

Red

Mercury was associated with generation and with gold; it also formed an essential component in the dramatic process for making a red pigment called vermilion. This pigment was an artificially produced imitation of naturally occurring cinnabar (fig. 3.18). In making vermilion, mercury and sulfur are heated together until they become a black substance,

3.18. Cinnabar, HgS. Staatliches Museum für Naturkunde, Karlsruhe. CC BY-SA 3.0.

which is stirred until it is dark blue on the outside and silver on the inside. After prolonged heating in a closed vessel, the mixture forms a vapor that condenses as a bright red cake on the walls of the vessel.[42] The putrefied black paste turns a series of colors, and finally is transformed to a splendid red hue. This cake around the walls of the vessel is then scraped off to form a painter's pigment. The saturated red of vermilion plays a pivotal role in the chain of connections among lizards, gold, and mercury, and a close examination of this color yields a further glimpse into the knowledge system of metalworkers. One of the most pervasive material correspondences in the early modern world was that between gold, blood, and the color red. The connections extended from pigments to alchemy to diet, linking artistic, alchemical, spiritual, and medical practices.

The color red could indicate the strength or perfection of a substance. An early seventeenth-century German collection of pigment-making recipes recapitulates the cosmic hierarchy of the planets and metals, beginning its list of recipes with red vermilion, composed of sulfur and mercury; followed by blue "lazur," composed of white sulfur, mercury, and sal ammoniac; and after that *spangrün*, green copper acetate, from copper.[43] Thus, red is associated with gold, blue with silver, green with copper, and so on through the rest of the metals. The *Rechter Gebrauch* includes instructions for an acid, a "Red water" that "belongs to" gold. The water, "red as blood," has a "virtue" and strength that makes it burn like *aqua vitae* (distilled alcohol) that is able to soften any substance from wood to iron.[44]

Red-colored foods also possessed powerful associations: red grapes were good for coitus because they made a full, rich blood, while foods that were black, the color of Saturn and lead, increased melancholy.[45] Blood was an ideal food—in blood sausage, for example—because it was predigested and tempered.[46] Moreover, blood was produced by the digestion and fermentation of food in the body, and was further refined into bodily spirits that tied body and mind together. The digestion and fermentation of wine was seen to operate on similar priciples, and wine could also be further distilled into *aqua vitae*, or spirits.[47] Blood was the carrier of life and heat, and its deep red hue was naturally associated with red-colored substances. Red stones were associated with blood and regeneration. Red coral, for example, was used to staunch blood flow, as Albertus Magnus noted: "And it has been found by experience that it is good against any sort of bleeding. It is even said that, worn around the neck, it is good against epilepsy and the action of menstruation, and against storms, lightning, and hail. And if it is powdered and sprinkled with water on herbs and trees, it is reported to multiply their fruits. They also say that it speeds the beginning and end of any business."[48]

Blood

In both vernacular practices and high theology, blood thus brimmed with overlapping and sometimes seemingly contradictory meanings. It signified vitality, fertility, the material of conception, and the spirit of life, but at the same time, blood poured out could signify death, and of course the blood shed by Jesus signified death, life, and redemption.[49] Michael of Rhodes includes a recipe "To staunch the flow of blood," which intones, "Blood is in you as Christ was in himself. Blood is fixed as Christ was crucified. Blood is strong in your vein as Christ was in his pain. And break one stone and immediately put it on your nose and take the breath of the stone deeply into yourself."[50] Blood was specified in recipes as an extremely powerful agent of transformation, as in recipes for the cutting of precious stones (to which we shall return in chapter 6).[51] In a substance intended to make metals melt at lower temperatures, one recipe includes cow's milk and goat's blood.[52] In *Rechter Gebrauch der Alchimei*, one of a number of recipes to prepare precious stones for cutting and casting calls for goose's and goat's blood, which is to be dried, pounded, and mixed with willow ashes, then boiled and combined with strong vinegar. The stone to be softened should be warmed in the mixture, which allows the practitioner to "cut or form it as you want."[53]

The material and symbolic significance of blood for artisans can also be seen in Benvenuto Cellini's well-known account of casting his 1551 *Perseus Beheading Medusa*, a sculpture that featured a spectacular flow of blood springing from Medusa's decapitated body. As Cellini told it, in the midst of casting this sculpture, he was attacked by a fierce fever and believed himself to be dying. At the same time, the metal for casting began to "curdle" in the furnace. Cellini proclaimed to his assistants, "From my knowledge of the craft I can bring to life what you have given up for dead."[54] As what he called the "corpse" of the metal came back to life, Cellini recovered from his fever, but still the metal would not flow, so he ordered all his pewter dishes thrown into the molten mass. He fell to his knees in prayer, calling out, "O God, who by infinite power raised Yourself from the dead and ascended into heaven!" On this, the mold filled in an instant.[55] Michael Cole has pointed out the manifold connections in this work between vivifying blood and the infusion of matter with spirit, as well as the act of incarnation emulating God's original creation.[56] Through his art, Cellini portrayed himself as able to resuscitate the dead metal to flowing, corporeal vitality by employing a vivifying force that sent life coursing back through the dead metal just as it had reentered his own veins. Other artists, too, notes Michael Cole, used "the medium of blood to emulate God's original acts."[57] While Cellini wrote about his feats of casting, sculptor Adriaen de Vries (1556–1626) embedded and proclaimed them instead in his works of art. He associated his artistic creation and blood in his *Hercules Pomarius* (1626–27), where he left the sprues (casting channels), concealed as vines, feeding directly into the figure's veins, thereby preserving, as the conservator Francesca Bewer has put it, "the very channels through which the master metalsmith infuse[d] the figure with life."[58]

The heat and vivifying power of blood could be stimulated by gold when drunk as potable gold or when worn on the body, just as mercury stimulated growth in the mineral and vegetable realms.[59] Gold was associated especially with Christ. The blood of Christ purified mortals of their sins, just as gold purified their bodies, and as ores were purified by the action of smelting and refining, an analogy made explicit in the depictions of the *Speculum metallorum* and the *Handsteine* discussed in chapter 2.

Vermilion

As gold was associated with Christ, so vermilion, also associated with gold, was connected particularly to the blood of Christ, and painters and scribes of illuminated manuscripts often made the mark of a cross to indicate where the red pigment was to be used in the text.[60] A fifteenth-century Psalter, for example, employs vermilion to spectacular effect to depict the blood of Jesus (fig. 3.19). Cennino insisted that vermilion should be used to represent the blood of wounds. In "How to Paint Wounds" he instructs the painter: "[T]ake pure vermilion. Make sure that you lay it in wherever you want to do blood. Then get a little fine lacca which has been well bound in the usual way and start shading this blood, either drops or wounds, or however it is."[61] He describes how one is to lay in the flesh tones of living individuals and specifies that this flesh tone is never to be used on dead faces. As we saw in chapter 1, he called this color "*incarnazione*," and regarded its use as akin to the incarnation of life in a body.[62] In the sixteenth century, *incarnatio* was referred to as a "mixture of white and red, like milk and blood," and artists often made an analogy in their paintings between the process of painting and the mystery of God's incarnation.[63]

3.19. Psalter and Rosary of the Virgin, c. 1480–c. 1525, England. 180 × 130 mm (120 × 90). MS Egerton 1821, fols. 1v–2r, British Library, London. This Psalter employs vermilion to depict the blood of Jesus. The worn state of the right page (fol. 2r) may show where devotees kissed and touched the pages. © British Library Board.

Gold

The incarnational red of vermilion was added to recipes for making gold pigments,[64] even when the vermilion was unnecessary to the process.[65] As noted previously, metalworkers and alchemical theory agreed that gold, like vermilion, was made of a mixture of sulfur and mercury. Cennino himself gives a recipe for mosaic gold—a sparkling golden pigment that imitated pure gold—and this recipe included mercury and sulfur, although only the sulfur is necessary for the production of the pigment.[66] Such a practice indicates that the color red (represented here by the mercury and sulfur—the ingredients of red vermilion) was viewed as essential in processes that sought to generate or transform, especially in those intended to produce gold, or gold pigment.[67] Conversely, gold was added to glass to create ruby-red glass.[68]

Material Imaginaries

Mercury and sulfur, the actual material ingredients of vermilion, and the metallic principles that gave gold its qualities, were viewed as crucial in processes of regeneration and transformation. In the 1531 collection of recipes for making gold by catching, feeding, and burning lizards, these creatures are associated repeatedly with generation and transformation, in part through their connection with mercury and sulfur, the generators of gold and blood-like vermilion. These recipes reveal a "material imaginary" that connects red, vermilion, blood, mercury, sulfur, lizards, and gold. Metalworkers explained processes of generation and transformation within this network of meaning-laden materials, which enhanced the significance of, and helped to organize, their practices.

Objects, too, reflected this material imaginary. The goldsmith Wenzel Jamnitzer, known for his life-casting of lizards, snakes, and flowers, made this material imaginary visible in his extraordinary depiction of the moment at which Daphne was turned into a laurel tree in order to escape the unsought and undesired attention of divine Apollo (fig. 3.20). In this work, Jamnitzer philosophizes about metamorphosis and transformation, moving from the metal ores at the base (fig. 3.21), through their refinement in gilded silver on the body of Daphne, into the blood-red coral arms, from which tiny green leaves, cast from life, begin to sprout (fig. 3.22). The sculpture bodies forth the multiple meanings of its materials: the tiny fragments of generative stones, such as sulfur, pyrite, and marcasite, at the base of the sculpture all suggest generation, as well as the transformation by human art of raw ores into precious metals. The coral of the arms, given a deep connection to processes of transformation by Ovid in his *Metamorphoses*, was likened in the sixteenth century to blood flowing in the veins. It was understood in natural philosophy to be formed from a "juice of the earth" that caused stones and metals to grow in the earth.[69] The sprouting leaves, cast from life, betoken the new life in which these processes of metamorphosis

3.20. Wenzel Jamnitzer, *Daphne*,
c. 1570–75. Silver, gilded silver,
coral, and semiprecious stones.
H 66.5 cm. Musée national de la
Renaissance, Château d'Écouen,
France (E.Cl. 20750). © RMN-
Grand Palais / Art Resource NY.
Photo: Mathieu Rabeau.

3.21. Jamnitzer, *Daphne*, detail of stone fragments at the base, including, from right, pyrite, silver ore, calcified marine worm tubes, malachite, sulfur, silver ore, pyrite or marcasite. Sulfur, pyrites, and marcasites were generative materials, and the calcified worm tubes were seen as evidence of growth. Author photo.

and transformation culminate. Depictions of Daphne, who was transformed by the touch of Apollo's hand, can also call to mind the hand of the artist who transformed the material of nature into the work of art.[70] Objects, as well as the books of art and their recipes, thus testify to the material, physical, and philosophical engagement of craft with the generative and transformative powers of nature.

Craft, or "Art," was a quintessential human activity, and it had the goal of making visible the hidden elements in nature. As Biringuccio noted, repeated experiences were a way to reveal Nature's hidden virtues and make manifest God's powers in the world. The practice of art thus revealed the hidden virtues and material connections in God's creation, like butter applied on the fragile wings of insects made possible their lifelike resurrection in metal. Through repeated encounters with natural materials—the experimentation

3.22. Jamnitzer, *Daphne*, detail, life-cast leaves, with green paint, sprouting from coral arms. Author photo.

evident in Ms. Fr. 640, for example—an artisan came to understand the "active ingredi-ents" (the "virtues") that could be identified in and extracted from natural materials by action of the human hand. At this interface of the human body and natural materials lay the power to create and transform.

Systems of Practical Knowledge

When read as records of artisanal practices, alongside objects and the techniques of their making, the recipes and *Kunstbücher* surveyed in part 1 of this book reveal an underlying framework by which craftspeople understood and oriented their working in the material world. As we have seen in this and the previous chapter, this framework had four general components: first, the material influence of the heavens over humans and the earth; sec-ond, the health worldview, with its core focus on balance and its view that the workings of the human body and its humors formed a model for other natural materials; third, art as an exploratory and experimental practice of making visible the hidden power in nature; and, finally, an underlying material imaginary, such as that employed by metalworkers, which connected blood, red, gold, mercury, and lizards. This material imaginary, or knowledge system, provided a more general framework for understanding materials and their transformation. It was a practiced "theory," rather than a discursive one expressed in conventional exposition or syllogisms, and, when it was expressed in writing, it was as a set of particular instances, procedures, and recipes. At the same time, however, the system incorporated and related different materials and processes, and influenced techniques.

It thus informed more general approaches to the material world. This material imaginary, with its connections among red, blood, gold, and lizards, sounds strange to us today, yet it stimulated persistent testing and trying—as Biringuccio expressed it, "oft-repeated experience"—to identify and employ the "hidden," "inner powers" of things. This imaginary thus allowed the craftsperson to investigate and engage in life forces, such as generation, and the relationship of matter to spirit, and even to consider through material engagement the most profound mysteries, such as resurrection and incarnation. As we have seen, such practices were oriented to the mundane production of goods and livelihood, but at the same time, they gave access to the greater powers of the universe.

Steven Shapin posits a similarly practical and flexible knowledge system that he calls the "proverbial economy,"[71] building on Clifford Geertz's proposal that common sense can be seen as a cultural system.[72] Shapin examines the function of proverbs, noting that they are oriented to action and judgment, but, rather than operating simply as rules of action, they instead function as a generalizing system that can be reflective, skeptical, and relativistic. At the same time, they are deployed in particular situations, and their meaning and use is determined by this particular context. They both represent themselves as embodying and, in some cases, actually do embody the fruit of accumulated human experience. Like the material imaginary I have set out in this chapter, they operate in general and particular registers at the same time. Artisanal knowledge is inherently particularistic; it necessitates playing off and employing the particularities of materials. As Biringuccio expressed it: "Although I am aware that you should already know how to make these [cannon balls] from the methods of moulding taught you before, still I shall tell you the method that is followed in making these, because in the practice of every particular thing there is some divergence from the general procedure."[73] The particularistic nature of experiential knowledge, as we have seen, does not preclude developing generalizations and knowledge systems through experimentation with materials, or producing proofs of that knowledge in objects. In part 2, "Writing Down Experience," we turn fully to the challenges of putting this type of simultaneously particular and general knowledge into writing.

Writing Down Experience

CHAPTER 4

Artisan Authors

Writing Artisans

As we have seen in part 1, many types of craftspeople, ranging from a galley oarsman to painters and metalworkers, began to write down their techniques and their experience beginning in the fifteenth century. While artisan writers may call to mind well-known individuals like Leonardo da Vinci and Benvenuto Cellini, many lesser known handworkers, including gunpowder makers, ships' pilots, fortification builders, masons, mariners, and dancing masters, all took up the pen from about 1400 on. These artisan authors practiced the mechanical arts, working with their hands to produce objects and make a livelihood. Their writing signaled a departure from the traditional modes of producing and ways of conceptualizing textual knowledge, in which those trained in schools and universities in the liberal arts (the arts of a *homo liber*, the freeborn man, not dependent on having to labor for a living) gathered knowledge from texts of the ancients, seeking to interpret, dispute, or to draw out general principles. Some artisan authors attempted to mold their writings to a learned pattern, while others wrote texts consisting mostly of recipe-like texts, a form of codifying knowledge that had little to do with the learned culture of the university scholars.

Writing of such "handbooks," "manuals," or "books of art" (*Libro dell'arte* in Italian; *Kunstbuch* in German) began in manuscript form, such as those of Michael of Rhodes, Lorenzo Ghiberti, Francesco di Giorgio, Leonardo da Vinci, and others, but it became a flood of texts after the invention of printing in the 1460s, with technical volumes appearing in large numbers by the 1480s and 1490s. As we saw in chapter 1, these texts—often based on collections of recipes—emerged out of the collective production of knowledge that could endure over centuries, and they often grew larger with each reprinting, as recipes were added and varied, and, sometimes further compiled by printers.[1] These works in the vernacular include *Distillirbücher* (books that instructed in distillation) in the 1490s, the *Kunstbuchlein* in the 1500s that encompassed many techniques and trades, and assaying and metalworking treatises (known as *Probir-Buchlein*) in the 1530s. Early works were

4.1. Johann Schönsperger the Younger, *Ein ney Furmbüchlein*, title page, printed c. 1525–29. Woodcut, 20 × 15.5 cm. Metropolitan Museum of Art, New York, 18.66.1(1r), Rogers Fund, 1918. CC0 1.0.

largely published without an author's name, but by the mid-sixteenth century, recipe collections began to appear under author names, although often pseudonymous ones.[2] The *Secreti del Reverendo Donno Alessio Piemontese* (The Secrets of the Reverend Don Alessio Piemontese), first published in Venice in 1555, was the most widely disseminated of these. It appeared first in Italian, then was translated rapidly into many other European languages and into Latin, going through more than one hundred editions by the end of the seventeenth century.[3] The *Bâtiment des recettes* appeared for the first time in 1539, composed of two texts: first, a translation of an Italian collection of household recipes for preserving foods, making perfumes and soaps, and producing wine, among many others, entitled *Dificio di recette* (first published in 1525); second, a shorter collection of recipes, entitled *Autres secrets médicinaux [. . .] expressément pour les femmes* (Other Medical Secrets [. . .] Expressly for Women). The *Bâtiment des recettes* continued to be enlarged, reprinted, and revised up through the eighteenth century. In her careful study, Geneviève

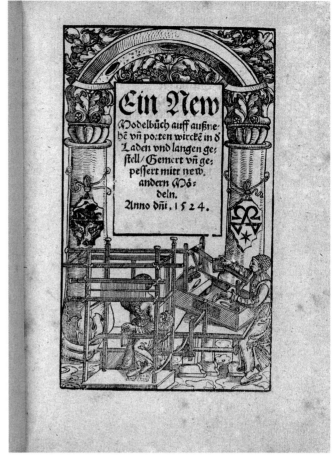

Deblock has illuminated the printer-entrepreneurs behind its publication.[4] Such printers, themselves practitioners of a trade, used such texts as sites to explore content, format, and marketing strategies. For example, Johann Schönsperger the Younger (active 1510–30) experimented with typefaces (fig. 4.1) and book formats, as well as various methods of representing patterns for weavers, lacemakers, and embroiderers (fig. 4.2). He proclaimed the novelty of his patterns and the new, "improved" state of his book, titling it *Ein new Modelbuch […] Gemert und gepessert mitt new andern Mödeln* (A New Muster Book […] Expanded and Improved with New, Different Models) (fig. 4.3). This boom in writing and printing books of art continued through the seventeenth and eighteenth centuries, which saw the publication of many more books on the making of all manner of objects and preparations, including remedies for humans and animals, beehive construction, hunting and fishing, gardening and husbandry, fencing, saltpeter making, embroidery, surveying, and even how to build a sundial in your garden.[5]

4.2. Schönsperger the Younger, *Ein ney Furmbüchlein*, p. 2v, embroidery pattern. Metropolitan Museum of Art, New York, 18.66.1(2v), Rogers Fund, 1918. CC0 1.0.

4.3. Schönsperger the Younger, *Ein new Modelbuch […] Gemert und gepessert mitt new andern Mödeln*, frontispiece, printed 1524. Woodcut, 18.5 × 13.6 cm. Metropolitan Museum of Art, New York, 29.71(1r), Gift of Herbert N. Straus, 1929. CC0 1.0.

Templates and Models

It may not be immediately obvious why this boom in technical manuals is unusual and intriguing, for artisans had committed technical information to various kinds of media and forms of codification before this time. Model books (fig. 4.4), templates, and drawings were passed down in workshops through many generations. One very early example is a well-known compilation by master mason Villard de Honnecourt (fl. 1225–50), consisting of drawings that apparently served as models for architectural details. Documentary evidence of drawings, models, and stencils being passed down can be found in a will drawn up by Bernardin Simondi in Aix-en-Provence in 1498.[6] Such drawings and templates helped to move motifs and images from one workshop to another, and made possible the rapid spread of some motifs and compositions—for example, those of the Flemish painters Robert Campin and Jan van Eyck.[7] Similarly, some printed sea pilots' books and charts seem to be a survival of workshop notes passed down by pilots.[8]

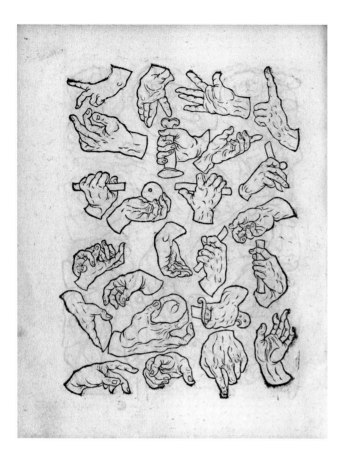

4.5. Heinrich Vogtherr the Elder (1490–1556), *Ein frembdes und wunderbarliches Kunstbüchlin / allen Molern / Bildtschnitzern / Goldtschmiden / Steynmetzen / waffen / und Messerschmiden hochnützlich zügebrauchen / Dergleichen vor nie keines gesehen / oder in den Truck commen ist* (A Strange [i.e., foreign] and Wonderful Little Book of Art, highly useful for painters, sculptors, goldsmiths, masons, and weapon and knife makers, the like of which has never before been seen or come into print). Strasbourg: Christian Müller, 1572. Bayerische Staatsbibliothek, Munich, Rar. 693#Beibd.1, http://daten.digitale-sammlungen .de/bsb00025720/image_18.

These paper materials functioned as tools of the trade, like the other tools necessary to any artisan, and they helped to build the reservoirs of collective, or distributed, cognition and knowledge. Model books continued to form the basis of many artisans' work into the sixteenth century and beyond, including Heinrich Vogtherr the Elder's *Kunstbüchlin* of 1538, reissued six times in twelve years (fig. 4.5), which, like that of Schönsperger the Younger, proclaimed its absolute novelty in its title, and the 1549 *Kunstbuch* of Nuremberg goldsmith Peter Flötner (d. 1546), with its forty pages of designs and not a word of text, used by generations of painters, cabinetmakers, goldsmiths and draftsmen.[9] Although today we think of books and writings as conveying codified information, constituting "cognitive prostheses"[10] by which knowledge is gained and transmitted, it is important to remember that artisans and practitioners of the mechanical arts did not need written texts in order to learn or transmit knowledge, for craft knowledge was efficiently transmitted by means of apprenticeship and disseminated relatively rapidly by the embodied knowledge moving in the artisans themselves.[11] Codification of artisanal techniques and processes took place not primarily in written form or on paper, but much more frequently in the ordinary training and tools of the workplace. Tools function both as extensions of bodily capacities and as embodiments of abstracted working processes and, as such, can be understood as codifications of certain techniques.[12] That craftspeople did not need writing to teach, transmit, and preserve often quite complex techniques and patterns bears stating explicitly, first because it is hard in our current text-centered society to remember this, and also because it continued to be true for centuries after 1400. One example is a Spanish tailor, Juan de Alcega, who published a collection of patterns in 1580 that had to be approved by two court tailors, one of whom had to ask the notary to sign on his behalf because he did not know how to write.[13] Yet tailors performed complex mental figuring to use a given length of fabric to maximum advantage, and, even today, they continue to size up their customers and perform expert calculations of this sort.[14] As we saw in the introduction, many practitioners made the point (paradoxically *in writing*) that they could not adequately transmit their practices in words or in writing.

The Book of the Art

In contrast to pictorial and embodied forms of transmitting knowledge, the "book of the art," as Cennino called the account of his art, and the *Kunstbuch* or *Kunstbüchlein*, as it was known in German, were a new kind of text. Around 1400, craft practitioners became authors, and brought into being a new genre to articulate a form of knowledge that had not had systematic written form. The appearance of these technical guides has been linked to the growth of urban culture and the cities' increased population of a "middling sort," who grew more isolated from familial sources of technical knowledge as their social mobility increased, and who sought out information not only useful to running their household, but also to assist in emulating their social betters.[15] This would appear to explain the publication during the same period of books of landed estate management (*Hausvaterliteratur*), how-to texts for conceiving children, and manuals of conduct, such as Baldassare Castiglione's *Book of the Courtier* (1528).[16] This general explanation, however, belies the many and varied—often singular—motives for, and aims of, these texts. Take, for example, the earliest ones, such as Cennino Cennini's more or less step-by-step exposition of the techniques of the painter, which contrasts with that of Lorenzo Ghiberti (1378?–1455), the celebrated sculptor of the Florence Baptistery doors, who decided late in his life (1448–51), when already a famed sculptor, to write a book. Ghiberti's book contained little about his actual work, and, instead, assimilated portions of antique authorities—in particular, Pliny and Vitruvius—as well as medieval optical theory transmitted from Arabic into Latin, in order to provide a history of the arts and to theorize about artistic practice. He titled his book *I Commentarii*, thereby imitating the commentary form of university teaching.[17] Ghiberti strove to prove his authorship by articulating his knowledge in scholarly, academic terms. Artisan authors seeking to philosophize in writing became more common in the next generation. Artists, such as Leonardo da Vinci (1452–1519), sought to engage on equal footing with the literati in philosophical debates of their day. Leonardo trained in the workshop of Andrea di Cione, known as Verrocchio (1435–88), but in contrast to Verrocchio, who philosophized only in his works of art,[18] Leonardo put his innumerable ideas down in words on paper. While none of Leonardo's writings appeared in print during his lifetime, his younger contemporary Albrecht Dürer (1471–1528), working at the heart of the German printing industry in Nuremberg, set up a printing press to to make certain that his own prints and books on architecture, perspective, and human proportions were disseminated. Such artist authors are well studied, but their writings are only the most visible of a deep reservoir of texts of practice. As we shall see in this chapter, even seemingly simple and straightforward collections of recipes and techniques turn out to have quite diverse forms and aims: they run the gamut from clear instructions to accomplish a specific task, to advertisement for a practitioner's abilities, to wonder-working promises, to competition for a patron or commission, to a desire on the part of printer-entrepreneurs to sell books, to administrators wishing to control a production process through making known its techniques, to legitimation of particular handwork practices, and various combinations of these aims. This chapter uncovers some of the diverse circumstances that gave rise to books of art and that contributed to developing this new genre.

The Prehistory of Artisan Writing

Before focusing further on texts produced by artisans in the early modern era, it is useful to briefly survey the long tradition of writing down technical procedures *before* 1400. Written instructions can be found far back in the historical record: a manual on horse training for chariot use written in cuneiform on a clay tablet from the fourteenth century BCE (fig. 4.6);[19] horse training manuals and perfume recipes in Middle Assyrian from the twelfth century BCE; and Assyrian glass-making recipes from about a half-millennium earlier. These Assyrian texts on glass-making, containing much local dialect, gave rise to "two written traditions which consolidated from forty to sixty recipes in specific sequences," which were preserved in multiple versions in the library at Nineveh among the tens of thousands of Assurbanipal's cuneiform tablets of the seventh century CE.[20] A corpus of recipes from this period for alloying and coloring metals, fabricating gemstones, and producing cosmetics, dyes, and pigments was remarkably long-lived and would appear again in Pliny's *Natural History*, written about 78 CE, as well as in papyri written in Greek in Egypt in the third and fourth centuries CE (known as the Leiden and Stockholm manuscripts for their present locations).[21] Many of these recipes continued to be integrated into recipe texts in much the same form throughout the Islamic and Christian lands, in some cases into the eighteenth century.

A study of the aims of these texts reveals that few of the pre-1400 texts were apparently written by or for practicing artisans. While some of the earliest surviving technical

4.6. Training program for chariot horses, allegedly written by Kikkuli, horse-master to the Hittite king Suppiluliuma. Cuneiform tablet, from Hattusa (now Boghazkoi, Turkey), fourteenth century BCE. Tablet no. 4 of a series. Front and back inscribed. Baked clay, 28.5 × 16 × 5 cm. Vorderasiatisches Museum, Inv. VAT 06693. Vorderasiatisches Museum der Staatlichen Museen zu Berlin—Preußischer Kulturbesitz. © bpk Bildagentur / Art Resource NY. Photo: Olaf M. Tessmer.

writings—the Assyrian glass-making texts from the library at Nineveh—contain recipes that read as if they were being used in practice at the time of writing, a different picture emerges when the texts are compared to the archaeological record. These recipes seem to be associated with a new type of glass that appears in excavated strata of the fifteenth to fourteenth centuries BCE, centuries *before* the recipes were recorded on clay tablets. At that time, this glass was new, part of a sudden change in glass-making practices that brought hot glass shaping to the fore, in place of cut and polished, monochrome-colored glasses set in mountings.[22] The recipes thus describe a type of glass already made for centuries, and scholars have hypothesized that the cuneiform tablets may have accompanied gifts of this antique glass between rulers. Thus, they were not necessarily associated with *producing* the glass, but rather with *consuming* it. These recipes may look as if they are meant to be about *making* and *knowing*, but in these cases, they were probably aimed at an audience more interested in *acquiring* and *owning*. A similar dynamic was at work in the publication of recipes apparently directed at connoisseurs in seventeenth-century China[23] and eighteenth-century Europe.

At the end of the fifth century BCE, cookery books were familiar to Athenians.[24] Cook-books also constituted some of the how-to literature of the Roman period, and the man-uscripts containing them continued to be copied into the sixth century CE. In addition, extensive technical writing, especially in the fields of medicine (both for humans and domestic animals) and in the arts of war, existed in the Roman period. The agricultural manual of Columella (4–70 CE), *De re rustica*, would become very important as a model for similar works in early modern Europe. The practitioner writer Vitruvius (c. 80/70–15 BCE), who produced a book of architecture, was the rare practitioner writer at this time,[25] and his book formed a model for the formulation and articulation of technical information and experiential knowledge in the Renaissance and beyond. Subsequent books that aimed to compile the wisdom of classical antiquity, such as Martianus Capella's fifth-century *Mar-riage of Philology and Mercury*, sometimes included sections on the mechanical arts, but these books often served to highlight the differences between the practical, earthly arts and those that could lead an individual to a higher form of knowledge.

The classical scholar Marco Formisano argues that the ancient texts are not concerned so much with actual practice, but rather are above all meant as a challenge within the Latin discursive field: because they claim to teach "how to *do* something through *writing*" (my emphasis), they in effect create a new discursive field that challenges the paradigm of ancient knowledge. Their goal of "doing in writing" puts new rhetorical demands and creates new opportunities for exploring nontraditional systems of expression, systems outside that of conventional Roman *eloquentia*.[26] These texts make claims to a new set of rhetorical "instruments," centered on *utilitas, sollertia, diligentia*, and *dissimulatio*. *Utilitas* is used "when the author wants to emphasize that the principal characteristic of his work is not an elegant and eloquent style, but the aspiration to direct applicability." *Sol-lertia* means the "careful collection, presentation, and arrangement of material collected from both the oral and written tradition." *Diligentia* and *dissimulatio*, already present in the classical rhetorical tradition, assume new meanings and values in these technical writings. *Diligentia* "denotes the great care and intellectual commitment of the author in inquiring into traditions of the past, searching for true and indispensable knowledge," while *dissimulatio* "refers to methodological error and bad faith in concealing knowledge out of self-interest or simply neglect."[27] This reevaluation of what constitutes *eloquentia* (and hence *sapientia*) and of the content of knowledge would subsequently be pursued by both Roman and Christian writers. While these four categories would appear in later technical writings,[28] Formisano does not argue for the influence of these late antique works on medieval reevaluations of technical and manual knowledge; rather, his point is that writing down technical information was done largely in order to advocate new ways of using language and expression—in part, as a reform and reconceptualization of what constituted knowledge, rather than as technical exposition. These examples give a sense of the diversity of motives for writing down technical procedures.

The scholarly culture of medieval Arabic writers incorporated a sustained interest in crafts. In the ninth and tenth centuries, the Arabic writers al-Fârâbî and Qusṭā ibn Lūqā created taxonomies of craft, only one part of a discussion of the crafts among scholars in the Islamic world.[29] The larger discussion also included an agricultural treatise written in Andalusia in the eleventh century, and books of secrets that included recipes for both mundane and magical practices that would go on to influence the Latinate corpus. One of the most influential of these Arabic books of secrets was translated in the twelfth century into Latin as the *Liber vaccae* or *Book of the Cow*, from a now almost entirely lost, ninth-century Arabic version. Recipes from the *Book of the Cow* continued to be included in books of secrets and alchemical treatises into the sixteenth century, including those involving lizards cited above.[30] Moreover, in the early thirteenth century, Arabic craftsman Badi'al-Zaman Abu al-'Izz ibn Isma'il ibn al-Razaz al-Jazari (1136–1206) penned *The Book of Ingenious Mechanical Devices*, which would also be influential in the Latin West.[31]

As for technical recipe books, there were a few survivals in Latin from the eighth to twelfth centuries: a recipe collection in low Latin from Greek sources (*Compositiones variae*) is preserved from the eighth century; a collection from tenth-century Italy in Latin, Eraclius's *De coloribus et artibus romanorum*; and a well-known compilation of recipes, the *Mappae clavicula* in Latin, that appeared in several versions from the ninth through twelfth centuries, and incorporated many earlier recipes.[32] The first entirely vernacular technical recipe collections—at least those that have been preserved—appear in Britain, written in French, in the second half of the twelfth century. In the twelfth century, there is also an apparent increase in interest among scholars, especially in monasteries, in collection and copying of recipes for metalworking and painting. On the basis of his *De diversis artibus*, which contains lucid recipes for metalworking, the twelfth-century writer Theophilus has been viewed by generations of scholars as a metalworking monk. This view has recently been contested, and the possibility has been raised that he was instead a participant in debates about pedagogy and the status of the arts—both mechanical and liberal—probably stimulated by contact with Islamic writings, and produced within the context of university debates.[33] Scholarly engagement with classical collections of recipes would combine, from the thirteenth century onward, both traditional techniques and recipes from contemporary practices. *De arte illuminandi*, by an anonymous fourteenth-century Latin author, and the Montpellier *Liber diversarum arcium*, composed between 1300 and 1400,[34] are recipe collections that contain both recipes and processes that go back to at least the first century, as well as those collected by the scribes from practicing artisans in their lifetimes. Recipes in German vernaculars appear in the fourteenth century, and also incorporate older, textually transmitted recipes copied down side by side with those dictated by practicing artisans.[35] In 1431 Jehan le Bègue redacted the collection of recipes of Johannes Alcherius, a schoolmaster who had likely gathered his recipes from painters between 1380 and 1411.[36] Jehan included a list of synonyms for materials, which shows him attempting to puzzle out a language that is foreign to him, indicating his own unfamiliarity with painters' practices.[37] The slightly later Tegernsee manuscript contains recipes from the *Mappae clavicula* as well as those in use by contemporary artisans.[38]

This brief survey of practical texts before about 1450 shows that most were written down by nonpractitioners. It is worth noting that the recording and compiling of recipes and practical information on the part of nonpractitioners did not by any means end in 1400 (as we shall see in the case of mining, below), but rather coexisted with the texts produced by artisans themselves; indeed, the texts by nonpractitioners peak between 1380 and 1450, at the time when Michael of Rhodes and Cennino Cennini composed their manuscripts. This indicates a more general interest in practical knowledge during the whole period.

We can trace the collecting of technical recipes by scholars and elites as it continued unabated through the sixteenth and seventeenth centuries: in the sixteenth century, for example, Hugh Plat (1552–1608), a brewer's son, was one of the most prolific compilers of recipes, publishing only a fraction of them in *Jewell-House of Art and Nature conteining divers rare and profitable Inventions, together with sundry new experimentes in the Art of Husbandry, Distillation, and Molding* (1594) and other books.[39] The very large recipe collections revealed by the Medici Archive Project make clear the scale of technical recipe collecting by rulers,[40] as do the tens of thousands of recipes in compilations found in the library of the Electors Palatine, whose library was sacked and carried to Rome in 1622. These compilations contain mostly medical recipes, often copied out by a scribe and then reordered by the compiler, but they also contain manuscript metalworking *Kunstbücher*.[41] Alisha Rankin and Elaine Leong's research has shown the importance of recipe collection within the domestic household of elite and lay folk alike.[42] Theodore Turquet de Mayerne (1573–1655), physician to royalty, collected painting and pigment recipes, and Nicolas-Claude Fabri de Peiresc (1580–1637) visited Parisian goldsmiths to collect accounts of their processes, just as Hugh Plat did in London, although Mayerne and Peiresc never published.[43] Most famously, the founders of the Royal Society inaugurated their "Royal Society of London for Improving Natural Knowledge" with a History of Trades project that sought out artisans in order to take down and collect their technical recipes, which they viewed as containing information essential to the pursuit of natural knowledge.[44]

The activity of compiling technical recipes resulted in ever larger tomes in the late seventeenth and eighteenth centuries. Some of these texts, with more schematic descriptions of processes, seem intended for amateurs and connoisseurs taking the measure (and learning the jargon) of the burgeoning world of material goods. For example, Marshall

Smith's *The Art of Painting according to the Theory and Practice of the Best Italian, French, and Germane Masters* (1692) was made available through a London art auction house, so eagerly snapped up there that it rapidly required a second edition.[45] Much-reprinted texts, such as Godfrey Smith's *The Laboratory or School of Arts* (1738), Robert Campbell's *London Tradesman, a compendious view of all trades, professions and arts, both liberal and mechanic* (1747), Robert Dossie's (1717–77) *Handmaid to the Arts* (2nd ed., 1764), and William Lewis's *Philosophical Commerce of Arts Designed as an Attempt to Advance Useful Knowledge* (London, 1748), all saw the collection of technical knowledge as useful to the "encouragement of arts, manufactures, and commerce," and Dossie's book was dedicated to the London society of that name founded in 1754. Other compendious, eighteenth-century collections of practical information, such as Ephraim Chambers's *Cyclopedia or an universal dictionary of the arts & Sciences* (2 vols., 1728), Johann Zedler's *Universal Lexicon* (1730s),[46] and Abbé Pluche's *Spectacle de la nature* (1732–51, one of the biggest bestsellers of the eighteenth century),[47] aimed to reform manufacture, craft, or common thought. The systematic exposition of the arts and trades by the Académie royale des sciences in the eighteenth century in the *Descriptions des arts et métiers* (1761–88) aimed to dispel the "secretive obscurity" of the tradespeople and to bring to craft the "torch of physical science and the spirit of observation."[48] Finally, the *Encyclopédie, ou Dictionnaire raisonné des sciences, des arts et des métiers* (1751–72) of Diderot and d'Alembert sought to reform the practice of craft by applying rational organization and consistent nomenclature. These authors viewed this as simultaneously an intellectual reform and a way to stimulate the production of goods and boost the national oeconomy.

To conclude this brief survey of technical writing by nonpractitioners and literati, we may note that none of these writings aimed solely to convey technical information, and many of them shared a goal to reform existing paradigms of knowledge, thus playing a more complex role within existing structures and hierarchies of knowledge than might be expected of so-called how-to literature—a point to which we shall return in chapter 7, when we focus on the readers of these texts.

After 1400

ARTISANS AND THEIR PATRONS

From the foregoing survey, it is apparent that a wide variety of technical information was put into writing long before (and after) 1400, so we must ask, what changed around 1400? The major difference is that craft practitioners themselves became authors. This sudden appearance of self-consciousness and self-assertion took place in the context of increasingly powerful territorial rulers and their need of artisans for war technologies, for the building and outfitting of ships for trade, and for the representation of power in the no-bility's new urban seats of rule. We have already discussed the Book of Michael of Rhodes in this context. Similarly, from the early fifteenth century, practitioners responded to the demand on the part of territorial rulers for experts in fortification and gunnery with manuscripts detailing the machines and fortresses of war. These self-styled engineers in-cluded the Sienese Mariano di Jacopo, called Il Taccola, who wrote about siege engines and guns in *De ingeneis* (1433) and *De machinis* (1449). Francesco di Giorgio Martini (1439–1501), trained as a painter in Siena, followed suit in two treatises written between the 1470s and 1490s on architecture, engineering, and military machines.[49] These practitioners responded to the rapidly changing war technologies in Europe, as cannons and other new types of guns came into use. Another product of the new warfare was the *Feuerwerkbuch*, a work on gunpowder, written down in the first decades of the fifteenth century, but not printed until 1529, then reissued three more times in the next ninety years.[50]

Practitioners authoring works on fortification and weaponry often emphasized the mathematical and geometrical basis of their crafts. Matthäus Roriczer, for example, men-tioned the many times he had discussed the "free art [i.e., liberal art] of geometry" with the Bishop of Eichstätt, to whom he dedicated a book on constructing spires, *Büchlein von der Fialen Gerechtigkeit*, in 1486. In 1488/89, Roriczer even published a *Geometria deutsch (Geometry in German)*.[51] On investigating such claims about a foundation in ge-ometry, particularly those of writers on weapons and fortifications, as well as surveyors and gunners, Steven Walton found that practical mathematics was used to various degrees in fortification and surveying, but that gunners used almost no mathematics. Walton finds that the gunners' use of a "mathematical posture" in the manuals was either aspirational or referred to the use of instruments.[52] Simon Werrett has similarly noted that gunners sought to raise their status by writing practical manuals about fireworks, not for their fel-low gunners but for the recreation of a noble audience. Indeed, the mathematics that did appear in these manuals was often not used by gunners for actual calculation, but rather was a way to relate to court and gentlemanly audiences, providing a sort of mathematical recreation.[53]

The same nobles who sought to defend themselves or consolidate their territory with the latest weaponry also strove to master and rule their growing courts and their urban populace by splendor and magnificence. Artisan authors, such as Antonio Averlino, called Filarete, contributed to this effort to create a "theater of state" by dedicating his treatise on architecture (c. 1460) to Francesco Sforza. This text, structured as a dialogue between architect and patron, sets out plans for the ideal city of "Sforzinda," and speaks not just to the relationship between ambitious artisans and centralizing territorial nobles, but also to the ways in which society in Europe had become increasingly urbanized, with con-centrations of artisans experimenting with different media and engaging in an intense exchange of skills and ideas with their fellow craftspeople and other social groups. This lively back-and-forth between artists and patrons emboldened Averlino to make the less than dignified proposal (as he called it) that Sforzinda would harbor a school in which pupils from all social classes could receive instruction not just in letters, law, canon law, rhetoric, and poetry, but also that "some manual arts should be taught here by their practitioners." The faculty would include a master of painting, a silversmith, masters of carving in both marble and wood, a turner, a master of embroidery, a tailor, a pharmacist, a glassmaker, and a master of clay, "that is of beautiful vases." Filarete makes clear the novelty of his proposition—this is "a thing that has never been done before."[54]

In the sixteenth century, the Medici rulers concentrated artists from throughout their territories in the workshops (or *uffizi*) in Florence, where the grand dukes could efficiently commission works and control the output and sale of goods by the artisans. The della Rovere family, Dukes of Urbino, also brought artists to live at their courts.[55] Artists worked on all sorts of commissions for these patrons: they produced objects, they scouted out and purchased works of art, they advised their patrons about the authenticity of the works of art they so avidly sought, and they assessed completed works of art by their fellow artists for adherence to contractual specifications.[56] As part of this same effort, nobles founded academies, such as the Accademia del Disegno for painters, founded in Florence by Cosimo I de' Medici (1519–74). With such a degree of aristocratic interest in painting, panel painters themselves founded fraternities in order to distance themselves from other trades, such as grocers, merchants, saddle makers, and simple house painters, with whom they had previously shared a guild or trade association. For example, artists founded the Accademia di San Luca in Rome in 1577, claiming the saint as their patron.[57] The congregations of artisans and the academies for training artists became "trading zones" or "agora" among scholars, princes, and artisans, and surviving contracts between patrons and artists show that much discussion between the two parties went into the creation of a work of art.[58] The concentration of practitioners of different crafts in close proximity brought about the sharing of techniques and types of expertise and, interestingly, resulted in transfer of goals and techniques from one group of expert practitioners into the materials and practices of another.[59] One outcome of this noble patronage of the visual arts was that by the early 1500s, Castiglione (1478–1529) advised in his manual for decorous conduct that a gentleman needed to know how to talk about art.[60]

Texts written by dance masters, beginning with the 1450 "Manuscript of Bourgogne" up through Louis XIV's founding of a Royal Academy of Dance in 1661, also resulted from courtly patronage. These texts often functioned as bids for patronage, but they also strove to establish the authority and authorship of dancing masters. They asserted the status of the craft and their individual identities as experts, and attempted to counter moral and religious criticism of their practice.[61]

Urban male artisans became increasingly literate, often attending local, informal schools, and they produced a variety of other texts, such as autobiographies, memoirs, family books, spiritual autobiographies, and plays, as well as account books, combined with recordings of all kinds, similar to the Italian merchants' *ricordanze*.[62] In the exchange that took place in the burgeoning cities and princely courts in the fifteenth and sixteenth centuries, many scholars took an interest in the arts as well. Francesco Petrarch (1304–74), for example, who advocated the new *studia humanitatis* as a means to re-create a world based on Greek and Roman antiquity, felt it was justified to praise individual artists, such as his contemporary, the painter Simone Martini (c. 1284–1344), in the words of Pliny's praise for Apelles in the *Natural History*. Petrarch's praise did not, however, extend to the mechanical arts in general, which he disdained.[63] Humanist Leon Battista Alberti (1404–72) viewed practice as useful to civic life, as he made clear in *De pictura* (1435) and *De re aedificatoria* (composed 1440s–50s, published in 1485). Increasingly, artists also began to advise scholars who wanted to understand and restore recently excavated antiquities.[64]

As has been discussed by many scholars, artists began to argue that they practiced liberal rather than mechanical arts, marked by the mental work of creative imagination —*poesis*, even genius—that went into their work. Painter Bernardino Pintoricchio petitioned the Sienese government in 1507 in these terms when requesting a tax exemption. He called upon the city officers to treat painting as a liberal art, in imitation of alleged Roman practices:

Most respected officers, it is argued with due reverence by Master Bernardino Pintoricchio as servant of Your Lordships and one of the most worthy painters, that the republic is accustomed to esteeming such artificers dearly who may be compared to those of ancient times and so should treat them as did the Romans. As Cicero wrote, at the earliest period they took little delight in painting, but as the empire grew after the oriental victories and the conquest of the Greek cities, the Romans took pains to gather in from all parts of the world the most famous works of painting and sculpture (not hesitating to plunder them). They held painting to be the most distinguished and similar to other liberal arts and to compete with poetry.[65]

The interest in establishing their craft as a liberal art spread to masons and carpenters, now sometimes calling themselves architects or engineers (trades for which, at this time, there was no established path of training or apprenticeship). In the sixteenth century, they increasingly began to have themselves portrayed with their instruments.[66] The mention of mathematics (sometimes only in the titles of these manuals) assimilated these mechanical arts to the higher status branches of mathematics, taught as part of the liberal arts in universities. Part of the problem of writing down experience lies in choosing what, out of the myriad parts of a trade or circumstances of practice, to record. Writing can never take the place of practice, thus artisans' books demanded a reflection on language and expression, with the result that the certainty of mathematics, or the illustrated diagram, could seem a welcome refuge for practitioners struggling to transfer their bodily gestures onto paper.[67]

Craftspeople who strove to ally their trade to the liberal arts include such ambitiously self-conscious workers as goldsmith Benvenuto Cellini and painter Giorgio Vasari (1511–74). Cellini's *Two Treatises* (*Due trattati, uno intorno alle otto principali arti dell'oreficeria, l'altro in material dell'arte della scultura* [Florence, 1568]) gives a compelling example of the function of such books to simultaneously explain a process and convince a patron. He ended his account of how he would reproduce a life-size statue of Hercules with the reflection that "there are many different ways of doing the thing and each master chooses the method to which his technical excellence or his fancy guides him [. . .] When I had delivered myself of these words to the King, *he said it was all so clear and he had understood it all so well that he very nearly thought he would himself be able to undertake such work.*"[68] Cellini's pride in his ability with words—to describe his techniques, to flatter, and to raise his credit in a crowded field of artists competing for the favor of a patron—reveals the rivalry among both artists and rulers as a central factor in producing artisan authors. The form in which these artisan authors expressed themselves drew upon older cultural hierarchies, such as the liberal and mechanical arts, and the importance of allying oneself with antiquity, and would come to include new cultural tropes, such as ingenuity (more on this in chapter 8).

We can recognize in these examples the dynamics still at work that we observed at the very start of the boom in artisan authorship: we saw in chapter 1 how the seaman Michael of Rhodes seems to have been spurred to writing by the intense competition for positions of command in the Venetian fleet, believing that a book gave him the edge in this competition.[69] A contemporary cookery book from the ducal court of Savoy in 1420 demonstrates even more vividly the structuring pressure of the court and its need for expert managers of spectacles, such as elaborate banquets. Court painters and sculptors worked alongside pastry makers and cooks to manage the increasingly elaborate festivities that displayed and broadcast magnificence and power. We see in this cookery book how the competitive court context brought out assertions of individual authorship and expert technique, as well as attempts at learned forms of expression. After an invocation of Jesus's name, the book begins:

Man's unretentive memory often reduces clear things to doubt. The foresight of worthy ancients therefore determined that ephemeral things should be rendered immortal by being written down, so that whatever the feebleness of the human mind cannot retain might survive by means of immutable writings […] Here follows […] this little compendium and booklet which has been compiled *On the Matter of Cookery* by Master Chiquart, cook of our most respected lord, the Duke of Savoy […] written out by me, John of Dudens, clerk, burgess of the town of Annecy. To begin with, the introit or preamble contains the four prime causes that ought to be found in every good composition, to wit, the efficient, material, formal and final causes.[70]

Every part of the text that follows is dedicated to materializing hierarchy at court by creating and serving food for the noble table. This is just one early example of the type of writing that accompanied the emergence of a court bureaucracy, embedded in social hierarchy, which would go on to engender many written texts of practice in the following centuries.

GEOMETRY AS PRACTICE

Geometry was one of the liberal arts taught at the university, associated with learned culture, and centered on the axioms and proofs of Euclid's *Elements* as well as computation of planes, lines, and bodies. From the late Middle Ages, references to Euclid and geometry were frequently also found in books of art. The earliest writing in the vernacular about the work of masons, a poem in Middle English, c. 1390, emerged from an attempt to create a textual genealogy for masons, making clear from the outset the dignity of the craft, remarking that many younger sons of lords and ladies took up masonry, the most honest work of any craft. Moreover, the masons descended intellectually from the "grete clerk […] euclyde," who in Egypt founded the craft of masonry on the basis of "gemetry."[71] This text, probably composed for a group of masons by a scribe, as Lisa Cooper argues, reflects the concerns of masons but also contains traces of the clerks and scribes thinking through their own craft *of writing* by engaging with craft practitioners.[72] Such a motive may also have impelled the Nuremberg writing master, Johann Neudörfer (1497–1563), to set down in 1547 a list of all the "*Künstlern und Werckleuten*" (artists and handworkers) active in Nuremberg during his lifetime.[73]

In the sixteenth century, practitioners increasingly set pen to paper with a new model of knowledge in mind, as they sought to formulate a theory of their art, most often simply by associating it with geometry, as the dancing masters did in 1581, when Fabrizio Caroso attempted to apply mathematical notation to describe a variety of dances.[74] The pattern book of the Spanish tailor, Juan de Alcega, is entitled *Libro de geometria, practica y traca* (Book of Geometry, Practice, and Dress).[75] The connection to geometry was asserted by many practitioners—engineers, for example, made it a cornerstone of their self-identity.[76]

But even as Alcega advertised his knowledge as connected to geometry, his practical guide to, in his words, the "economical layout of patterns on fabrics of different widths for a range of clothing and textiles" showed no trace of Euclidean geometry[77] In his *La piazza universale di tutte le professioni del mondo* (Venice, 1585), which listed four hundred types of manual work, Tommaso Garzoni wrote that tailors exhibit *"gran giudizio"* (excellent judgment) because they need to know which styles suit different types of people; indeed, "the last excellence of the tailor is this, that he shows himself to be the best Geometer, because in one cast of the eye, in one glance only, he sizes up the measure of the whole person from head to foot." And, in 1650, Giovanne Pennacchini asked in his treatise on tailoring whether tailors were not comparable to architects, as both applied theory and practice, for tailors needed "to understand measure, design, and calculation, and to study the science of geometry accompanied by practice." "[O]n account of this science and doctrine," who could deny that a tailor "does not transcend the confines of the ignobility of the manual arts?" Tailors must possess so much knowledge, he noted—including the meaning of colors, the relationship of colors to humors, to the elements, the metals, the virtues, to social ranks, the days of the week, and the months—that they must be considered both natural and moral philosophers.[78]

Throughout the sixteenth century and into the seventeenth, artist artisans produced books of many kinds that often incorporated geometry. Sibylle Gluch argues that until Albrecht Dürer's *Unterweisung der Messung* (Instruction on Measurement, 1525) and *Vier Bücher zu menschlicher Proportion* (Four Books on Human Proportion, 1528), craft geometry did not overlap with the theoretical and computational geometry of the universities.[79] The geometry of craftspeople, such as Roriczer, mentioned above, was constructive, using and manipulating simple geometrical forms to solve—by means of arithmetical or proportional computation—real-world problems of design and building. Dürer, too, was concerned with these problems, but he strove to integrate elements of university geometry into his *Unterweisung* in order to create a more certain "science," and to elevate the craft to the status of liberal art. Dürer had many artisan emulators whose geometrical manuals followed his lead; however, the connection to university geometry was more tenuous in these texts, and they continued to be concerned with solving problems of construction by means of measuring tools.[80] As Andrew Morrall concludes in his insightful case study of Augsburg craftsmen who made luxury measuring instruments and clockwork-driven mechanical wonders for noble patrons in Prague, Vienna, and Dresden, a book alluding to skill in "geometry" could advance claims of expertise, giving an artisan the edge in the fierce collaboration and competition among goldsmiths, brass makers, clockmakers, and *Freikünstler* (artisans outside the guild system) promising to create novel "inventions" for wealthy patrons.[81]

The Nuremberg goldsmith Wenzel Jamnitzer displayed great skill both in his craft and in his creation of self as author. In the *Perspectiva corporum regularium* (Perspective of Solid Bodies, 1568), engraved by Hans Sachs, Jamnitzer provided a theoretical and natural philosophical introduction to what was essentially a model book and a proof of his skill.[82] He prefaced this virtuoso representation of polygonal solids with a discourse on the place of the Platonic solids in a philosophy of nature.[83] Then, at age seventy-eight, he

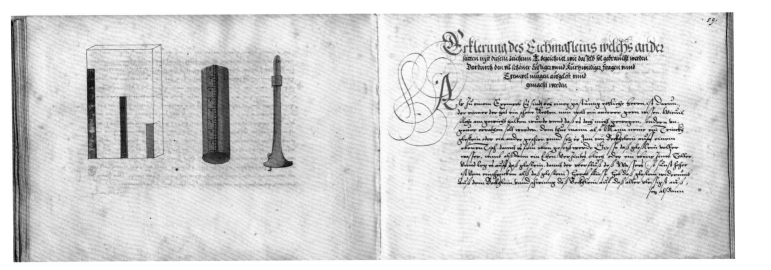

4.7. Wenzel Jamnitzer and Jost Amman, "Ein gar kunstlicher und wolgezierter Schreibtisch sampt allerhant kunstlichen silbern und vergulten newerfunden instrumenten," "Erklerung des Eichmasleins" (Explanation of the Little Gauging Vessel), 1585. Vol. 1, fol. 58v. National Art Library Special Collections, MSL/1893/1600–1601. © Victoria and Albert Museum, London.

composed a two-volume manuscript book of instructions for a collection of luxury measuring instruments apparently manufactured on spec for a noble patron. It is not surprising that this collection eventually found a buyer in Duke Johann Casimir of Saxon-Coburg, for its title alone, set within a rich cartouche painted by Jost Amman, was designed to make a prince salivate: "A thoroughly artful [*Künstlich*] and well-ornamented desk with all kinds of artful silver and gilt newly invented instruments. Of use to the geometrical and astronomical, as well as to the liberal and useful, arts. All newly manufactured by Wenzel Jamnitzer, citizen and goldsmith in Nuremberg." The extended subtitle referred as well to "hidden and pleasurable [*verborgen und lustigen*]" elements of weight and measure, "which may all be learned and resolved" through his instructions.[84] The social context in which hidden things were to be "learned and resolved" is clearly indicated in an entry that begins, "Often in a gathering of some lords, one has a beautiful chain that another wants to know the weight of […]," and then goes on to explain how to place the chain in a glass of water, and to measure the amount of displaced liquid with a specially invented gauging vessel that indicates the weight in different types of gold (fig. 4.7).[85] In this lavish manual, Jamnitzer made frequent reference to the new, inventive, and artful (*künstlich*) nature of the measuring instruments for weighing, gauging, scaling up and down, comparing sizes and weights of artworks constructed with different metals (fig. 4.8); casting cannonballs to fit specific cannon calibers (fig. 4.9); instruments for surveying lands, fortifications, seas, and mines (fig. 4.10); ingenious time-keeping instruments that enabled multiple calendrical calculations; and specially designed tools for constructing and copying ornamental designs. To top off these wonders of measurement, the desk contained a diverse collection of cut and carved stones, lenses, and ornamental objects. All in all, the instruments *and* the written manual together were a stunning display of the novel artfulness in the mechanical arts so avidly desired by European nobility for their collections. In such

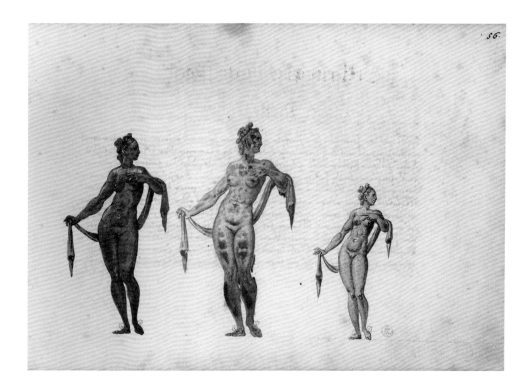

4.8. Jamnitzer and Amman, "Ein
gar kunstlicher und wolgezierter
Schreibtisch," drawings representing a sculpture cast in the same
weight of the metals copper, silver,
and gold. Vol. 1, fol. 56r. National
Art Library Special Collections,
MSL/1893/1600–1601. © Victoria
and Albert Museum, London.

4.9. Jamnitzer and Amman, "Ein
gar kunstlicher und wolgezierter
Schreibtisch," Gentleman measur-
ing the bore of a cannon. Vol. 1, fol.
27r. National Art Library Special
Collections, MSL/1893/1600–1601.
© Victoria and Albert Museum,
London.

advertisements of artisanal skill, necessary within the fiercely competitive atmosphere, artisans sought to gain a foothold by allying their ingenious abilities and novel products with the liberal art and "science" of geometry.

MINING AND AUTHORSHIP

Just as court culture was a powerful incubator of artisan authors, so too the burgeoning industries of mining and metalworking produced much writing on these newly visible industries, crucial to the financing and equipping of rulers and their armies (figs. 4.11 and 4.12). Not all such texts were authored by artisans, and hence they form an important exception to artisan authorship after 1400. They illustrate instead the exchange between practitioners and scholars,[86] and show how experiential knowledge was valorized within established cultural hierarchies and values. Generally, mining and metalworking books sought to transform smoky, dirty work into something more noble (and textual). One of the earliest mining texts, the 1492 *Iudicium Iovis* by Paulus Niavis, took the form of a legal suit in which the Earth played plaintiff against humankind because of the damage they had caused her in mining. All sides, including those who claimed the nobility of mining, aired their views.[87]

In his numerous works on mining and metals published between 1530 and 1556, the physician Georgius Agricola (1493–1555) sought to ennoble the work of miners and mine investors among elites scornful of nonagrarian pursuits. As silver production increased fivefold in this period, and mountain towns grew from small hamlets to tens of thousands of inhab-

4.10. Jamnitzer and Amman, "Ein gar kunstlicher und wolgezierter Schreibtisch," fold-out leaf, showing the surveying of a mine. Quadrants hang from a series of lines, illustrating methods of calculating angles and distances. Vol. 2, fol. 54. National Art Library Special Collections, MSL/1893/1600–1601. © Victoria and Albert Museum, London.

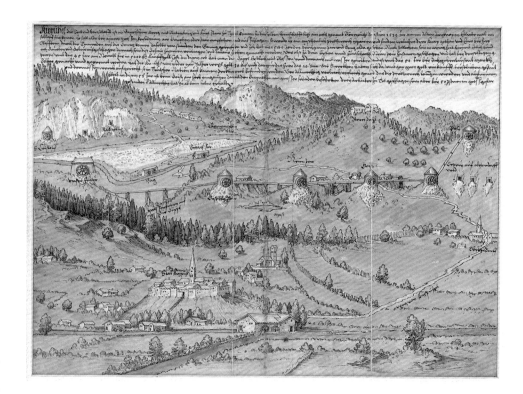

4.11. Silver mining district near Kitzbühel in the Tyrol, showing a linked system for draining water in five different mines, from the final version of "1556 Perkwerch etc.," known as the *Schwazer Bergbuch* (1556), compiled by Georg Ettenhart. The mining boom after the 1530s, when rich silver veins had been discovered, brought great technical and environmental changes to this region. Ferdinandeum Library, Dip. 856, Tafel 14, "Kitzbühel." Tiroler Landesmuseum, Innsbruck.

4.12. Falkenstein silver mining district. From another copy of the final version of the *Schwazer Bergbuch* (1556), compiled by Georg Ettenhart, showing the thousands of mines in this area, and the technical infrastructure that supported the extraction, smelting, and transport of metals. Ferdinandeum Library, FB 4312, Tafel 6, "Falkenstein." Tiroler Landesmuseum, Innsbruck.

Tigna ſtrata A. Pali cuneati B. Tigna tranſuerſaria C. Aſſeres D. Ca
ua E. Venti F. Operculum G. Puteus H. Machina carens operculo I.

O 2   Secunda

4.13. Georgius Agricola, *De re metallica: libri XII* (Basil: Froben, 1561), book 6, p. 159. Aboveground fans intended to bring air into the mine shafts. Bayerische Staatsbibliothek, Munich, Res/ BHS II B 5#Beibd.1. Notice the analogy between the older classical structures in the distance and the new mine ventilation towers in the foreground. http://mdz-nbn -resolving.de/urn:nbn:de:bvb:12 -bsb10199044–2.

itants in a matter of years, Agricola was confronted, in his medical duties as city physician and in the material conditions of his life in these boom towns, with a world far different from that of ancient languages and classical medical authors in which he had been trained. As Chemnitz city physician, he dealt with the newly emerging illnesses of miners, and as a municipal official and a wealthy mine shareholder, he was personally interested in the discovery of rich ores as the basis of both civic and personal wealth. Agricola was also deeply committed as a humanist to trying to square his own new world with the world, culture, and material conditions of antiquity, and he taught Greek to schoolboys in a boom town. He justified mining as a moral and honorable activity by modeling his best known work on mining, *Twelve Books on Mining* (*De re metallica*, 1556), on the Roman writer Columella's advice to landed gentlemen, *Twelve Books on Agriculture* (*De re rustica*) (fig. 4.13).[88]

Agricola's book was produced in collaboration with the artist Basilius Wefring, who made the drawings, and woodcutters employed by the Froben press, including Rudolf Manuel, known as Deutsch, and Zacharias Specklin.[89] Like his other books, *De re metallica* shows the traces of his own interaction with miners and other local informants. Agricola has one of the humanist physician interlocutors in his book *Bermannus* (1530) note that he will take home the rocks and minerals that the miners gave him in order to study them, just as when he was studying at university in Italy, he took home and studied the flora and fauna brought to him by practitioners.[90]

The recording and valorizing of experiential knowledge in the mining towns also occurred by positioning mining within a religious framework. Johannes Mathesius was pastor of St. Joachimsthal from 1542 to 1565, where he preached sermons to the miners that contained minute descriptions of miners' diseases, mining techniques, tools, and machinery. He published his sermons in 1562 as *Sarepta oder Bergpostill*. At one point, his detailed description of a new watermill ends in a prayer:

For this machine, we thank God and the inventor, and all those who daily help to improve such art. Many hands make light work, as the saying goes, but clever heads make light work and prevent much damage. God reward all such who use their heads and ideas for something better and thereby serve our dear mines. Amen.[91]

Like the *Speculum metallorum*, compiled and commissioned by Christoff Hofer, a silver smelter in Schwaz, in 1576, the *Bergpostill* of Mathesius was primarily a work of religious devotion, but also contained much technical information.[92]

In contrast, the *Schwazer Bergbuch*, a compilation of texts relevant to mining, emerged from the economic dictates of mining, put together at a moment of crisis in the mining areas of the Tyrol in the 1550s. With productivity declining and fierce rivalry occurring among the various mining works, mine shareholders sought new investors, especially among the nobility. The manuscript, drafted in 1554, included mining ordinances from the 1490s, and detailed technical information about the processes and people involved in mining, framed as a dialogue between a scholar and a miner (like Agricola's *Bermannus*). It was initiated with the intention of bringing together the principal Tyrolean mining works, the mining officials, and the nobility in a general meeting, a *Bergsynode*, on the model of meetings that had taken place from 1490 to 1513. The 1554 draft was revised and copied out in several versions in 1556 in preparation for this meeting. One of the copies was ornamented as a presentation version for Ferdinand I (1503–64). This version portrayed the miners going about their work in Sunday garb, while the other three 1556 copies showed the miners in torn clothes and barefoot. The *Bergsynode* took place in 1557, and several further copies were made after that.[93] This compilation thus can be aligned with diverse other technical writing on mining that was produced with the aim of informing and attracting investors to an activity that required large capital investments.

Another field requiring capital investors was overseas voyaging, as Gabriel Plattes transparently proclaims in *Discovery of Subterraneall Treasure, viz. Of all manner of Mines and Mineralls, from the Gold to the Coale; with plaine Directions and Rules for the finding of them in all Kingdomes and Countries. And also the Art of Melting, Refining, and Assaying of them is plainly declared, so that every ordinary man, that is indifferently capacious, may with small charge presently try the value of such Oares as shall be found either by rule or by accident [. . .] Also a perfect way to try what colour any Berry, Leafe, Flower, Stalke, Root, Fruit, Seeed, Barke, or Wood will give: with a perfect way to make Colours that they shall not stayne nor fade like ordinary Colours. Very necessary for every one to know, whether he be Travaler by Land or Sea, or in what Country, Dominion, or plantation soever hee shall Inhabite* (London, 1639). In this book, Plattes sets out the tools that will be needed by the man who wants to prospect on his voyages or plantations: "pipkins, urinalls, iron pickaxe, spade, crowbar, take a little piece of each kind of ore." And to do more complicated assays: a grate of iron, some bricks, hand-bellows, tongs, lead, salt-peter, sandiver, borax, Flanders melting pots, ring of iron for the test, a hatchet, a handsaw, aqua fortis, weights and scales. And, says Plattes breezily, "if a man be not active handed, he may have a man for a trifle to shew him the Manuell practice in a day before he goe his voyage."[94] While Plattes's book seems to be written for the man in the service of a state colonial enterprise, other writers, such as Ithier Hobier in his 1622 *De la construction d'une galère et de son equipage*, claimed to have written for commanders, officers, and royal commissioners, who would thereby be provided the means to regulate more efficiently, "without the trouble of inquiring of individuals," that is, artisans, who, he disparagingly noted, often did not understand their profession well.[95]

Conclusion

This chapter has surveyed only some of the great variety of artisan authors and their books of art from the deep past through the eighteenth century. Historians of science, such as Allison Kavey, see the significance of these books in giving readers, not actual experience with nature, but rather the sense that nature was accessible to and manipulable by Everyman (and, in the case of some subjects, by Everywoman).[96] These texts became important to the new experimental philosophers in the early modern scientific societies, including the Royal Society and the Académie royale des sciences, and scholars agree they formed one of the principal paths (along with collaborations with practitioners) by which experiential knowledge and techniques came to underpin the new modes of producing natural knowledge that emerged in the period of the so-called scientific revolution. Yet, we should also acknowledge the varied roles these books played for those who wrote them. While these texts seem to record or to teach how to conduct a technical procedure, their ultimate purpose might be as a defense of the art, proclaiming that "doing" had dignity, and could be a pious and noble undertaking, even an activity of high status, which was capable of expression in written form. As we have seen, many of these texts, describing the practice of an art, involved an effort to establish the theoretical underpinning of the author's expertise.[97] Other texts sought to advertise the skills of the author, or to attract attention or investment to a trade or field of activity, or to provide diversion to an elite audience. Some technical writings functioned simultaneously as "how to do" and "how to be" books,[98] while others were compilations put together by entrepreneurial printers. Like some celebrations of craft and craftspeople in the present day, many texts about practice aimed to reform existing paradigms of knowledge and knowledge-making. This survey of technical writing makes clear that such texts must be examined on a careful, case-by-case basis to understand their genesis and their intent. Their varied origins and functions help to explain why the information in them frequently seems less than useful as instruction and why such books of art cannot easily be classified by type.

CHAPTER 5

Writing *Kunst*

Writing and Embodied Knowledge

As we saw in the last chapter, books of art do not invite their readers to a single kind of reading or use. Even those texts that sound more akin to straightforward instruction books can suffer from the problem of conveying information in writing. We see this in the manuscript of an anonymous pupil of Nicholas Hilliard (c. 1547–1619), the painter of miniature portraits favored by Elizabethan nobility. Hilliard had penned an account of his practice around 1600, and his pupil followed in his footsteps. Initially, this pupil seems to be laying out a step-by-step instruction book:

Haveing furnished yourselfe with all necessary Implements […] & 1st, learn to Draw by a drawing or a Print […] or some part of a Drawing, or Print, as a Bird, a Flower, a Hand, a Foot, a Face, or the like. And […] there are divers such Prints to be sold, for a Young Practisioner to draw after at the first. But afterwards harder Drawings or Prints are better for you: & when you can draw thus reasonable well: then practice by some good imbossed Picture, ether of Wax, Mettall, or Plaister of Paris, & then by the Life observeing exactly every part that doth foreshorten according to the Rules of Art. But take heed that you doe not learn, or practice in colours, before you can draw very well.

But, before long, just like many other practical texts, this one undermines the very premise of writing down art, saying, "The best way of learning to draw well, is to be taught by a good Artist that is able to direct you, & shew you, where you err, than by yourselfe alone."[1] This point is one made over and over again by practitioners when they sit down to write about their techniques.[2] Like other artisan authors, this writer acknowledges that technical writing can be descriptive only; it points to bodily activity, but cannot accomplish that activity, teach it completely, or even describe it fully. This chapter focuses on the paradoxical and ultimately impossible effort to put skilled practice fully into writing.

Craft and Secrecy

In his *Ten Books of Surgery with the Magazine of the Instruments Necessary for It*, Ambroise Paré (c. 1510–90), a barber-surgeon who had trained, as was common in the sixteenth century, by apprenticeship and practice on the battlefield, wrote, "You may find this manner of practicing rather obscure and difficult to understand, but you must consider that it is a very difficult thing to put manual surgery clearly and entirely in writing, for it is rather to be learned by imagination and by seeing good and experienced masters perform, if you have the means, or, indeed, to try it on dead bodies, as I have done many times."[3] This is a familiar expression of the frustrating process of conveying practice in written form. In his *Apologie and Treatise*, Paré recounts how he came by some of his most prized and secret techniques. In Turin, he writes, he found a surgeon famed for curing gunshot wounds, "into whose favour I found meanes to insinuate my selfe, to have the receipt of his balme." Although this master would hold Paré off for two years, Paré was ultimately able to "draw the receipt from him" by gifts and presents. This secret, "to boyle young whelpes new pupped, in oyle of Lillies, [with] earth wormes prepared with Turpentine of *Venice*," aroused great pleasure in Paré, and he wrote, "See then how I have learned to dresse wounds made with gunshot, not by bookes."[4]

Paré's writings point to two common characteristics of books of art: the frustration expressed at not being able to transmit art adequately in writing, and a rhetoric of secrecy. Secrets, such as Paré's "balme," were matters concealed in order to retain them as proprietary knowledge. *Kunstbücher*, recipe compilations, and other books of art often made reference to "secrets." As with Paré's remedy, the secrets of the books of art, however, turn out to be more or less straightforward processes.[5] In these books, "secrets" were often synonymous with "arts" or "techniques."[6] In the 1604 "Goldsmith's Storehouse," an English assayer's book of techniques, the author uses "sciences" and "mysteries" as synonyms—meaning specialized, expert knowledge—and states that a "grounded experience in this Science or mysterie" is necessary to the practitioner of the craft.[7]

Craft knowledge as "secret" thus encompasses several different senses of "secrecy" that need elaboration: "craft" is related to "crafty" or "craftiness," which in English have overtones of deceptive practice. In German, *Kunst* (art)—and, in English, "cunning"—derive from *können*, meaning "to be able." *Metis*, used by Greek writers to mean practical efficacy that could also be effected by magic and trickery, is the root of the French *métier*, or trade.[8] Craft knowledge was thus "secret" insofar as it was hidden from the understanding of those not experienced in the craft, or because its efficacy was "hidden" in the things of nature or in the material objects craftspeople made.[9] Thus, as this etymology shows, craft was perceived as powerful and potentially deceptive.

In addition, crafts involved initiation into both a technical body of knowledge and a social group, so they possessed elements of esoteric knowledge and the rituals of "mystery" cults. Craft, as bodily knowledge, could be linked to the mystery of union with the divine, which was often experienced through the body. In the Christian ceremony of the Eucharist, the body and blood of God is ingested. Moreover, God was beyond human reasoning, thus spiritual mysteries might be better apprehended through the senses. The Muslim mystic Abu Hamid Muhammad ibn Muhammad al-Ghazālī (1058–1111) stated that true knowledge of the divine cannot be expressed in words but can only be experienced as a kind of sensory "tasting."[10] Similarly, medieval Christian mystics, especially female mystics, described union with God through the language of bodily sexual union, hence the extraordinarily erotic—to our ears—language of some medieval religious writings.[11] According to Paracelsus, the artisan had to "sound out" his materials, to be attuned to them, to taste them through the bodily senses, and to "overhear" matter.[12] The artisan's or religious mystic's sensory engagement with the world is also a radically particular and personal type of knowledge, very hard to convey to others in words. In this light, the secrets of craft knowledge thus seem less a matter of deliberate concealment than one of being—like the experiential knowledge of the religious mystic—ineffable. Sensory apprehension—tasting, smelling, seeing, feeling and even hearing God—formed an alternative type of knowledge known in the body and, like craft knowledge, was often referred to as "secrets" or "mysteries."

Another meaning of "secret" was associated with so-called "empirics," vernacular medical practitioners who displayed the "tricks of their trade" and sold their special secrets on public squares.[13] Books of secrets could be one component of astute marketing strategies practiced by self-described "professors of secrets," such as Leonardo Fioravanti (1518–88).[14]

Artisanal associations often called their techniques "secrets" or "mysteries." A guild might enjoin its members to hold close the secrets of the trade, or even include ritualized "mystery" in its initiation procedures. For example, a general meeting of masons' lodges in the German-speaking lands in 1459 proclaimed that "no workman, nor master, nor parlier [master mason in charge of a building site], nor journeyman shall teach anyone, whatever he may be called, not being one of our handicraft and never having done mason work, how to take the elevation from the ground plan." Despite such injunctions, within a generation (in 1486), the German mason Matthäus Roriczer published this secret; however, no one except an expert mason could have seen that his text and illustrations presented a special case of the method of producing a plan of elevation from a ground plan.[15] As Anne-Françoise Garçon has made clear, such practices of "secrecy" were part of the culture and socialization of craft. They could be a means of holding on to the position of expert, and were often formulaic, even employed as a cover to hide officially prohibited processes, such as subcontracting.[16] Craft "secrecy" eventually came to be seen as opposed to an ideal of "openness" proclaimed by economic thinkers, wishing to destroy the corporate power of guilds, and by eighteenth-century writers, especially those of the *Encyclopédie*, who sought to organize and publish craft techniques.[17]

Historians of science and technology adopted this binary that opposed secrecy to openness, and economic historians long maintained that craft secrecy was about the proprietary protection—even monopoly—of technical secrets. However, over the last couple of decades, economic historians have shown that, while such "secrecy" might give the practitioner an economic advantage for a short time, there is little evidence of processes actually being kept secret for very long and much data to the contrary showing just how rapidly technical innovations, such as Murano glassworking and the arcanum of Meissen porcelain, spread.[18] Indeed, in an eighteenth-century Habsburg investigation of whether to retain the journeyman *Wanderschaft* among craft guilds, a report from the Duchy of Krain emphasized that journeymen sent abroad could learn much of use to the territory that they would bring back as innovations when they returned, concluding that, "it is known that in this way very frequently many secrets are discovered and new arts and techniques are introduced."[19]

John K. Ferguson, in *Bibliographical Notes on Histories of Inventions and Books of Secrets*, noted that secrets could refer to expert knowledge, but the "secret" component of them was usually due to the nature of craft labor and bodily work:

It is hardly to be expected that a practical art can have any literature worth speaking of. The man who is busy practicing it can have little time to write about it, and he who wishes to learn it must put to his hand and work at it, and that under the supervision of a master, and not by merely reading books. This is the apprenticeship that everyone must serve. No amount of reading will make a sculptor, or a gardener, or a shoemaker, or a surgeon, or a musical executant. The arts must be acquired by practice, and they are extended and improved by practice. Everyone who exercises them comes to have special power and certain ways of doing things, which may enable him to surpass others who are similarly engaged. These are his "secrets", which very often he cannot or will not, reveal to others. Rapid insight into a particular case, power of overcoming physical obstacles, ingenious adaptations of means to ends, exhibition of due care at the right time, enable one man to effect what others cannot.[20]

Kunst as Knowledge

Having considered in the foregoing section all the various ways in which craft can be "secret," we can now draw together the characteristics of craft as a form of knowledge that makes it prone to being called secret. At the most basic level, descriptions of craft procedures cannot capture workshop experience because such work involves unpredictable qualities of materials, always changing workshop conditions, and rapidly transforming matter, all of which the craftsperson had to respond to in real time. The acute observation and attention to the circumstances of the ephemeral moment necessitated by such conditions makes much of craft knowledge tacit, scarcely able to be captured or codified in writing. As chapter 4 showed, books of art, or *Kunstbücher*, could be written for all kinds of reasons, and in some cases, as I show in this chapter, these books of art sought to convey this tacit knowledge and other essential hallmarks of *Kunst*. The specific historical context of each *Kunstbuch* thus provides one mode of studying these texts on *Kunst*, but we can also investigate them from an ahistorical or phenomenological perspective, in order to define the type of knowledge they contain. "Tacit knowledge" is often used to refer to knowledge that is not written down—as Michael Polanyi says, "we can know more than we can tell."[21] Sometimes this is because no one has ever attempted to write it down, but sometimes it is simply impossible to reduce such knowledge to writing or to transmit it only through written texts. Such is the bodily knowledge involving "muscle memory" that can only be acquired by repeated practice, like that needed by a pianist or a bicycle rider.

In his 2010 book, *Tacit and Explicit Knowledge*, sociologist of knowledge Harry Collins calls this muscle memory "somatic tacit knowledge." Collins does not see this knowledge as in principle tacit because it is often possible to explicate it in physical and mechanical terms.[22] This will not necessarily make it possible to employ in practice, but Collins's differentiation of this individually embodied tacit knowledge from the type of knowledge that he sees as irreducibly tacit—namely, knowledge that is learned and practiced within society—makes a useful distinction. How we move through the world of our fellow human beings cannot be explicated in the quantitative language of the natural sciences, and cannot be incorporated materially into machines, such as robots. The particularity of each human interaction in the social sphere is analogous to the irreducibly particular circumstances of craft processes in an early modern workshop. This chapter proposes that we use the term *Kunst* to refer to this sometimes unverbalizable knowledge of the particulars—what previous ages called "secrets." Although *Kunst* is now translated into English as "art" and is generally understood as fine art, this is a relatively recent development that began in the period of the Renaissance. When Cennino used "art" in the title of his "book of the art," he meant handwork in general. This narrowing of the meaning of art is in fact partly a result of the long process described in this book, through which the explication of knowledge in writing became self-evident as the more powerful form of containing and transmitting knowledge, while the tacit knowledge of particulars came to be seen as an inferior mode of knowing. As ambitious artists, claiming a higher status as practitioners of the liberal arts, stressed the conceptual, imaginative, and theoretical dimensions, rather than the bodily aspects, of their productive capacity, craft came to be viewed as "mere" practice, not rising to the level of "theory." As we saw in chapter 1, in his book of the art, Cennino Cennini already praised "imagination," and held up "theory" as of higher status.

As a result of such long-term developments, *Kunst* (and art), in the broad meaning of handwork, has, like the knowledge of particulars, become devalued: anthropologist Tim Ingold points out that "artificial" once meant "full of deep skill and art," but now it means "shallow, contrived," and almost worthless.[23] Rachel Maines recounts the ways in which activities that previously played a role in survival, such as hunting, fishing, or gardening, have evolved into hobbies, engendering a new attitude to craft.[24] The place of handwork and craft in global capitalism today, in which mass-produced items are routinely called "artisanal," is captured in a *New Yorker* cartoon that depicts a large industrial building, smokestacks filling the air with fumes, topped by a sign "Artisanal, Inc." At the same time, however, handwork and craft produced by knitting collectives and used in "yarn bombing" are performed as subversive acts that intend, through their everyday handmade quality, to critique mass consumption and the status quo. Such contradictions speak to the continued power of the idea of craft in the present era.

Among artists today, conceptual art often privileges form, or the "design," over skilled process of creation. So-called fabricators of contemporary artworks often remain anonymous as their artifacts are exhibited under the name of the artist, who may have provided a design on paper or a verbal description, and left it to the fabricator to bring into being. In contrast, the essence of *Kunst* in the early modern period lay in its origins in *können*, "to be able to do," and *Kunstbücher* evoked this ability to "do," to make, and to produce. Like its English counterpart, "craft," which itself is derived from the Germanic root meaning "power" (*Kraft*), *können* connoted power, as the medical and intellectual reformer Paracelsus indicated in the sixteenth century: "through practice and experience comes power/knowledge [*künden*]."[25] In the early modern period, a *Kunststuck* meant "masterpiece," as a proof of expertise that might in fact seem a deceptive trick to someone unfamiliar with the "tricks of the trade"—the *Kunst*—through which an object could be brought into being.

Nature/Culture and Matter/Form

Theorists of craft, such as Howard Risatti, see craft as springing "directly from our confrontation, as a species, with nature; it is grounded in physiological necessity and in our struggle for survival. This means purpose in craft objects has a primordial dimension that goes beyond culture." Moreover, "craft objects must be seen as nothing less than a physical manifestation of human subjectivity [meaning "that capacity of the human mind to cognition, knowledge, and awareness"] in confrontation with nature. They are a concrete expression of human subjectivity's worlding capacity, of human subjectivity's potential to create a world of culture out of the realm of nature."[26] Ultimately, Risatti views crafts as human beings creating culture "out of the raw substance of nature itself."[27]

While Risatti's view of the material and cultural dimensions of craft is powerful, is it really unique to human subjectivity? Animals also engage in making: beavers build dams to admirably precise effect, honey bees create hives with sophisticated geometric forms, weaverbirds weave regular nests that function like netted bags. Further, new forms of tool use have been observed to be transmitted among communities of wild chimpanzees.[28] Tim Ingold begins, like Risatti, in nature and biology, viewing skilled practice as central to being alive in the material world, whether for nonhuman or human animals,[29] but he does not view this biological sphere as "primordial" and separate from culture. Instead, Ingold sees a continuum between organism and environment, with "skill" being developed at the interface of the two in the organism's perceptual engagement with its environment. As organisms engage with their environment to subsist, skill emerges.[30] Thus, experience does not mediate between cognitive processes and nature, but rather is intrinsic to the process of being alive in the world.

Animals and humans transmit skills in similar ways: through observation and imitation of expert practitioners' bodily movements and gestures. As Ingold puts it, "the progress from clumsiness to dexterity […] is brought about not by way of an internalization of rules and representations," on the model of knowledge derived from books, but rather "through the gradual attunement of movement and perception."[31] Ingold sees imitation and then the coordination of perception and action through practice as giving rise to skill. He argues that male weaverbirds create their woven nests by learning a habitual pattern of movement while young. We may think of weaving as a quintessential result of human culture, but Ingold points out that both for human makers of string bags and for weaverbirds, it is the pattern of regular movement, rather than a prespecified design, that generates the form of the woven object: "the fluency and dexterity of this movement is a function of skills that are developmentally incorporated into the *modus operandi* of the organism—whether avian or human—through practice and experience in an environment."[32] Such movements are not, as is often assumed, instinctual in birds and acquired in humans. Instead, for both, their woven artifacts emerge from the interaction between organism and environment. For Ingold, humans differ from other animals in that they are "peculiarly able to treat the manifold threads of experience as material for further acts of weaving and looping, thereby creating intricate patterns of metaphorical connection. This interweaving of experience is generally conducted in the idioms of speech, as in storytelling, and the patterns to which it gives rise are equivalent to what anthropologists are accustomed to calling 'culture.'"[33]

Early Modern Theorists of *Kunst*

We can use Risatti's and Ingold's insights about human interaction with the material environment for survival, and the drawing out, or emergence, of form from the materials of nature to help us read early modern accounts of *Kunst*. When Albrecht Dürer set pen to paper to define *Kunst* and the actions of the *Künstler*, at the end of the third book of his *Four Books of Human Proportion* (1528), he memorably described art as something embedded in nature, which the artist must extract. Art makes visible what is invisible in nature.[34] In a felicitous double signification, he wrote that the artist must "draw" art out of nature—meaning, as it does also in English, both "to extract" and "to draw on paper," thus adding a deeper dimension to Dürer's repeated admonition in this excursus to develop *"Brauch"* by much drawing. Understanding, says Dürer, must begin and grow with practice (*"gebrauch"*).[35]

As Dürer noted, form and matter are not separate in this process of manipulating the materials of nature. Today, we tend to view design as preceding making, with a preexisting form in the maker's mind, ready to be impressed upon the material. But materials are not freely available everywhere, nor can they assume any form a craftsperson dreams up, and coming to know the properties of materials by working them is part of the necessary testing and trying of craft. In many cases, every material a craftsperson uses is different from the rest. Not only are there differing types of clays and glazes, but also they depend for their properties on the environment inside the kiln during firing. Or, as woodworker Robert Tarule has observed, every tree has different patterns of growth, necessitating varying responses from the woodworker. Thus, as woodcarvers work their way through their blocks of wood, encountering the history of the tree's growth as they go, their works of art emerge only gradually from this sensory encounter with the idiosyncrasies of their material.[36] Indeed, the characteristics we regard as human, such as intention and agency, emerge gradually in the process of making things. Anthropologist Lambros Malafouris regards the directed action of making—stone knapping, in his example—as not *"executing an intention, but actually bringing forth* the maker's intention." In other words, humans make things, and making those things makes humans.[37]

The constraints and the affordances of the materials teach the handworker; indeed, they "grow" the artisan's skills, as Ingold has it. It is the reciprocal interaction between the materials and the body of the maker that brings into being an object; thus, a work of art emerges through a process in which the human body sounds out, and to a certain extent intermingles with, natural materials. Ann-Sophie Lehmann has eloquently shown how, in opposition to the conventional view, materials actually help shape and impose form on the work of art.[38] As Ingold points out, even when a basket maker begins work with a clear idea of the form they wish to create,

the actual, concrete form of the basket, however, does not issue from the idea. It rather comes into being through the graduate unfolding of that field of forces set up through the active and sensuous engagement of practitioner and material. This field is neither internal to the material nor internal to the practitioner (hence external to the material); rather, it cuts across the emergent interface between them. Effectively the form of the basket emerges through a pattern of skilled movement, and it is the rhythmic repetition of that movement that gives rise to the regularity of form.[39]

Form and matter, as Dürer noted, are not distinct in the work of *Kunst*.

This interaction of practitioner and materials is conducted not through an internal process of representing the object in the mind and carrying it out with the hands, but rather through acting, sensing, observing, and adjusting; acting, sensing, observing, and learning; acting, and so on, repeated over and over. This complex process of "mind-body" work is very difficult to convey in words—an established lexicon is not to hand—a problem confronted by the writers of *Kunstbücher*. Cennino talks about "sounding out" his materials, Paracelsus talks about "overhearing" the virtues of natural things, while Biringuccio writes, "Even though to advise you better might require that I should have told you more of these things than I have, you will learn many things by yourself while working and practicing it, such as how to make a choice of clays, stone, moulds, furnaces, seasons, weather, and the like, and it would take too long if I should wish to tell you all of them."[40] Like al-Ghazālī's analogy of "tasting" the true knowledge of God, historically the learning of artisanal skills often involved processes of embodiment and incorporation. Dürer titled his first sketch for a painter's manual *Speis der Malerknaben* (Food for Painters' Apprentices).[41] Paracelsus wrote, "The art of medicine cannot be inherited, nor can it be copied from books; it must be digested many times and many times spat out; one must always rechew it and knead it thoroughly, and one must be alert while learning it."[42] In *Idiota de sapientia et de mentia (The Layman on Wisdom and the Mind)*, Church reformer and cardinal Nicholas of Cusa (1401–64) has the layman say to the scholar, "[Y]ou will hear and taste the truth," and he points out that *sapere* encompasses both "knowing" and "tasting" (carried into English *sapient*, meaning "wise," and *saporous*, meaning "tasty").[43] "Tasting" conveys the merging of subject and object in the bodily engagement of craftspeople in matter, as well as the habituation through experience, and the fine abilities of discernment that it produces. In their focus on materials, bodily gestures, and sensory signs, books of art reveal a type of knowledge that derives its efficacy from being literally in touch with the world, engaged in matter through the body and its senses.

Bodily Knowledge

The human body was an integral part of early modern making; it functioned as a tool in myriad ways—for warming, blowing, handling, manipulating, sensing, tasting, and providing force and dexterity. It also formed a source of materials used in manufacture—including urine, bones, blood, ear wax, saliva, and even excrement. Manuals of art provide many indications of the body in the workshop: a recipe for soldering works of gold or silver calls for the "white thick saliva that is found on the teeth" to be applied, "& next, with a little fat earth wetted with saliva, make a small layer on the opposite side to hold the solder better."[44] The human body was also a conceptual model for natural processes: the fermentation, digestion, purging, and excretion of the body provided a theoretical framework for the transformation of materials in nature. Moreover, the quotidian stuff that sustained growth in the human body—including bread, butter, eggs, milk, cheese, honey, and garlic—was also employed on a daily basis in workshop practices. And, finally, it was by means of the body and its gestures and techniques, learned through interaction with materials, that the embodied knowledge of craft was produced and reproduced, passed on from one generation to the next.

Craft recipes expressed measures in bodily terms such as "four drops of spittle," or "two-fingers wide."[45] A lock of woman's hair was used to measure the temperature of material being heated,[46] and human touch could measure whether an object was "cool enough to be held for a short time in your hand," before being subjected to the next process.[47] All five bodily senses were fully employed in the workshop: vitriol could be identified by its biting, sharp-to-the-taste, pungent-to-the-tongue, and astringent nature, while rock alum had "a bitter taste with a certain unctuous saltiness."[48] Other measurements relied upon hearing: "You will recognize that they are dry enough when […] they cry & crackle once brought near the ear."[49] In a process for hardening mercury, the material in the crucible was supposed to sound a loud bang to sign that it had had enough of the fire.[50] And, in another, "if the tin cries loudly, it is a sign there is not enough lead. If it cries weakly, that means that there is too much."[51] The purity of tin was tested by biting to see whether it made cracking sounds, "like that which water makes when it is frozen by cold." Good iron ore could be indicated by the presence of a red, soft, fat earth that made no crackling noise when squeezed between the teeth.[52] In a dramatic account of casting bells, Theophilus advises the caster to "lie down close to the mouth of the mold," as the metal is poured into the bell mold, "and listen carefully to find out how things are progressing inside. If you hear a light thunder-like rumbling, tell them to stop for a moment and then pour again; and have it done like this, now stopping, now pouring, so that the bell metal settles evenly, until that pot is empty."[53] A goldsmith's treatise advises the assayer to make certain that an acid bath has dissolved all available silver in an alloy by listening carefully to see if the glass vessel makes a "bott, bott, bott" sound when tapped,[54] also demonstrating just how challenging the attempt to render sensory knowledge into words could be.

The workshop and its tools functioned as an extension of the capacities and products of the human body. But the body was more than a tool in production—it was also implicated in the work. The bodies of metalworkers and the very matter upon which they labored interpenetrated each other: bad breath could prevent the adhesion of metal gilding, while metal fumes were known to shorten the lives of metalworkers, necessitating the consumption of butter, as discussed in chapter 2.

The Apprentice's Attention

Working with natural substances entailed bodily labor in the material world, struggling with unpredictable forces and qualities that always had the potential of transforming in unexpected ways. Like living beings, materials came to be known through intimate acquaintance. Materials had needs and desires, and behaved in idiosyncratic ways, which artisans had to learn—and respect or master—through experience. A craft apprenticeship in the workshop educated the body in its interaction and entanglement with materials through training the senses to discern the character and differences among materials (an understanding that can be discerned in the material imaginary, discussed in chapter 3), and, at the same time, transforming the body—its "muscles, attention patterns, motor control, neurological systems, emotional reactions, interaction patterns, and top-down self-management techniques," as one neuro-anthropologist has put it.[55] Today, we forget how highly discerning and effective trained sensory apperception can be. As Raymond Tallis points out, the hand is not a dim, groping object, but an organ of exploration and cognition. The ridges on the hands and fingertips, individual to each person—the fingerprint—register and amplify small changes in the surface being touched.[56] In daily life, these stunning abilities of sensing and motor control become second nature to most people, but robotics has reminded us that stacking blocks, which utilizes the sensitivity of the hand, and placing one foot in front of the other to take steps are some of the hardest actions for robots to master.[57]

Consider again the recipes involving lizards, discussed in chapter 3. These recipes include many naturalistic details, for example, about approaching the lizards carefully and preventing them from suffocating by perforating the vessels in which they are contained. Might these be understood as inculcating certain habits necessary to craft practice? The concentration of the lizard hunt, for example, reminds one to "stay alert!" Many recipes make reference to being wide awake to the changes in the state of the material and the task at hand. In a similar vein, the injunction not to suffocate the lizards seems to say to the practitioner, Put yourself in the position of the material—enjoining the practitioner to come to know the materials through complete bodily absorption.[58] The burning of lizards to a temperature at which the lizards have turned to ash but brass is not yet melted might indicate the importance of judging temperature and knowing the behavior of materials on heating, for, as Biringuccio noted about metalworking, the regulation of fire is "a very important thing at the beginning, middle, and end, because it is the principal agent."[59] These instructions perhaps can be understood as conveying the necessity of "feeling" and "sensing" in coming to "know" matter.[60] Such attitudes, necessary to a craft, may be an important component of the "mystery" or "secret" of the craft that could be reinforced by an apprentice's initiation.

Collections of written recipes, such as the lizard examples, might encourage attentive meditation on materials and help to develop habits. Moreover, by their very repetition, often listing different variations of ingredients or different methods of doing an operation—sometimes simply titling these variations as "another way"—recipes could encourage constant testing and experimentation. The recipe collections' repetition with variation teaches that materials are something through which to explore challenges and "resistances," in order to seek out the characteristics of a material in different situations. As Biringuccio put it about smelting: "[i]t is necessary to find the true method by doing it again and again […] to have a superabundance of tests […] not only by using ordinary things but also by varying the quantities, adding now half the quantity of the ore and now an equal portion, now twice and now three times," for the "ores must be so tormented that the obstinacy of their hardness is overcome."[61] In *L'Arte vetraria* (The Art of Glass Making, 1612), Antonio Neri advised readers to add colors to the glass melt very gradually and methodically, repeatedly pulling it out of the furnace to check and adjust the color. Only by experimentation with different quantities and types of pigments could a practitioner come to know the materials.[62] A thorough grounding in the habit of exploring methodically the behavior of matter, like Michael's computational practice, developed a capacity sometimes called "intuition," an ability to make split-second responses to the variable behavior of the materials. Books of secrets could not bring about that ability, but they could lay out the importance of educating the attention, of overhearing or thinking with matter, and of repeated exploring and responding to the contingencies of the workshop.

Education of the attention necessitates being observant and attentive in the face of particular instances. In perception and cognition, the senses recognize, assign, sort, and classify external phenomena; in a word, they "discern," as the body literally takes in the world.[63] Ingold describes perception as based in attentive "listening." He chooses "listening" rather than "seeing" because listening implies a resonance between listener and the world, while seeing detaches the observer from the world, setting up a relationship of distanced "objectivity."[64] Seeking to move away from thought to the activity that better describes the sensory engagement with the world, he notes that perception has been described as starting with seeing and hearing, that is, where things are "translated or 'cross over' from the outside to the inside." Because sight starts from light entering the eyes of a perceiver and hearing from the point where sounds begin to reverberate in the ear—"the interface […] between inside and outside"—this is usually described as the beginning of perception. But, Ingold goes on,

the notice beside the railway tracks did not advise the pedestrian to 'stand, see and hear'. It advised him to 'stop, look and listen', that is, to interrupt one bodily activity, of walking, and to initiate another, of looking-and-listening […] In what, then, does this activity consist? […] It consists, rather in a kind of scanning movement, accomplished by the whole body—albeit from a fixed location—and which both seeks out, and responds to, modulations or inflections in the environment to which it is attuned. As such, perception is not an 'inside-the-head' operation, performed upon the raw material of sensation, but takes place in circuits that cross-cut the boundaries between brain, body and world.[65]

Craft apprentices' training, then, is not about transmission of information, but rather about educating, or drawing forth, habits of attention. Early modern writers, such as Paracelsus and Bernard Palissy, wrote that the practitioner must be wide awake,[66] he must learn to pay attention with all the senses, to "overhear" matter, as the metal caster learns to seize the moment at which bronze is ready to pour and the mason begins to recognize the signs of the vein at which the slab in the quarry can be broken away.

Learning to attend is the essence of skills taught in craft apprenticeships, which, in the early modern world, began by teaching humility. Beginning with the meanest of tasks— picking up chicken bones under the table, for example, as Cennino recalls—attending to the menial tasks of the workshop creates vulnerability and opens a person up to attending to and learning from the world. Nicholas of Cusa wrote that true knowledge, the kind that comes from "tasting," will of necessity be humbling.[67] Dürer's so-called *Aesthetischer Exkurs* can be read as a manifesto of humility, a stance to being that emerges from engaging with the particularity and variety of the world. He repeatedly contrasts the power of God to create to the puny capacities of the individual: "thy might is powerless against the creation of God." Anyone who is sure he has certainty can only deceive himself.[68] And there is no human perfection. Man is the measure of all things, wrote the "sophist" Protagoras, expressing the same uncertainty as Dürer, if from a radically different religious standpoint.

Doing chores around the workshop put the apprentice in a liminal position, but it also began the process of learning by unobtrusive observation and imitation. A survey of apprenticeship indentures from Antwerp in the sixteenth to eighteenth centuries reveals that learning was taken for granted in these contracts as soon as the apprentice was active in the workshop. In this study, Bert de Munck has found that the only systematic procedure for learning recorded by these documents was that apprentices were assigned simple repetitive tasks first, and only began to carry out more complex assignments as they gained experience in these simple procedures. Even a 1674 silversmith's apprentice contract, which specified that the master would teach the apprentice "the science of silversmithing," did not specify any modern didactic procedures, or a separation of empirical and theoretical knowledge that we might consider necessary today.[69]

Sociologists and anthropologists label this form of learning "legitimate peripheral participation." Jean Lave contrasts such an "understanding in practice" with older views in which learning "entails an internalization of collective representations or, in a word, *enculturation*." Instead, "understanding in practice" is "a process of *enskilment*, in which learning is inseparable from doing, and in which both are embedded in the context of a practical engagement in the world."[70] Scholars of pedagogy regard this as an extremely effective mode of learning because it is "situated."[71] As Paracelsus put it in the sixteenth century, "To teach and do nothing is little. To teach and do, that is great and whole […] A stonecutter who teaches his apprentices with his hands rather than with his tongue teaches and acts at the same time."[72]

In the natural sciences, too, this kind of knowledge and teaching is central to experimentation. In his treatment of tacit knowledge, Harry Collins points out four components in which natural scientists learn by such apprenticeship: (1) "mismatched salience," in which the scientists, unknown to themselves, focus on different salient variables in a new experiment, and only learn about their differences by watching others; (2) "ostensive knowledge," which can only be conveyed by pointing, demonstrating, or feeling; (3) "unrecognized knowledge," in which aspects of an experiment are performed without their importance being recognized explicitly, but rather absorbed implicitly; and (4) "uncognized/uncognizable knowledge," which involves skills like language that can "be passed on *only* through apprenticeship and unconscious emulation." Through repetition and the social dynamics of the lab, they can also move out of the realm of tacit knowledge as "procedures that were once esoteric and difficult […] become routine." For example, unrecognized knowledge can become obvious and explained as the procedure becomes routinized. Moreover, "mechanical and 'turnkey'" methods for packaging the experiment are worked out.[73] Such turnkey methods might have been called "secrets" in early modern Europe.

Collective Cognition

The fundamentally social dimension of handwork, whether in the trades or in the sciences, has often been overlooked because our prevailing models of knowledge acquisition are top-down, with ideas born in the heads of individuals, later to be applied through imposition of a design. In contrast, in considering craft knowledge, it is helpful to bring in the concept of "distributed cognition," in which "people appear *to think in conjunction or partnership* with others and with the help of culturally provided tools and implements."[74] In this view, cognition is situated in certain practices, and intelligence "emerges" from group processes.[75] Early modern artisanal texts note this process: Dürer wrote that an artist must see the paintings of many famous masters, copy them, and hear the masters speak—an apprentice must learn from many different artists, and listen to everyone.[76] Biringuccio discussed the necessity of observing the methods of other craftsmen, for "seeing the methods that others use is a wide gateway toward proceeding confidently by other paths in order to arrive at desired ends."[77] And, when Cellini worked in Paris, he witnessed this dynamic of knowledge creation: "I had in my employ many workmen, and inasmuch as they very gladly learnt from me, so I was not above learning from them."[78] Michael Baxandall's discussion of Herbert Gerhard's making of the Fugger altarpiece, in which the metalsmiths consumed butter to ward against the "evil smoke," makes clear this collective cognitive process. Baxandall recounts how sculptor Hubert Gerhard (1540/50–before 1621) modeled the figures, then a Florentine stuccateur translated the models into molds, then the molds were "filled with great difficulty by a native copper-smith of Augsburg," and the casts "worked up by a local goldsmith of not very great reputation." Baxandall concludes, "Gerhard's models were the basis for a cumulative development within the capacities of a succession of craftsmen."[79] Dürer theorized this cognitive process in his remarks about the practices of imitation, that is, first copying after other masters, and then copying after nature, which resulted in a bodily storing up of experience. But Dürer emphasized that out of this bodily practice, *Kunst*—by which he meant a kind of collective, higher order knowledge, larger than an individual's skills—comes into existence. Furthermore, these practices—transformed into art—are manifested in an object. In the 1520s, Dürer articulated this development of knowledge:

no man can ever make a beautiful image out of his private imagination [*eygen sinnen*] unless he have replenished his mind by much painting from life. That can no longer be called private but has become "*Kunst*," acquired and gained by study, which germinates, grows and becomes fruitful of its kind. Hence it comes that the stored-up secret treasure of the heart is manifested by the work and the new creature which a man creates in his heart in the shape of a thing.[80]

Craft knowledge not only required a community to create and train, but also a community of experts to discuss, compare, and judge expertise.[81] As craft was not practiced in words, so expertise and mastery were not demonstrated in words, but rather, craft knowledge was demonstrated in public. Artisans proved their mastery of a craft after their years of apprenticeship by producing a masterpiece, judged by the other guild members. In the early modern period, when most artworks were made in accordance with a contract or sold on an emerging open art market, the proof was in the product. Often contracts included a clause that specified that "good artists" would judge the final work before a maker could be paid the final installment. A contract involving the Tyrolean artist Michael Pacher (c. 1435–98) stated that in the case of dispute, patron and artist would appoint equal numbers of experts; a sixteenth-century contract dictated that the artists "will make and complete for said city, as well as they are able, according to the standards of craftsmen and people who understand this matter, all and each of the works of painting and sculpture."[82] Isabella d'Este directed her agent Francesco Malatesta to consult Leonardo da Vinci about the value of some vases to be sold from the Medici collection.[83] The merchant and go-between, Philipp Hainhofer, in writing to his princely patrons about works of art, assured them that the authenticity and value of the works had been judged by knowledgeable artisans.[84] Skill was always practiced and had to be judged within a community of practitioners and other members of their particular cultural field. As Peter Dormer in *The Art of the Maker* expresses it, "Public scrutiny is a key element in the development of tacit knowledge."[85] Skills are built, craft is demonstrated, and things are made within a community of experts, and proof and demonstration of knowledge are rendered through material objects.

Transmitting Culture

Anthropologists, including Michael W. Coy, have employed apprenticeship as fieldwork in cultural anthropology because learning a "culture" is also a process of enskilment that involves ways of knowing and perceiving that are experiential. As he writes, apprenticeship is the means of imparting specialized knowledge to a new generation of practitioners. It is the rite of passage that transforms novices into experts. It is a means of learning things that cannot be easily communicated by conventional means. Apprenticeship is employed where there is implicit knowledge to be acquired through long-term observation and experience. This knowledge relates not only to the physical skills associated with a craft, but also to the means of structuring economic and social relationships between oneself and other practitioners, between oneself and one's clients.[86]

Apprenticeship consists in becoming familiar through constant exposure with a subject, through testing and trying bodily techniques and various materials over years, resulting in a gradual tutoring of one's senses. This brings an ability to see patterns, to discriminate between similar but different qualities, and to recognize nuances. It involves not just looking and listening, and trying and testing, but also talking and conferring, both about different experiences of matter and about judgments of artifacts.[87] Discriminating can also bring about purposeful improvement in techniques.[88]

Repetitive practice, emphasized again and again by early modern artisanal writers, is the key to developing the necessary abilities of discernment. Artists' manuals record the constant replication of bodily practice necessary: Cennino, for example, wrote that the apprentice and journeyman must keep "drawing all the time, never leaving off, either on holidays or on workdays,"[89] and reiterated this advice to "make sure that every day, without a break, you draw at least something."[90] Albrecht Dürer, son of a goldsmith, stressed the need for copying, from one's master and from life.[91] Michelangelo, too, saw repetitive bodily practice as the key to knowledge; on a study sheet over some attempts by an apprentice, he scrawled: "draw Antonio, draw Antonio / draw and do not waste time."[92] In lawsuits about apprenticeships gone wrong, Bert De Munck finds evidence that the model for apprenticeship presupposed that "the work of apprentices consistently underwent corrective adjustments, presumably in their [the master's] presence. Teaching and learning meant demonstrating and emulating, with the understanding that apprentices were informed of their mistakes and received encouragement and advanced through trial and error."[93]

Michael Polanyi seeks to describe the development of these abilities. Repetition in acquiring a skill is not "mere" repetition, but "a structural change achieved by repeated mental effort."[94] Although he uses the term "mental effort" here, his account of how this happens makes clear this is also a bodily process. Polanyi draws a distinction between subsidiary awareness and focal awareness, giving the example of the pianist who shifts her attention from a subsidiary awareness of the movement of her hands in relation to the notes and music to a focus on the individual movements of her fingers. Such a shift in attention can lead to a breakdown in performance. This kind of focal awareness on the particulars of a skill, whether the hammering of a carpenter, the skilled handling of a tennis racket, or the abilities of the average car driver, moves in the course of repeated practice from a focus on particular components of the skill to an increasing unconsciousness of the particular actions, and finally results in attainment of the ability to hold the particulars in subsidiary awareness while performing a series of integrated movements and procedures to bring about a whole skilled performance or result. Polanyi says, "In the exercise of skill and the practice of connoisseurship, the art of knowing is seen to involve an intentional change of being: the pouring of ourselves into the subsidiary awareness of particulars, which in the performance of skills are instrumental to a skillful achievement, and which in the exercise of connoisseurship function as the elements of the observed comprehensive whole."[95] In another example, as described by the anthropologist Gladys Reichard, the Navajo weaver "must keep the composition of the entire rug surface in her mind, but she must see it as a huge succession of stripes only one weft strand wide. It matters not how ideal her general conception may be, if she cannot see it in terms of the narrowest stripe, meaning a row of properly placed wefts, it will fail of execution." The weaver's focus shifts between the entire composition and the weft strip being woven.[96]

In considering the stages of proficiency in glassblowing, Erin O'Connor further develops the embodied dimension. Expert action, she says, takes on a "*lived* character" that flows out in an "arc of embodied technique." Instructions by experts often involve admonitions to "pay attention" to particular components of technique, yet these are then meant to slip back into the unconscious of the novice, to become, as Pierre Bourdieu called it, "habitus." This is a gradual process of the "attunement of movement and perception," rather than the internalization of rules.[97] The apprentice absorbs—or better, incorporates—this by bodily imitation and mimicry, and much practice results in "non-reflective anticipation," an ability to anticipate what needs to be done without reflecting upon it.[98] Cognitive scientist Cecilia Heyes sees imitation as crucial in this process because it calls upon "associative sequence learning," in which "sensory representation is activated by observing another person's behavior; this activation is propagated to the motor representation via the associative link between the sensory and motor representations; and the resulting motor activation enables […] the observer to perform the observed action." These "matching vertical associations are forged by learning, predominantly social learning."[99] Neuroscientists today see suggestive connections between brain regions responsible for spontaneous speech and those for skills such as musical improvisation, which may indicate that neural processes of learning a language are similar to learning a technical skill. It is worth noting that both are socially immersive processes.[100]

In the constant practice of the workshop, much experience in the sometimes idiosyncratic properties and varieties of materials is gained, and this working through the resistances of materials in order to develop this active anticipation is part of the journey that an apprentice makes from clumsiness to dexterity. Cognitive neuroscience calls this process "motor learning," the study of which promises to shed light on the neural basis of a process that can be found expressed in artisanal accounts from early modern Europe to the present day: The author of "The Goldsmith's Storehouse" included a clear description of the bases for this process: "a p[er]fitt Assay Master, whose perfection [is] grounded upon Artificiall Exercise, for these thinges doe rather consiste in doing then in referringe, for they are not easelye reduced to matter of Argument unless exercise bee ioyned with speech." In his words, the trade "asketh a good Judgment, gotten rather by years & experience, th[a]n by speculation & dispute." The assay master must possess "grounded experience in this Science or mysterie, [having] a perfect Eye to vewe, & a stedye hand to waye for other mens senses cannot serve him."[101] As a modern Japanese swordsmith trained in traditional methods put it, "Remember, our work is not done by measuring and talking. The hammering, the forging, all the processes are performed by intuition. It's the split-second intuitive decision to remove the iron from the fire, when and how to bring up the flame, to immerse the blade in the water now—it is these acts of intuition that produce."[102] In other words, the handworker must know when to seize the moment, discerning patterns through long experience, despite the varying circumstances and state of the materials. Such "intuition" could only come through practice: it was not an "instinctive" reaction, but one developed by the long process of what modern investigators have called "reflective practice"[103] or sequenced motor learning.

Judgment and Discretion

The goal of a practitioner's repeated trial and error was "skill," that is, a capacity of "judgment" that made him able to improvise in response to the contingencies of the workshop conditions and the materials. There were no fixed rules in such an environment, especially in the metalsmith's workshop. Biringuccio wrote of the metal founder's judgment born of experience in designing vents and gates.[104] He notes that it is the requirements of the work that will determine the correct alloy of tin and copper,[105] and warns that "no other rule can be given in this except to tell you that in regard to the weight you must use discretion according to your judgment and experience."[106] Such ability to discern, built on long experience, can only inspire admiration, and he praises the working of iron, which he notes has more secrets, "perhaps more ingenious secrets," than the art of any other metal, especially "when I consider that the masters make their works without moulds or pattern, letting only the eye and good judgment suffice for it."[107] In his repeated references to such capacities, he uses *judicio* (judgment) to express this proficiency in knowing what is "right," and *ingegno* to mean the capacity to improvise in response to new circumstances. Edward Norgate calls this facility "discretion," when he notes that "in a word all must be left to discretion and practice which will infallibly lead you to the right middle temper [of pigment and binder]."[108] We might call this "intuition" or "intuitive response," as Nicholas of Cusa has the craftsman say to the scholar in trying to explain that God is both simple and incomprehensible: he uses the verb *intuetur*, that is, to apprehend by "direct, non-deductive, non-discursive understanding."[109] In 1306, Henri de Mondeville called this capacity of expert problem solving "foresight," suggesting that physicians should follow the example of masons on holiday: "These workers of Paris, coming and going through the streets on Sundays and holidays, examining and criticizing the mechanical structures such as walls, houses and other similar work, under way or completed. They are very useful to other workers in their foresight, and to the bourgeois for the construction of their houses."[110] Paracelsus distinguished between "practice," by which a person learns to make a thing so the work comes out as planned, and "experience," which teaches how to carry out the work in all situations.[111] All these writers were attempting to define and convey the practitioner's knowledge, gained by engaging bodily and sensorily with materials, which lay in the ability to respond to uncertainty by drawing on experience. This skill, based on many iterations of practice, allowed the practitioner to work with different (nonstandard) materials and in varying environments, yet to produce with certainty a desired result. As we saw in chapter 3, Biringuccio believed "those things that have such inner powers, like herbs, fruits, roots, animals, precious stones, metals, or other stones, can be understood only thru oft-repeated experience."[112] More recently, Tim Ingold put it this way:

acting in the world is the skilled practitioner's way of knowing it. It is in the direct contact with materials, whether or not mediated by tools—in the attentive touching, feeling, handling, looking and listening that is entailed in the very process of creative work—that technical knowledge is gained as well as applied. No separate corpus of rules and representations is required to organize perceptual data or to formulate instructions for action.[113]

This facility, often referred to today as "expertise" or "skill," which enabled the practitioner to develop the capacity to intuit and improvise in the face of contingency, was, and is, the essence of craft knowledge—*Kunst*.

From Training Skill to Inborn Talent

The view of *Kunst* as a powerful form of knowledge is so different from current views of art that it is useful to consider how changes in the structure of artistic training set the path for conceptions of knowledge and skill that are still generally held today. In the mid-seventeenth century, academies of art began to be established to train individuals in the "fine" arts of painting and sculpture. These academies were just one aspect of a competition within artisanal communities that elite artisans had been waging since the late 1400s in an attempt to differentiate themselves, as practitioners of the liberal arts, from others in their craft. This drive to distinguish themselves also led to the breaking off of "fine" painters from house painters, apothecaries, and pigment sellers. This stratification grew stronger over the subsequent centuries, reinforced by the new academies of art, as stated baldly by the Secret Council of the Academy of Arts in Antwerp in 1770: "it would be improper to confuse fine arts with mechanical arts; the most astute painters' brushes with the coarse brushes of workmen, applying colours to a wall or a door."[114] We can observe in the structure and methods of the academies the transformation in attitudes to manual activities, indeed, to the meaning of "art" itself: to enter the academies, artists had to present a work of their own creation at the very beginning of training, in order to exclude those with "so little talent that they did not merit this distinction."[115] In a guild, artisans had no evaluative entrance test, but rather had to prove that they could be socialized and acculturated into a family or a workshop and were able to work to prevailing standards of the craft and value systems of the guild. If a guild required a masterpiece of prospective masters after apprenticeship, it was to test whether they met the standards of the guild in performing techniques, not whether they could express an individual creative or stylistic talent.[116] The emphasis in the guilds was on learning through observation, imitation, repetition, and experience, underpinned by the assumption that with sufficient practice most individuals could learn the necessary skills. The academies, in contrast, stressed inborn creative talent, competition for places, and a hierarchy that made manual technique inferior to an innate quality of intellect.[117] Academy training became standard for fine artists over the course of the eighteenth century, and notions of art as the preserve of the talented few grew ever stronger.

Conclusion

In *Ways of the Hand: The Organization of Improvised Conduct*, anthropologist David Sudnow sets out to describe improvisation from the standpoint of the practitioner. He asks whether it is possible to describe in words "the body's improvisational way," not through an introspective consciousness but by a close examination of concrete problems "posed by the task of sustaining an orderly activity."[118] In answer to this task he set himself, he recounts his five-year apprenticeship in learning jazz improvisation (he was already a competent pianist, but could not improvise), and his description, with its many evocative passages and neologisms, demonstrates the challenges of writing down the *Kunst* of improvisation: he describes his breakthrough after five years as "learning to sing" with the body, hands, fingertips, and everywhere. As he recounts, "from the middle of the piano, the beginner gradually acquires an incorporated sense of places and distances, 'incorporated,' for example, in that finding the named, recognizable, visually grasped place-out-there, through looking's theoretic work, becomes unnecessary, and the body's own appreciative structures serve as a means of finding a place to go."[119] And he concludes, "I learned this language through five years of overhearing it spoken. I had come to learn, overhearing and overseeing this jazz as my instructable hands' ways—in a terrain nexus of hands and keyboard whose respective surfaces had become known as the respective surfaces of my tongue and teeth and palate are known to each other—that this jazz music *is* ways of moving from place to place as singings with my fingers. To *define* jazz (as to define any phenomenon of human action) is to *describe* the body's ways."[120]

At the close of his account, Sudnow considers trying to "sketch an orchestration of descriptive writings that seems to more closely 'point to' possibilities for the study of bodily activities, in terms of which such inquiry might receive mundane grounding in accounts of accomplished paced-place achievings."[121] While experts might quibble with Sudnow's description of jazz improvisation, I quote it here to suggest that such an "orchestration of descriptive writings" is what we confront in the writings of craftspeople around 1400: their descriptions of observing, imitating, practicing, gaining experiences, and achieving skill—articulated in the inadequate mode of the written word—in order to point to the bodily activities and the embodied knowledge of *Kunst*. This chapter has provided an account of that extremely complex amalgam of knowledge-in-practice and thinking-through-making that modern cognitive scientists study and that artisan authors such as Biringuccio and Dürer sought to convey. This complex form of knowledge can also be unearthed in a much more quotidian and humble written form in early modern Europe: the recipe, which is the subject of the next chapter.

CHAPTER 6

Recipes for *Kunst*

Recipes

Early modern recipe texts often seem simple records of practice or straightforward instruction in specific techniques. They possess frustrating characteristics: they set us down in the middle of action, mostly starting with the term *recipe* (often abbreviated to "Rx"), Latin for "take," sometimes followed by a list of ingredients, signifying to a reader to "take these ingredients." Recipe collections often contain many slightly varied—and sometimes contradictory—recipes for the same process. Moreover, recipes are often contained in compilations that seem to have no central organizing principle, as they move, for example, from steel-making to digestive remedies on the same page. They seem to form scattershot collections of particular instances.

Although recipes and recipe collections appeared simply as how-to or "technical writing" to a previous generation of scholars, recent work on recipes has made clear their much broader dimensions. Elaine Leong, Sara Pennell, and Alisha Rankin have made clear that recipes possessed social functions and exchange value in Europe in the sixteenth through eighteenth centuries. As Leong and Pennell note about medical recipe collections,

recipes can be seen as analogous to particular forms of early modern financial transaction, notably bills of exchange, in that their realizable value was tied up with the trustworthiness of the relationship on which the exchange was based. But recipe exchanges also at times involved recipes as a variety of gift, where the values placed on the texts donated and received were framed by social relations, as much as any inherent "value" in the recipe itself.[1]

Pennell and Michelle DiMeo have captured this socially connective function of recipes in the early modern period in labeling them "choreographies of connection."[2] Wendy Wall has used recipes to open up "household culture," to understand attitudes to gender and domestic work, and hierarchies of taste and knowledge,[3] and Elizabeth Spiller has explored the intersections among cooking and medical practices.[4] In examining the first known illustrated cookbook, compiled by Bartolomeo Scappi, Deborah Krohn revealed the role that culinary recipes played in the emerging culture of antiquarianism and the more general interest in practical knowledge.[5] Alisha Rankin and Elaine Leong have shown how recipes were connected to the making of natural knowledge.[6] These scholars, among many others, have illuminated the multiple functions and registers of meaning of recipes and recipe collections.

Francisco Alonso-Almeida has described the recipe as a "discourse colony" (a term used by Michael Hoey to describe encyclopedias and inventories), in an analogy to a hive in which a "queen bee" oversees the "colony" but could never be considered the creator of the whole. He sets out the characteristics of a "discourse colony," which makes clear its flexibility as a textual form:

· Meaning does not derive from sequence
· Adjacent units do not form continuous prose
· Framing context may or may not be present
· No single author, and/or anonymous
· One component may be used without referring to the others
· Components may be reprinted or reused in subsequent works
· Components may be added, removed, or altered, and many of the components serve the same function.[7]

Enumerating the ways in which recipes are open-ended and flexible enables us to see their accretive nature as a strength of the genre rather than a weakness.

Recipe Varieties

Most people associate recipes with step-by-step instructions for cooking and preparing food; however, recipes are extraordinarily diverse. Some recipes do contain step-by-step exposition, such as a recipe for a "beautiful fine" blue color from a collection of recipes for pigment-makers and dyers written down at the very beginning of the fifteenth century in Old Middle German and known as the Strasbourg Manuscript:

If you want to make beautiful fine (*blue*) clothlet colour. Then take during the first eight days after Whitsun, 7 (*12 hand*) fulls of cornflowers gathered in the morning before midday. Break the blossoms at their top into a clean dish. Crush the blossoms of the flowers in a clean mortar until they become as a mush. After that put them (*from the stone*) in a clean twill cloth and wring the juice very well through the cloth into a glazed vessel. Take a *settin* of sal ammoniac and lay it in the colour, so it will dissolve directly. Then take a well clean washed cloth from an old veil or an old tablecloth. Dip the cloths into the colour until it absorbs all the colour and the cloths should become neither too wet nor too dry. Boil the colour until they have received (the colour) everywhere. Let them well dry. After that the next morning you have to gather again fresh flowers—the same quantity as before. You should again break them at their top and again make a mush like before and wring them through the twill cloth in the (*vessel*). Then take gum Arabic which must be very clear and which has softened before (*with water*). One should levigate the gum with a finger together and mix the dissolved gum under the (*juice of*) the flowers. Stir it all together with the stick. Take a *settin* of ice alum finely ground to powder. Lay the powder in the aforesaid cloths and press them in the (*blue*) colour again. Leave them in the colour until they have completely absorbed the colour and they are well coloured. Then hang up the cloths again in the wind and let them dry completely. Then wrap the cloths into a convenient paper and put them in a clean new wooden box for keeping. Place it somewhere high, exposed to the air, so that [it] will not moisten.[8]

Although the aim of this recipe is unfamiliar to us today—the blue clothlets are in effect concentrated dye "tablets" that will later be finely ground and the blue extracted for dyeing of new fabric—this recipe contains a step-by-step exposition that is easy to follow, in a form still familiar today.

Similar to the lizard recipes discussed in chapter 3, other recipes provide step-by-step instructions in entirely implausible actions:

If you want to carve a piece of rock crystal, take a two- or three-year-old goat and bind its feet together and cut a hole between its breast and stomach, in the place where the heart is, and put the crystal in there, so that it lies in its blood until it is hot. At once take it out and engrave whatever you want on it, while this heat lasts. When it begins to cool and become hard, put it back in the goat's blood, take it out again when it is hot, and engrave it. Keep on doing so until you finish the carving. Finally, heat it again, take it out and rub it with a woolen cloth so that you may render it brilliant with the same blood.[9]

There is a remarkable specificity and detail about these unlikely techniques—why a two- to three-year-old goat or a woolen cloth, for instance? The tone of this recipe seems to indicate that the practice comes out of experience even when, in written form, it goes back millennia more or less unchanged, to at least Pliny. Pliny consulted technical manuals available to him, probably including an early version of the third-century BCE recipe collection now known as the Stockholm Papyrus, which contains a recipe for softening crystals with goat's blood.

Recipes such as this one using goat's blood can be found repeated in recipe collections over many centuries, even though they could scarcely have functioned as written. Such ancient recipes were not necessarily copied mindlessly by a scribe; instead, slight variations appear to indicate that authors, scribes, and recipe compilers integrated information from other sources and in some cases from their own experience.[10] To explain this, scholars have speculated that such recipes originated in practical techniques—in the goat-blood recipe, the "quench-cracking" of rock crystals, in which the rock was heated, then plunged into a cold liquid to crack it and allow dye to penetrate and color the stone.[11] It may be the case, however, that such recipes owe their durability to the way they reinforced the Hippocratic-Aristotelian-Galenic view of nature and health, in which opposites must be combined (or "tempered") in order to bring the four elements/humors/qualities into a balance that results in good health, as described in chapters 2 and 3. As noted there, books of secrets often indiscriminately mix medical recipes for human health with instructions for pigment, dye, and metallurgical recipes. This should not necessarily be viewed as a random combination, for all these recipes operate on the basis of the same coordinates of the four humors and qualities, and the same principles of tempering, in order to bring about balance (and thus health). This is further demonstrated in Albertus Magnus's comments about goat's blood, in which he notes that the diamond—by which he means three different types of material (diamond, magnetite, or very hard metal)—cannot be softened with fire or iron, but it can be destroyed by the blood and flesh of a goat, "especially if the goat has for a considerable time beforehand drunk wine with wild parsley or eaten mountain fenugreek; for the blood of such a goat is strong enough even to break up a stone in the bladder, in those afflicted with the gravel."[12] Albertus moves seamlessly from the stones to be worked by an artisan to those to be cured by a physician.

The 1604 "Goldsmith's Storehouse" contains an even more striking mix of firsthand experience and material imaginary with regard to goat's blood: "some Aucthors doe wryte that the Dyamon cannot be broken, butt with the new warme bludd of a goate, but it is not soe, for Dyamon Cutters have dayly experience to the Contrary, who doe continually use the powlder of Dyamons." It goes on to make the point that nothing will cut or polish a diamond except the diamond itself. While modern scholars might seize upon this "correct" observation that "must" represent firsthand experience, it is important to notice that the anonymous author of this text has not actually contradicted the goat's-blood recipe, but rather has stated that it is not the *sole* means of cutting a diamond.[13] Indeed, in other sections of this extensive manuscript, the author provides a curious (to our ears) combination of material, medical, and monetary values inhering in precious and semiprecious stones. Coral, for example,

is a wede betweene a herbe and a tree, that groweth in the bottome of the sea, smale & braunched like hornes […] There is two sortes of it, the one reddish like old Ivorye, the other whyte, & is figured in manner like the braunches of trees, the Redder sorte being polyshed, as is first fyled, then polyshed with Sande and after with Tripole & oyle; is most excellent redd but the whyte coullor never taketh any other collour but whyte. It is experienced to prevaile against the fluxe of bludd, and being hanged about the necke it is good against the falling sicknes, and the worke of Menstne which is woemens termes, & against tempests of thunder & lightninge

/ The fayer redd coullor being well polyshed, and fairely braunched is worth 2- [?] the oz but unpolyshed not above 2 […] the oz by reason of the waste & labor in the fyling & polyshinge.[14]

Monetary value is just one of the important active virtues of these powerful stones.

The author exhibits a similar mix of values when describing the agate: he notes that because the agate contains the color red, it preserves eyesight. Moreover, it "cherisheth the hearte of man[,] being holden in the mowthe, it quencheth thurste, It is good against the stinging of scorpions, and against any other kind of venom or poyson." He concludes with a financial note that the value of agates has gone down, not because the virtue has been diminished, but rather because they are more plentiful in his day.[15] Again, the health, medical, and monetary virtues of the stone are not regarded as separate. He praises the magnet both for its usefulness to sailors in direction-finding, and "it is sayed" for learning if a woman is chaste by putting it under her pillow.[16]

Today the mixture of descriptive, experiential, and textual authority contained in these entries on stones might sound dissonant, but it demonstrates the importance of both first-hand experience and textual authority at that time. This is even clearer in the author's comment on unicorn horn, in which he reports on a textual authority's experimental practices:

And this Experiment hath bene proved (by my Authore) for vertue of Unicornes horne. That he hath caused two doggs to [be] poysoned, to the one he given double quantity of poyson, & to him he gave good quantitie of Unicorn's horne in Powlder as scraped with water, & the dogg hath bene well againe. To the other he gave less quantity of poyson and no unicorn's horne and the dogg presently dyed.[17]

These passages show that these collections of techniques can cover a wide spectrum from firsthand experimentation to "proof" by citation of textual traditions. The combinations of these elements in recipe texts are quite diverse, but a common denominator among them is *action*, either impelling to action (including by the "Rx" or "Take" with which they often begin), or in their chronicling of practical activity.

The Storehouse of Experience

Recipe collections are characteristically quite repetitive, containing multiple recipes for the same process. These iterations of recipes often contain only small variations. The *Kunstbuch* entitled *Artliche künste mancherley weise Dinten und aller hand Farben zubereiten* (Arts to Prepare All Kinds of Dyes and Colors in Many Ways, 1531), for example, gives a recipe for a pigment and follows it with up to seven recipes labeled *"Ein andere"* (Another) that offer different ways to achieve the same result. In painting recipes compiled by Theodore de Mayerne, physician at the court of Charles I, recipes for making lapis lazuli are repeated numerous times under many headings: "*Azuro in altro modo*," "*Azurro Bello*," "*Azurro perfetto*," "A more compendious way," "Another varrey faire one," "Azurro made of Quicksilver," "Another experimented to be excellent good and to abide the fire," "Another," "Azur made of silver," and on and on.[18] These repeated recipes replicate the training in a diversity of methods that Dürer, Biringuccio, Cellini, and others celebrated, as discussed in chapter 5. In *The Art of the Maker*, Peter Dormer sees iterative process as "a spur to conceptual reflection: it prompts questions such as 'what happens if I try it this way?'"[19] Problem solving is a major cognitive activity of craftspeople, and creative procedures and techniques, or "secrets," are the result of systematic and reasoned problem solving.

Just as Francisco Alonso-Almeida's view of recipe collections as "discourse colonies" underscored their capaciousness as a strength, we can see their repetitive character as a core component of their active nature. Even the name "recipe" indicates their character as an invitation to repeated practical action: "Take these ingredients . . ." A collection of recipes could thus form a repository of particular cases, by which a person could gain or record experiences. These experiences, either extrapolated or recorded on paper, were stored up to be recalled when needed. No wonder that collections of recipes often carried the title of "storehouse." Recipe collections acted as storehouses of possible ingredients, or categories of ingredients, of trials, of past experience, but like the mathematical exercises of Michael of Rhodes, they also modeled the process of gaining experience and the journey to expertise. The individual recipe constituted an invitation to action, and the recipe collection's serial format enacted on paper the repeated testing and trying in order to work through the variables and resistances of materials and eventually to bring about a successful (and replicable) result.

Are Recipe Texts Knowledge?

The recipe format thus functions as an invitation and framework for action, but does it constitute "knowledge"? Since Plato's Athens, the abstract Euclidean proof has formed a standard of certainty, and universal knowledge has been regarded as of higher value than particular experiential knowledge. In many ways, a recipe text is the opposite of a Euclidean proof. Recipes list particular ingredients and specific techniques out of which they aim to create replicable material results or objects. They thus seem to be directed toward particular instances of creating, rather than toward making generalizations about universal cases. As Edward Norgate in his *Miniatura, or, The Art of Limning* (1650) noted in relation to making the pigment "carnation" (flesh color), any effort to put such knowledge in the form of propositional knowledge was futile: "To prescribe an absolute and general rule is both impossible, and a little ridiculous, *Nature herself* so infinitely full of variety, in the shadows and colors of faces and all so differing from one another that when all is said that can be, your own observations, practise and discretion, must be your best director."[20]

But other early modern artisan authors, such as Ambroise Paré, the sixteenth-century surgeon, asserted that experience was a form of certain knowledge. For Paré, experiential knowledge of singular and particular things was more certain than the generalizations of universal knowledge:

because the sayd experience is a knowledge of singular and particular things, and science on the contrary is a knowledge of things universall. Now that which is particular is more healeable than that which is universall, therefore those which have experience are more wise and more esteemed, than those which want it, by reason they know what they doe. Moreover I say, that science without experience, bringeth no great assurance.[21]

Paré is just one of many early modern practitioners who regarded recipes as an integral part of "science," by which they meant knowledge that was certain. Alisha Rankin has demonstrated this in her examination of medical recipes traded among the networks of sixteenth-century noblewomen. For the London practitioners chronicled by Deborah Harkness in *The Jewel House*, this "science" involved testing, criticism, demonstration of accountability, and a dedication to the goal of useful knowledge.[22] The work of both these scholars on recipes provides a sense of what kind of "science," or knowledge, *Kunst* can be, but the foregoing consideration of the nature of recipes and recipe collections illuminates further the ways that recipes, and the knowledge of *Kunst* that they codify, form a "science," and, more specifically, a "science of particulars."

A Science of Particulars

Kunst, or practical knowledge, differs from propositional knowledge, which can be captured and fixed in the words of a syllogistic proof. Following a recipe, in contrast, is different every time because the materials, workshop conditions, and practitioner's skill level, among other factors, can change the process and the outcome. Practical knowledge thus deals with emergent phenomena for which it is necessary to manage a process of material transformation. In his work on geometry and medieval masons' practices, Lon R. Shelby compares a Euclidean proof to a master mason's proof:

For Euclid the construction of a geometrical figure with compass and straightedge was merely a part—and not an absolutely necessary part—of his mathematical exercise; there remained the more difficult and important task of demonstrating the mathematical correctness of the construction. For Roriczer and his fellow masons, such a construction was, geometrically or mathematically speaking, the end of the exercise; the next task was not to prove its mathematical correctness, but to transform the geometrical construct into an architectural form in stone.[23]

Proof of practical knowledge consists in producing the right result by managing material processes successfully. In instructions to his son, master mason Lorenz Lechler emphasized that a practitioner needed to manage social as well as material processes in producing a successful result: "[t]herefore if you give proper attention to my teaching, you can meet the needs of your building patron and yourself, and not be despised as the ignorant are, for an honorable work glorifies its master, *if it stands up*."[24]

Kunst as "knowledge," then, must codify this emergent character from particular circumstances, as well as the materiality of the proof—"if it stands up"—that Lechler underscored, but, most of all, it must capture the characteristic of managing *process*. A 1696 compilation of recipes, *Curieusen Kunst- und Werck-Schul*, collected technical recipes from at least the previous two hundred years. It conveys clearly both the emergent nature of *Kunst* and the attainment of skill by which process is managed, a stage at which one no longer needs the rules of technique and can work without hesitation or conscious thought.[25] This recipe is for various ways of making "incarnatio" or flesh color for imitating the variety and particularity of lifelikeness of individual people, and it ends by noting that the color contains

so many and varied pigments, it would be difficult to give general rules about so many and varied particulars. But a person no longer turns to rules when he has attained dexterity and expertise through practice [*einen Handgriff und Fertigkeit*]. Indeed, those who are so far advanced that they either work from their original or simply from their imagination or ideas, and do not themselves know how they do it, so that the most skillful—who can produce without reflection and hesitation and with far less effort than others—must make much more of an effort to make clear the reasons and ideas [for their actions] when they are asked which colors they used than [the effort] they put into the painting.[26]

This passage makes explicit that the ultimate goal of the processes it records is the development of "foresight," or the skill to be able to respond to new situations intuitively and manage the process of material change, an ability attained through repeated practice. It also emphasizes that this intuitive response does not necessarily result in or correlate with any facility in describing the successful process in words.

Conclusion

As the written codifications of experiential knowledge, recipe texts provide paths of action by which to manage the particular and variable processes of material change. But, more than this, they implicitly set out the fundamental methods for managing these processes successfully, that is, to do it over and over, to fail, to extend, to try and experiment again. Recipes thus indicate, in abbreviated form, the essential character of making as coping with emergent phenomena, and they set out a pathway to the expertise of skill by which emergent and ever changing processes are channeled, harnessed, and managed.

Instead of seeing recipes as a frustrating and unforthcoming genre, then, we can understand them as a strong and flexible form in which to describe and transmit the core of practical knowledge. Recipe texts put into writing—self-consciously inadequate as that might be—experiential knowledge, the mind-body, thought-action work of handwork. In their early modern guise, recipe collections encouraged active testing and hands-on work, and they fostered working through material resistances; indeed, more than this, they laid out a path or a process by which to learn about and overcome the resistances of matter by the repetition of "another way." In this, they express, to the extent that it is possible in writing, a form of knowledge that has been difficult to theorize because it is made up of particular cases and is emergent from the material and social fields in which it is produced but, at the same time, can be transmitted as a corpus and powerfully employed to produce knowledge and objects. In their books of action and practice, then, artisan authors used writing to model the process by which "judgment" and "discretion"—the ability to respond to new situations on the basis of repeated experience—could be attained. As such, these practical manuals and collections of recipes can be viewed, in addition to all else, as a consideration of embodied practical knowledge and how it is acquired.

PART 3

Reading and Collecting

CHAPTER 7

Who Read and Used Little Books of Art?

The three chapters of part 2, "Writing Down Experience," have shown that practitioners wrote down their working methods and recipes for a variety of reasons neither aimed solely at providing utilitarian practical knowledge, nor necessarily at teaching their trade to others. Judging by the many editions of these books and their frequent compilation and recompilation, printers apparently viewed technical writing as best-selling material. Who bought these texts, and, more intriguing, who actually read this swelling tide of books of art? It was once assumed that the audience for these books was made up of aspiring practitioners of a trade. But the evidence about the audience for such books, as for any early modern publications, is scarce because it relies upon the vagaries of survival: first, the survival of the books themselves, then of copies that contain notes or marginalia, or the survival of inventories that tell us about the ownership of books (although, as we all shamefacedly know, owning a book does not guarantee reading it). Despite a wide range of scholarship devoted to reading, book history, bibliography, and popular and didactic literature, we still know far more about the production of books than about their reception. As book historian Roger Chartier made clear, a combination of close textual analysis, bibliography, and cultural history is necessary to understand how a book would have been read, and what it would have meant to its various readers.[1]

Marginalia

Statements about using practical manuals are rare. One occurs in Samuel Pepys's (1633–1703) diary when he records using and experimenting with a slide rule and its accompanying instructions, mightily pleased with himself that he "found out some things myself of great dispatch, more than my book teaches me."[2] Some of the most telling evidence for how these "books of art" were used consists of marginalia, or note taking by active readers. For example, a weaver entered notations on five pages of a 1529 pattern book published

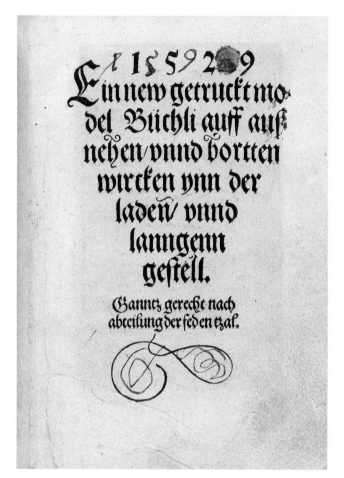

7.1. Johann Schönsperger the Younger (active 1510–30), *Ein new getruckt model Büchli* (Augsburg, 1529), frontispiece, with hand copying in the date and the ornamental line. 20 × 15.5 cm. Metropolitan Museum of Art, New York, 18.66.2(1r), Rogers Fund, 1918. CC0 1.0.

7.2. Schönsperger the Younger, *Ein new getruckt model Büchli*, pattern for weaving. 20 × 15.5 cm. Metropolitan Museum of Art, New York, 18.66.2, Rogers Fund, 1918. CC0 1.0.

by printer Johann Schönsperger the Younger (active 1510–30) and perhaps doodled on the frontispiece (figs. 7.1–7.3). Scholars have fruitfully explored marginalia as a source of evidence for reading practices, such as the wonderful account by Lisa Jardine and Anthony Grafton of how Gabriel Harvey read his Livy—in the process supplying himself with a "notebook of secrets." But locating such sources has mostly been the result of serendipity.[3]

We can also make inferences based on book ownership. For example, a manuscript of around 1458–85, written in Middle English, possibly by a scribe for a merchant patron in London, includes a taxonomy and defense of the mechanical arts, which suggests a mercantile audience (of at least one) for vernacular texts on *Kunst*.[4] By combing through inventories, scholars have also discovered that the libraries of some European nobility contained books of secrets.[5] In short, the audience for these books can only be suggested by rather fortuitous and somewhat anecdotal case studies. Moreover, once we know who owned a book or even marked it up with notes, we still do not know how or why that individual read and used that book—did they read it from start to finish, or dip into it for recipes, or perhaps use it to acquire a vocabulary with which to haggle about purchasing works of art or objects they desired? Mostly we can never know the answers to these questions, although, as we shall see, some marginalia suggest surprising clues.

In chapter 6, I argued that artisans produced collections of recipes as tools for fostering and modeling the embodied skills of *Kunst*. Some evidence suggests that printed books could be bought and used in this way. Elaine Leong and Alisha Rankin, for example, have shown that some copies of Girolamo Ruscelli's (d. 1565/6) *Secreti nuovi* have copious marginal commentary indicating that readers were trying, testing, and correcting the recipes in them.[6] Others might have been used by artisans, although not necessarily in the ways intended. An often reprinted book of architecture, *Regola delli cinque ordini d'architettura* by Giacomo Barozzi da Vignola (1507–73), first published in 1562 and going through hundreds of editions, was used by masons and joiners as a model book, from which they traced templates.[7] These were far from the only possible uses for the books of art, however. A wide range of readers acquired these books, coming from a variety of social strata and doing so for a variety of purposes. This chapter first surveys what incomplete evidence we do possess about such readers, then goes on to consider how these readers reflect a more general surge of interest in the productive potential of art and the reform of knowledge.

Digesting Texts

A fundamental set of questions that might not occur to a modern reader must be broached at the outset: what was a book, what did it "do," and what were the expectations of the person who sat down with it? These issues are surprisingly complicated. In *The Book of Memory*, Mary Carruthers elegantly showed that medieval texts must be understood within an entirely different regime of practice than our own. Engaging with a book did not connote what we today mean by "reading"—scanning over its lines, seeking information, sense, and meaning. Medieval texts were intended not as "a definitive statement of fact or experience but an occasion for rumination and meditation."[8] Books were bound up with a culture of memory in which memorization possessed ethical and piety- and character-building dimensions, because memory training was seen to be the source of developing judgment, prudence, and wisdom. "A 'book' was only one way among several to remember a 'text,' to provision and cue one's memory with 'dicta et facta memorabilia.'" A book could thus function as a mnemonic device,[9] which made "reading" a process of "chewing," ruminating, and digesting the text,[10] in order to lay it away in the memory to be pulled out again when needed. Such ingestion had the aim of making the words of a text the reader's own, to be used in appropriate and decorous ways that gave honor and greater authority to the text. The writing of commentaries, the collection of florilegia, and the reuse of texts in new settings were part of this process of digestion and rumination. Thus, when a reader opened a book, they expected not an easy or difficult read, but a kind of "unending collocation" with the text.[11] Such an understanding of, and expectation from, a text makes more sense of the Book of Michael of Rhodes and of recipe collections.

These expectations about the function of a book changed over the course of the period from 1200 to 1400. Up through the 1300s, with books tied to *lectio* (lecture or reading aloud) in the universities and to church sermons, many (even most) people probably did not have a concept of reading a book, but only of hearing one read aloud. They might be able to write a bit, but would never expect to read a book. As urban life grew, a more intense circulation of written products took place, and schools and professions began to produce new kinds of books for the private use of scholars and students. Books became articulated for readers to make them easier to divide up into shorter texts arranged in relationship to each other, with titles, illuminated initials, and subheadings. In the fifteenth century, increased literacy of lay people (including artisans) and humanist interest in the content of the books of antiquity brought about a transformation in attitudes to books and reading. Indeed, the buying and selling, collection, and curation of books, which began before the advent of printing, but increased with the humanist search for the works of antiquity, and then again with the greater availability of printed books, resulted in new expectations about what a book "was."[12]

The design of books indicates this transformation in expectations: books came to contain a variety of sizes, fonts, pagination, indices, and registers of the contents, among other aids for readers, and they also dropped in price—printing brought down the price of a book by a factor of eight.[13] These changes made books more useful for consultation, which presumably led to various kinds of interactions with books in the sixteenth century,[14] a shift that scholars have characterized as a move from "intensive reading" of very few texts to "extensive reading" of a broad range.[15] But we should not assume that older ways of "reading" fell out of use. We already saw the mnemonic structure and accretive nature of the Book of Michael of Rhodes. In the 1530s, as we have seen, Paracelsus referred to medical learning as something that must be chewed, spat out, and digested many times over.[16] Similarly, in 1612, Francis Bacon wrote that "[s]ome books are to be tasted, others to be swallowed and some few to be chewed and digested."[17] Even as late as 1711, the compiler of a lexicon of mining terms and practices felt he had to justify his use of alphabetical order, which, because it did not connect knowledge into an organic whole, encouraged a scanning, extensive type of reading. His ideal was what he called "systematic" order, that is, laying out the topics in an interconnected system (which would have fostered rumination), and he lamented a future day when alphabetical order replaced systematic order altogether, for then the world would no longer contain true scholarship.[18]

Varieties of Readers

Scholars who study the history of books and reading now recognize that people did not just *read* books; they *used* them.[19] For example, they could use a book for compiling information drawn from their reading, for note taking, or for commonplacing. They might practice their penmanship or their drawing skills in the margins. A book that might seem practical to us, such as health advice, could form the basis not just for bodily care but also for spiritual reflection (which could, of course, be bound up with health of the body), and its contradictory directives on healing, compiled out of a variety of authorities, could even invite the reader to reflect upon and challenge medical advice in general.[20] Indeed, the evidence of readers' engagement with books often seems at odds with what we would assume from a book's title and intent. We see this in the printer Robert Copland's preface to his English edition of the *Secreta secretorum*. This book of "secrets" was a handbook on statecraft for princes, attributed to Aristotle, but probably written first in Arabic sometime in the tenth century, then translated into Latin and vernacular European languages, and printed and reprinted frequently. Copland considers the book not only "very profitable for every man," but "also very good to teach children to learn to read English."[21] The traces left in the margins of a 1556 book intended by its gentleman author, Leonard Digges, as "most conducible for Surueyers, Landmeters, Ioyners, Carpenters and Masons," also upends expectations of the use of the book.[22] Digges clearly meant his *Boke Named Tectonicon*, which contained instruments, arithmetical rules, and numerical tables, to be used by mathematical practitioners of all kinds, especially land surveyors, but also those who called themselves architects. Digges expected the book to have direct impact on practitioners, and while it seems at least one craftsman joiner left his jottings in the margins, the most extensive note making still extant was carried out by a scholar deeply interested, not in the how-to content or procedures, but in the words themselves, glossing many of the terms, noting their Greek and Latin roots, and speculating about their etymologies.[23]

The philological approach to practical texts was widespread,[24] emerging out of the reading techniques of scholars trained in the *studia humanitatis*. In early modern Europe, these humanists strove to model their own civic spaces and intellectual practices on those of the ancients, and, in reconstructing a picture of the material world of antiquity, they also encountered the craft workshop. Scholars exploited the rich detail of Pliny's *Natural History* for all kinds of information about material life in the ancient world, and also drew out of it a vocabulary for the practice of crafts going on around them in workshops of their own day. The Padua-trained humanist scholar, Pomponius Gauricus (1481/2–1530), for example, claimed in his *De sculptura* (1504) to practice sculpture himself and to be friendly with founders, but acknowledged that he derived much of his technical information about alloys and patinas from isolated comments in Pliny's books 33–34.[25]

Where Gauricus went *ad fontes*, to the ancient sources, to articulate his observations about sculpting and founding processes, the German humanist, Georgius Agricola, by contrast, had to take a different approach. In writing about mining, Agricola found that he had to abandon his classical authorities for actual miners. Agricola wrote his first work on mining, *Bermannus, sive de re metallica dialogus* (Bermannus, or dialogue about metals, 1530), as a dialogue between a "learned miner, *Bermannus*" and two traditionally trained physicians, one expert in Latin and one in Arabic and Greek. The miner teaches these two physicians mining terms and concepts, which gives rise to conversation between the scholars about the relationship of these new terms to ancient texts, practices, and medicines.

For a learned humanist, then, a practical book could provide vocabularies of practice and of specific trades. Similar concerns about language appear also to have formed a motivation for the rationalization and reform of the arts attempted by the authors of the *Encyclopédie ou Dictionnaire raisonné des sciences, des arts et des métiers* (1751–72). In the entry on "Art," the author (probably Diderot himself) laments the regional proliferation of different terms for tools and processes in the arts, and advocates setting out standard terms to be adopted by all. This concern shaped the form of the *Encyclopédie*, with its multitude of plates that pictured and named artisans' tools.[26]

7.4. John Bate, *The Mysteries of Nature and Art* (London: Ralph Mabb, 1635), Robert Dent's "Phisicall observations," recorded in blank pages appended to the start of the book. National Gallery of Art Library, Washington, DC, Rare Q155.B32 1635.

Another kind of reader was interested not so much in etymology and ancient texts as in spiritual enlightenment through the consideration and perhaps practice of concocting recipes. A copy of John Bate's 1635 printing of a book of secrets, *The Mysteries of Nature and Art*, was used by two separate readers in quite different ways. This copy was bound with several blank pages at the start, in which one Robert Dent wrote down a selection of what he titled "Phisicall observations" and "Extravagants" (meaning wandering out of bounds, roaming, vagrant), amounting to a short collection of diverse recipes, dated 1675, including for metalworking, gilding, and medicaments (fig. 7.4). He filled the last of his bound-in blank pages with the "virtues of hearbs," including seven herbs, each listed with its planet, along with the ten "virtues of beasts." He concluded by enjoining the practitioner to work under the favorable planets of Jupiter and Venus, mentioning the effects and virtues "in the aforesaid things as I have proved & seene often times together with our Brethren in our time." He ended the blank pages with a verse entitled "Astrologicall Hieroglyphicks 1675" in which he prophesied "Great thinges" for England up through 1683, then fifty years

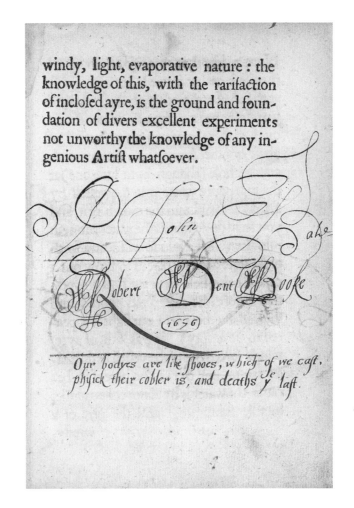

7.5. Bate, *The Mysteries of Nature and Art*, Robert Dent marginalia: "Robert Dent Booke 1676. Our bodyes are like shooes, which of[f] we cast, phisick their cobler is, and deaths the last." National Gallery of Art Library, Washington, DC, Rare Q155.B32 1635.

of vague and enigmatic tragedy, in a rhymed verse, whose "hieroglyphick," Dent tells us, would only be revealed in 1702. In 1676, he saw fit to add a last admonition to the world: "Robert Dent Booke 1676. Our bodyes are like shooes, which of[f] we cast, phisick their cobler is, and deaths the last" (fig. 7.5).

In the same copy of Bate's book, a Benjamin Day signed his name in book 3, which is devoted to instructions for "Drawing, Limming, Colouring, Painting, and Graving," and he or someone else practiced drawing on the reverse sides of pages that contain illustrations for drawing practice.[27] He sometimes traced the illustrations from the back, sometimes copied them on facing pages, with some changes (figs. 7.6 and 7.7). Here we have at least two different users of the book, one who used it to practice drawing, and another who speculated about the connections between heavenly and earthly things and sought spiritual consolation in it. For both, such a book of instruction made good on the genre's

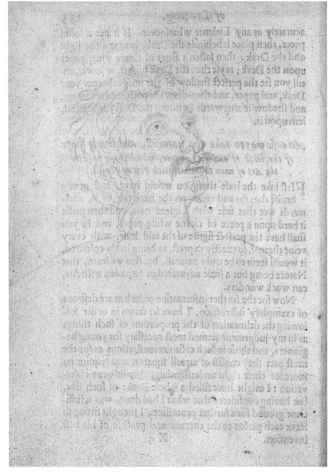

open-ended promise of (co-)producing the book's meaning by their active use.[28] All of which indicates that we must not assume that readers of practical manuals used them as we might expect them to have done.

Ownership of Books

What we do know about printing and book ownership in early modern Europe? Detailed studies have been carried out since the nineteenth century on the output of printing presses in Europe, and, while different regions have particular characteristics (more vernacular works on religious reform in some German territories than in Italian polities, for example), most scholars agree that from the invention of the printing press until the 1520s, clerical publications predominated, with those intended for secular scholars and

7.6. Bate, *The Mysteries of Nature and Art*. National Gallery of Art Library, Washington, DC, Rare Q155.B32 1635.

7.7. Bate, *The Mysteries of Nature and Art*, Benjamin Day use of the book, showing his copying of images on the facing page (see fig. 7.6). National Gallery of Art Library, Washington, DC, Rare Q155.B32 1635.

scholars with humanist interests increasing significantly from the 1490s. Professional books, especially for lawyers, came off the printing press with increasing frequency from around 1500, and this period also saw an uptick in works of "science," such as astronomy, natural philosophy, and mathematics, as well as in editions of the ancient authors who wrote on these topics. Vernacular and popular books took off in the 1490s, and continued to increase at a steady rate through 1540; by the 1580s they had come to hold the largest market share.[29] In Strasbourg, with a population of 20,000 and approximately 77 printing shops, there was a larger percentage of books in the vernacular than any other type by 1549.[30] These included medical books, herbals, recipe collections, technical writing, popular literature, and other books explicitly aimed at the "common man and woman."

Who read the books coming off the printing presses? In the first years of printing, it was only those who had learned Latin who could read the majority of these texts, so one might assume that anyone reading Latin could and did read the vernacular books. But this was not always the case: one scholar commented in 1529 that children were taught Latin but then had to teach themselves their native tongue. Only a few editions of a widely used Latin grammar, the *Rudimenta grammatices*, included some translation into regionally inflected Italian of the Latin phrases, but most such grammars simply used the classical language as a starting point.[31] From 1480 to 1520, there is an intriguing increase in vernacular books about teaching oneself to read the vernacular or Latin, as well as a greater number of books instructing on speaking and writing, such as a work by Heinrich Geissler that taught how to express oneself before a *Landrichter* (a judge). Hanseatic merchants used instructional phrasebooks for language learning.[32] Other books, published in both Latin and German, instructed the reader in how to write a form letter or petition (formularies), but such basic instructional texts tapered off after the Reformation of Strasbourg, perhaps because parish schools were established.[33] In Italy, such vernacular works continued to appear through the sixteenth century.[34] These instructional manuals were probably aimed at the same clientele attracted by the signboard of a schoolmaster, painted by Hans and Ambrosius Holbein in 1516, advertising the schoolmaster's teaching of reading and writing. The signboard portrays two grown men working with their teacher, advertising the master's wares to a working audience (figs. 7.8 and 7.9):

If someone wishes to learn to write German and to read in the shortest way that can be conceived; someone who cannot even recognize the alphabet can quickly grasp this principle through which he can learn to write and read by himself. If anyone has not the aptitude for this and cannot learn, I will not charge him anything. So whether you be a burgher, an artisan's apprentice, a matron, or young woman, if you wish to learn, enter here. You will be conscientiously taught for a fair price. But young boys and girls should come after the ember day fasts, as is customary.[35]

Wer jemand hie der gern welt lernen dütsch schriben vnd läsen vß dem aller kürzisten grundt den jeman erdencken kan do durch ein Jedr der vor nit ein büchstaben kan der mag kürzlich vnd bald begriffen ein grundt Do durch er mag von jm selber lernen sin schuld vff schriben vnd läsen vnd wer es nit gelernen kan so vngeschickt were Den will ich vm nüt vnd vergeben gelert haben vnd ganz nüt von jm zů lon nemen er syg × wer er well burger Durch handtwerckß gesellen frowen vnd Junckfrouwen wer sin bedarff Der kum har jn der wirt drüwlich gelert vm ein zimlichen lon ◦ Aber die jungen knaben vnd meitlin noch den fronuasten wie gewonheyt ist ◦ Anno ◦ m cccc xvi

Wer Jemandt hie Der gern welt lernen Dütsch schriben vnd läsen vß dem aller kürzisten grundt den Jeman erdencken kan Do durch ein Jedr der vor nit ein büchstaben kan Der mag kürzlich vnd bald begriffen ein grundt do durch er mag von jm selbs lernen sin schuld vff schribe vnd läsen vnd wer es nit gelernen kan so vngeschickt werr Den will jch vm nüt vnd vergeben gelert haben vnd ganz nüt von jm zů lon nemen er sig wer er well burger oder hantwercks gesellen kouwen vnd junckkrouwen wer sin bedarff der kum har jn der wirt drüwlich gelert vm ein zimlichen lon ◦ Aber die junge knabe vnd meitliu noch den kronualten wie gewonheit ist ◦ 1 5 1 6 ◦

7.8. Hans Holbein the Younger (1497/98–1543), *School Master's Signboard* (adult side). Mixed media on fir panel, 5.3 × 65.5 cm. Amerbach-Kabinett 1662, Inv. 310, Kunstmuseum, Basel. CC0 1.0.

7.9. Ambrosius Holbein (c. 1494–c. 1519), *School Master's Signboard* (children's side). Mixed media on fir panel, 5.3 × 65.5 cm. Amerbach-Kabinett 1662, Inv. 311, Kunstmuseum Basel. CC0 1.0.

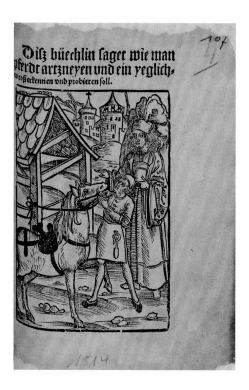

7.10. Meister Albrant (Master Albrecht), *Disz büechlin
saget wie man pferdt artzneyen und ein yegliches roß
erkennen und probieren soll* (Strasbourg: Matthias
Hupfuff, 1514). Stadtbibliothek, Trier, VD16 ZV 16614,
urn:nbn:de:0128-6-1180. dilibri Rheinland-Pfalz (www
.dilibri.de).

As a consequence of the Reformation, many schools were established, and literacy rates among both men and women rose as a result.[36] But the relationship between the ability to read and write was never direct: some people could read without writing, and some could write without reading. Generally speaking, there was improvement in literacy in Europe between 1500 and 1800, but it was never steady, or linear, or universal. Literacy in urban areas was much higher than in rural ones. For example, it has been estimated that in seventeenth-century urban England, 70–80 percent of skilled artisans and yeoman could write, whereas 30–40 percent of village artisans and merchants, and less than 25 percent of laborers and agricultural workers could do so.[37] Records from fifteenth-century Venice suggest that "some male artisans read for instruction and pleasure and that few were completely illiterate."[38] In sixteenth-century German cities after the Reformation, books are found in 65 percent of artisan households.[39]

Some manuals address themselves directly to an audience new to books and reading. An early book on horse care, ostensibly written by a Master Albrecht, the Stall Marshal of Frederick of Hohenstaufen of Constantinople in the thirteenth century, went through an extraordinary number of versions, including more than 200 manuscripts, 8 incunables, and multiple printed editions from the early sixteenth through the eighteenth centuries.[40] The edition published in Strasbourg in 1514 seems to be aimed at such an audience, proceeding in a way that imitates verbal instruction. It features a descriptive picture on its title page (fig. 7.10) and clarifies for the reader that "this little book says how one should

Von distillierung

Ar nach müstu haben blyßen ring in der mitten ingesencket mit frer durch gelöchert oren groß vnd clein licht vnd schwer die mittel messig võ zehen pfunde/die cleinen võ acht pfunde die grosse von .vii. oð .viii. pfunde, also disse figur zeigt, des glichen hültzin bretter als wyt die cappeln oð der offen ist/also das das glaß durch das mittel loch gon mag . Dar nach mancherlei öffen als ich hie vnden zeige will zü brennen distillieren in dem balneu marie/vff das dz glaß nit über sich stige mag vor der schwere des anhangende blyßes so es dar an gebunden vnd geheffret ist.

Ar nach ð helm võ wysser erde gebrät võ über glasurt innen vñ vssen/oð kupffer. blyßyn oð zynne/deren form also wirt. Auch etlich mit zweien faltz vnd zwo rören/also dz der öber faltz ein röz haß/gond in die vnder röz die helm vast vil wasser gebent.

Ar vnder müstu haben von wysser erden wol über glasurte pfanne oð blyßen oð küpffern nach dynem vermögen oder begeren, Darnach müstu haße

gleser genant cucurbit/võ den tütschen kolben gemacht vonn Venedischen scherben glaß/vff das sie füer erliden mögen/deren form also ist.

Ar nach müstu haben ettlich gleser als dz man zwei vff eynað stürtzen mag/deren form also ist/dar in zü distillieren an der sunne als ich in dem nünden capitel diß erste büchs leren will.

Ar nach müstu habe krüme gleser forman wie ein storck schnabel gnat retort also gesompt vnd der gleser mit zweien armen genat pelican/deren form also ist.

7.11. Hieronymus Brunschwygk, *Liber de arte distillandi. de simplicibus. Das buch rechten kunst zu distilieren die eintzigen dinge* (Strasbourg: Johannes Grüninger, 1500). Ornamented by hand with red initial letters and underlining, this copy was acquired for the monastery of Tegernsee in 1501. Bayerische Staatsbibliothek, Munich, 2 Inc.c.a. 3867, n.p. (19v), https://daten.digitale-sammlungen .de/bsb00031146/image_42.

know and test horse medicaments and each type of horse" (*Disz büechlin sagt wie man pferdt artzneyen und ein yegliches roß erkennen und probieren soll*). It then explains the "register," or table of contents: "The following register of this little book, in which you find the Title, also its corresponding number of the page on which the number stands that you want to read." Each recto side of the page is clearly labeled at the top with a large-print roman numeral, for example, *"Das. XI. Blat* [page 11]." After that, the author gets down to business, starting with the signs by which a good horse can be recognized, and telling how to prepare medicaments for horses. His reader is assured that Master Albrecht "tried and assayed everything" in his book.[41]

This tone of oral instruction also comes through in Hieronymus Brunschwygk's *Distillierbuch*, first published in 1497 and reprinted very frequently thereafter: "Here begins the book called Liber de arte distillandi—about the art of distilling—gathered together and set down by Hieronymus Brunschwygk / all that has been experienced by many experienced masters of medicine [*ertzny*], and that has been found out and taught by daily hands-on work [*hantwürckung*]." Brunschwygk also feels compelled to explain how to use what we would call the table of contents "in which you find according to the number on each page what book you desire," then provides explicit instructions for the use of each part of the book. The book's third part was constructed around individual herbs, their use and method of distillation, and the fourth part contained a "head to toe" list of ailments and their corresponding medicinal distilled waters. For these sections of the book, Brunschwygk experimented with a system of labeling by which "you easily and rapidly find what you need."[42] But our expectations of the readership for this book are once again belied by the signature on the end paper of the Bayerische Staatsbibliothek copy of the Strasbourg (1500) edition of this book. It records that it was obtained for the monastery of Tegernsee in 1501, and much of the book has been painstakingly ornamented with red initial letters and ornamental underlining, vestiges of the kind of working through and digesting with mnemonic aims, for which bright red ink (made with vermilion) and historiated initials were particularly effective (fig. 7.11).[43]

Practitioners as Readers

Although many practical books turn up in learned environments like the Tegernsee monastery, artisans themselves also appear to have begun acquiring books in this period. If we turn from prefaces printed in Strasbourg to the city's testate inventories, we find that, of those leaving wills in Strasbourg between 1500 and 1580, 40 percent left books, and 25 percent of those who did were artisans or working people.[44] This is a higher rate of book ownership among this social group than that found in many other places, such as Paris and Amiens, but it is not obvious that these artisans owned any technical writing, as the books listed are bibles, devotional books such as Psalters and prayer books, almanacs, and calendars. Some of the "assorted old small books" listed in more than one inventory may have included vernacular how-to books, but there are no specific mentions of practical manuals among the wills that Miriam Usher Chrisman examined for Strasbourg. Among inhabitants of Amiens from 1503 to 1576, meanwhile, many of the craftspeople possessed model books at their death; for example, a town crier (*hucher*) had "a book concerning the trade of crying in which are printed many models," and three goldsmiths each had collections containing a large number of images and models, as did a painter and an armorer. A carpenter had "a book with many pictures serving the trade of carpentry," as well as six more books of "pourtraicture." A few other artisans' wills record model books, but among all craftspeople, as among all the other Amiens residents,[45] the great preponderance of their books were devotional.[46] This seems also to have been true for northern Italy.[47] While an executioner in Heilbronn possessed Adam Lonicer's *Kreuterbuch* (Herbal) and the popular cookbook *Koch- und Kellermeisterei*, the majority of his seven books were devotional works.[48] In 1561, a cloth cutter from Schwäbisch Hall left behind a variety of books, including an herbal and two vernacular medical texts, while a ribbon or lace cutter from the same town left fifteen small devotional books in 1619, and a late sixteenth-century mason in Braunschweig owned two *Kunstbücher*.[49] In German cities after the Reformation, cobblers had remarkably well-stocked libraries, from which little books of art seem to have been almost entirely absent. In Valencia from 1474 to 1550, average book ownership among artisans in the textile trade rose from one to four books.[50] But, as in other wills and inventories among artisans, many of the books listed are without titles, and many inventories probably omitted works of little value as well.

From a detailed study of death inventories in German artisanal households, it appears that the great majority of the books were in fact recorded without titles, with reference being made simply to "small books."[51] This same study found that the only trade whose members clearly owned books correlated to their occupation was that of medical practitioners; masters of the bath, barbers, apothecaries, bonesetters, and oculists in German cities often owned substantial numbers of herbals, books of distillation, and vernacular medical texts, as well as works by Paracelsus.[52]

Well-known painters, sought after by elites in their own day, often owned books, but very rarely recorded books of practice in their reading. Among twelve books, Filippino Lippi's 1504 inventory includes "*uno libro di geometria* [one book of geometry]."[53] Gillis van Coninxloo (1544–1607) owned several devotional books, as well as the chronicles of Josephus and the mythology of Ovid, and a single practical book, "*het boeck van alber Duyr* [the book by Albrecht Dürer]."[54] Rosso Fiorentino (1495–1540) owned a copy of Vitruvius and Baldassare Castiglione's *Courtier*, as well as devotional texts, while Jacopo Pontormo (1494–1557), like many other artists, kept a notebook, called in Italian a *zibaldone*, or "jumble," filled with sketches and notes on work, which contains mention of his study of books in other artists' workshops.[55] This reminds us that books were often shared (thus not appearing in individual inventories), and some municipalities established libraries that loaned out books; thus, artisans did not have to own books to read and use them. The library of an artisan at the pinnacle of the social elite is indicated by the inventory of the *"sculpteur et fontainier du Roy,"* Jean Séjourné, who died in his apartments in the Louvre in 1614, leaving behind an enormous library of eighty-two books, all in the vernacular. The inventory includes descriptions of illustrated books only, probably because these were of higher value, with books of architecture by Serlio and Alberti, works that included ancient iconography, such as *Discours de la religion des Romans*, as well as Biringuccio's *Pirotechnia* in French, Guillaume Rondelet's *Histoire entière des poisons*, the *Songe de Poliphile*, and the *Roman de la Rose*, among others, with no further books of practice.[56] The famed teacher of Rembrandt, Pieter Lastman (1583–1633), owned 150 books shortly before his death; unfortunately, the inventory includes no titles.[57] There is no doubt that many artists read, and that some even became the *"artifex doctus"* idealized in sixteenth-century literature,[58] but they did not necessarily read books on their own craft. As noted above, however, any illustrated book could become a source from which an artist derived material for practice. This included instructional drawing books, which, like many other illustrated texts, were used as templates for practicing.[59] Jaya Remond has identified pricking in a copy of engraver Crispijn de Passe's (1595–1670) *'t Light der Teken en Schilderkonst* (The Light of Drawing and Painting, 1643–44), by which a user of this book transferred the outlines of the book's many illustrations to another working space, probably a notebook, such as the one Pontormo carried with him.[60] Elizabeth Merrill has demonstrated that the notebooks of self-styled architects and engineers also allowed them to experiment with working process considerations on paper (akin perhaps to the experimentation that artisans did more generally in authoring texts of practice).[61]

Civic Cultures of Writing

As James Amelang makes clear in *The Flight of Icarus*, some cities, such as Valencia, Strasbourg, Nuremberg, Antwerp, and, slightly later, Amsterdam, fostered a culture of writing in which artisans participated. In these cities, many books of art were authored, aimed at "the common man," and a ready audience for works in the vernacular emerged.[62] Many practical manuals were published by the entrepreneurial printers of Strasbourg, among them sixteenth-century German translations of manuals by the self-styled *medicus* and author, Walther Hermann Ryff (c. 1500–1548), on surgery, female medicine, medicaments, anatomy, and architecture, as well as his reworking of Hieronymus Brunschwygk's book on distillation. In Nuremberg, artist Jost Amman (1539–91) and cobbler and *Meistersinger* Hans Sachs (1494–1576) produced an illustrated *Ständebuch* (Book of Estates) in 1568, which depicted a variety of social ranks, with artisans as the most populous and productive estate, associated with diligence, industry, economic flourishing, and social order. This book illustrated numerous trades, each with an image of a tradesperson in the workshop, surrounded by the requisite tools and products, and a short caption that often described the work or process the figure was carrying out (fig. 7.12). Venice, too, had a lively

Der Kandelgiesser.

Das Zin mach ich im Feuwer fliesn/
Thu darnach in die Mödel giessn/
Kandel/Flaschen/groß vnd auch klein/
Darauß zu trincken Bier vnd Wein/
Schüssel/Blatten/Teller/der maß/
Schenckkandel/Saltzfaß vnd Gießfaß/
Ohlbüchßn/Leuchter vnd Schüsselring/
Vnd sonst ins Hauß fast nütze ding.

V iij   Der

7.12. Hans Sachs and Jost Amman, *Eygentliche Beschreibung Aller Staende auff Erden Hoher vnd Nidriger Geistlicher vnd Weltlicher Aller Kuensten Handwercken vnnd Haendeln* (Frankfurt am Main: Sigmund Feyerabend, 1574), "Der Kandelgiesser" (The Pewterer). Interior of a pewterer's workshop, with various pewter vessels, and men working at a wheel, at an object on a turning lathe, and in the background at a fire. Staatsbibliothek zu Berlin—Preußischer Kulturbesitz, Yg 9902.

vernacular culture, and it was here that Tommaso Garzoni published *La piazza universale di tutte le professioni del mondo* in 1585, in which he listed four hundred types of manual work. Garzoni both celebrated the power of print and illuminated one potential use of practical manuals—as guards against deceptive practices of artisans and pretenders. He noted that, with the printing of trade techniques, "it is no longer possible to sell lies and to pass black off for white. Now all pass judgment on infinite things, and if it weren't for the press, they wouldn't be able to open their mouths to speak, let alone pass judgment. [Printing] is the art which makes the made known, that manifests the arrogant, that clearly shows the educated, that brings death to ignorance, that gives life to virtue and to science."[63] Such an assertion presupposes his imagined audience as consumers of made things rather than the makers of things themselves. Garzoni's survey of the mechanical and productive trades was reprinted twenty-nine times in the following century.[64]

Even Garzoni's celebration of *Kunst* contained hints of a more long-enduring ambivalence toward manual labor and craft, here expressed by Garzoni as distrust of craftspeople and their ability to deceive until printing makes "the made known." This combination of lively interest and continuing low regard for handworkers comes through clearly in the preface of a book on the diseases of craftspeople by a late seventeenth-century physician from Carpi, Bernardino Ramazzini (1633–1714), who visited workshops and worksites, even venturing to visit sewer cleaners and gravediggers. "I, on my part, have done everything which I thought right," he wrote "and I did not feel belittled when, in order to observe the characteristics of manual work, I entered into very humble craftsmen's workshops; on the other hand, in this period also medicine uses observation derived from mechanics."[65]

Citing Books of Art

Another means to measure the readership of technical manuals can be found in how they are incorporated into contemporary writings. Artisans who wrote manuals drew upon ancient models. For example, Cennino Cennini paraphrased Horace's treatise on poetry, and the sculptor Ghiberti incorporated whole sections of Vitruvius into *I Commentarii*.[66] The treatises on architecture and painting of the scholar Leon Battista Alberti were mentioned by artists such as Leonardo da Vinci and Francesco di Giorgio. While these examples show artists striving to take part in the culture of learned humanism, the best evidence for the use of such manuals comes from the noble courts. Lorenzo de' Medici reportedly often read Alberti's *On Building*.[67] That work seems to have been used by patrons in church design in Italy, but perhaps the most important influence of such manuals was to disseminate a vocabulary and a set of critical categories with which to talk about works of art. Catherine King cites a soldier in 1509 who used Albertian categories in admiring paintings.[68] In 1506, Urbino humanist Baldassare Castiglione advised that a gentleman should know how to converse about art for pleasure and discernment:

it is fitting for our Courtier to have knowledge of painting also, since it is decorous and useful and was prized in those times when men were of greater worth than now. And even if no other utility or pleasure were had from it, it helps in judging the excellence of statues both ancient and modern, vases, buildings, medallions, cameos, intaglios, and the like, and it also brings one to know the beauty of living bodies.

Castiglione even suggests that the courtier should actually learn the art of painting:

And do not marvel if I require this accomplishment, which perhaps nowadays may seem mechanical and ill-suited to a gentleman; for I recall reading in the ancients, especially throughout Greece, required boys of gentle birth to learn painting in school, as a decorous and necessary thing, and admitted it to first rank among the liberal arts; then by public edict they prohibited the teaching of it to slaves. Among the Romans, too, it was held in highest honor […] besides being most noble and worthy in itself, [painting] proves useful in many ways, and especially in warfare, in drawing towns, sites, rivers, bridges, citadels, fortresses, and the like; for, however well they may be stored away in the memory (which is something that is very hard to do), we cannot show them to others so.[69]

Castiglione's concern with the mechanical arts in his *Courtier* is connoisseurial and utilitarian, but, as Jessica Wolfe has argued, his interest in artifice also derives from the analogy he and his fellow authors made between an artificer's skill, which makes virtuoso feats look easy, and the *sprezzatura* needed by an expert courtier. Both relied upon practice and bodily control, and could be used to deceive: "like mechanical *virtù*, courtly *virtù* is an adversarial and dissimulatory power, one that relies upon trickery rather than force to overturn the natural course of things."[70]

While Castiglione says nothing about how-to books, his humbler emulator, Giovanni della Casa (1503–56), in his *Galateo, or The Rules of Polite Behavior* (1558) even models his "how to be" book on the model of a sculptor who

wrote a treatise in which he gathered up all the rules of his craft with the authority of someone who knows his art. He demonstrated how the limbs of the human body ought to be measured, each by itself and in relation to the others so that they should be in proper proportion. He called this volume *The Rule*, meaning henceforth every master of the craft should shape and design statues according to it, just as beams, stones, and walls are measured with a standard ruler […] In order to demonstrate even more clearly his own expertise, the above-mentioned valiant man, aware as he was of the lack of talent, acquired a fine block of marble, from which, after long and difficult labor, he carved a statue proportioned in limbs and symmetrical in parts as his treatise recommended. And, just like the book, he called the statue *The Rule*.[71]

Della Casa claims to be achieving an aim similar to that of the sculptor in putting "together in this volume the proper measures of the art I treat," but he cannot enact them in the way the sculptor displayed his art, because "in matters dealing with the manners and customs of men, it is not enough to know the science and the rules, but necessary to put them into effect through use."[72] Della Casa claims he is getting too old to perfect his own conduct. Continuing in this vein, he also emphasizes that by doing, one becomes a better teacher, because by making mistakes yourself, you are better versed in how *not* to do something. In order to learn from reading, you must follow the rules carefully, he says, but it takes experience and a long time to truly learn something. Intriguingly, Della Casa's story ends with the sculptor himself following his *own* instructions, demonstrating the extent of his skill by reenacting the method he has written down in his book. Della Casa seems to see the act of writing down technical expertise, including disciplining the self, as a personal and technical exercise for the writer rather than necessarily directed toward a larger audience.

Books of Art and Reform of Knowledge

These remarks about books of art do not just indicate attitudes to artisans, but, more immediately, they reveal their function as agents of reform. This could be the civilizing reform of how to comport oneself at court or in life, as it was for Castiglione and Della Casa, or it could be the reform of tradespeople. This was the aim of the gentleman Leonard Digges (d. 1571?) in his *Boke Named Tectonicon*, who saw his book as useful for tradesmen, who, as he imagined, would "first confusedly reade them thorow, then with more iudgement, and at the thirde readinge wittely to practise [...] Note, oft diligent reading ioyned with ingenious practise, causeth profitable laboure."[73] He believed that the craftsperson would need these three readings because the traditional practice of the crafts needed to be corrected by mathematical reasoning. He denigrated the rules of thumb that craftspeople used to calculate materials. As Anthony Gerbino and Stephen Johnston write, "Digges was not providing merely technical support but a prescription for a new relationship, in which the mathematician instructed the artisan from above."[74] They show that some artisans willingly accepted this vision, for example, the joiner Richard More, whose *The Carpenter's Rule* (1602) repeated Digges's criticisms and recommended further mathematical authors to his woodworking brethren. John Symonds, in contrast, saw it as enabling practitioners "to discourse equally with captains, statesmen, and governors." Digges offered an image of the mechanical arts ennobled by mathematics, for mathematics, he argued, would raise crafts out of the realm of rote, uninformed experience. Some craftspeople, however, challenged the authority of the gentleman author preaching reason to them, protesting that allowances could be devised for inaccuracies in the actual practice, and that their methods made them better equipped to work in the contingencies of the real world. The joiner Richard More recorded these protests dismissively, seeing them as holding fast to old, erroneous rules.[75]

Such craft disputes mirrored conflict in the Casa de Contratación established by the Spanish crown to equip and train navigators for its expanding empire. In this house of trade, scholar cosmographers sought to reform navigation by focusing on the construction of sea charts by mathematical means, while the pilots argued that their firsthand experience at sea was more effective in practice. In this debate, as in others throughout the early modern period, the scholars focused on "general and systematic knowledge in place of local craft knowledge."[76] A similar conflict erupted in Spain under Philip II, when the crown attempted to formalize gunners' techniques by applying mathematical methods. As with the ship pilots, this made possible an examination and certification process that eventually came to certify and police access to the profession. At the same time, schools to teach these methods proliferated, and they became a new audience for manuals teaching the new mathematical methods.[77]

While these examples demonstrate the reform of practitioners, practices, and institutions, a study of scholars' remarks about craft at this time reveals a deep interest in the mechanical arts as part of a more general reform of knowledge. The survey of technical writing in chapter 4 showed that discussions of practice and handwork were bound up with diverse reforms of knowledge in Roman antiquity, in the Middle Ages, and in the Renaissance. A second generation of humanist scholars, such as François Rabelais, Juan Luis Vives, Petrus Ramus, Paracelsus, and others used the arts in their efforts to reform the curriculum of the university. In François Rabelais's (1483–1553) satirical but useful overview of the humanist training of the fictional Gargantua in his pseudonymously published series, *Gargantua and Pantagruel* (first installment published 1532), the giant spent rainy days in observing practical activities:

They either went to see how metals were drawn or how artillery was cast; or they went to watch the lapidaries, goldsmiths, and cutters of precious stones; or the alchemists and coin minters, or the makers of great tapestries, the weavers, the velvet makers, watchmakers, mirror makers, printers, organists, dyers, and other such kinds of workmen; and, always treating to wine, they learned and observed the skill and inventiveness of the trades […] and instead of botanizing, they visited the shops of the druggists, herb sellers, and apothecaries, and considered attentively the fruits, roots, leaves, gums, seeds, exotic unguents, also at the same time how they were adulterated [note, again, craft cunning and deception].[78]

Rabelais hilariously satirizes technical manuals when he has Gargantua, with his massive strength, helping out the citizens of Orléans who had tried without success to unearth an enormous bell by employing precepts of Vitruvius, Alberti, Euclid, Theon, Archimedes, and Hero to no avail. Rabelais pokes fun at such texts again when he compiles a list of practical writings: "*On the manner of making black puddings*, by Mayr […] *On the practice and utility of skinning horses and mares*, written by Our Master de Quebecu […] *The Fart-Puller of the Apothecaries*," and so on.[79]

Like Rabelais, although in an altogether different tone, Juan Luis Vives (1493–1540) also lays out a new course of study in his encyclopedic *De disciplinis libri XX* that was oriented away from the scholasticism of university training, and based on the model of the arts. Vives, from a New Christian family in Valencia, studied in Spain and Paris, and then taught for most of his life in the southern Netherlands at the University of Leuven and in Bruges. In *De tradendis disciplinis* (On the Transmission of Knowledge, 1531), he considers the great variety of the arts, and the impossibility of drawing them into an orderly system. He attempts to classify them, dividing them into the contemplative arts, the goal of which is knowledge, and those that are active, which only aim to produce action, "as in music when, after the action, nothing is left."[80] Vives believes that once a man has become riper in knowledge and experience of things, he should begin to "consider more closely human life and to take an interest in the arts and inventions of men: e.g., in those arts which pertain to eating, clothing, dwelling." First, he should look to husbandry; then, to nature and strength of herbs and living animals; then, to architecture; then, to travel and conveyance, including navigation.[81] The pupil, he writes, "should not be ashamed to enter into shops and factories, and to ask questions from craftsmen, and get to know about the details of their work. Formerly, learned men disdained to inquire into those things which it is of such great import to life to know and remember, and many matters were despised and so were left almost unknown to them. This ignorance grew in succeeding centuries up to the present."

The fruit of the arts, writes Vives, would be practical wisdom, so necessary to the learned man in his active life, which brought the prized virtue of humanists, prudence. Vives defines prudence as "craftiness and astuteness," and a type of "carnal wisdom," thus associating it with the body:[82]

Practical wisdom is born from its parents, judgment and experience. Judgment must be sound and solid, and at times, quick and clear-sighted. Experience is either personal knowledge gained by our own action, or the knowledge acquired by what we have seen, read, heard of, in others. Where either of these sources is lacking a man cannot be practically wise. For in matters which are connected with any practical experience, unless at some time or other you have yourself gone through the experience, however much precepts may be expounded to you, if you never duly seek it yourself, of a surety, when you apply your hand to the work, there will not be much difference between your coming to it quite as a novice, and never having heard of it before.[83]

Vives did not believe that the mechanical (or "inferior," as he called them) arts were directly useful to the scholar for developing prudence—a scholar should not practice as a craftsman. Rather, the arts exemplified how judgment and practical action were cultivated by much experience: "[j]udgment such as is inherent in wisdom, cannot be taught." In order to acquire judgment, learned men must do much "reading of authors good in judgment—ancients, church writers, by dialectic, logic, art of right speaking." Experience, the second part of practical wisdom, which "brings a very great mass of detail to the power of thinking," was gained by being active in practical affairs and, most important, by studying histories,[84] those collections of human experience, most analogous to the case histories collected by medical practitioners: "[f]or out of how many practical experiences on all sides has the art of medicine to be built up, like rain-water composed of drops!"[85] In all, a scholar should not live his life behind the walls of a monastery or university, but should turn his knowledge "to usefulness and employ it for the common good."[86]

Vives's ideal education not only provides insight into his goals and those of his fellow reformers, but also sheds light on the type of pupil they were no doubt seeing more of in European cities. Vives goes on to describe the schoolmaster of the Lilian Gymnasium in Louvain, Charles Virulus. Whenever a father or relative of a pupil came to visit, Virulus

made a point of inquiring, some hours before the time fixed for dining, in what topics any coming guest was best versed. One was perhaps a sailor, another a soldier, another a farmer, another a smith, another a shoemaker, another a baker. In the meantime before their arrival, he would read and meditate upon his visitor's particular kind of work. Then he would come to the table prepared to delight his guest by conversing on matters familiar to him, and he would induce him to talk on his own affairs, and give him information about the most minute and secret mysteries of his art. He would thus hear in the briefest time detail which he himself could scarcely have gleaned from the study of many years [. . .] How much wealth of human wisdom is brought to mankind by those who commit to writing what they have gathered on the subjects of each art from the most experienced therein![87]

Other authors, such as Paracelsus, engaged in a more antagonistic attempt to reform university teaching by lecturing in the vernacular and by training practitioners, such as barber surgeons, alongside physicians. Paracelsus's effort to transform the university by including practitioners had roots in religious reform: God's first revelation in creation was primary, argued Paracelsus, preceding his revelation in the words of the Bible. Thus, true reformed religion would be gained by coming to know nature; artisans who engaged with the matter of nature would be guides in this reform of knowledge and society.[88]

While Paracelsus preached a sweeping reform, humble recipe collections could also function as vehicles of reform. The pseudonymous Alessio Piemontese's *Secreti* was one of the most reprinted recipe collections of the sixteenth century. In his posthumously published *Secreti nuovi di meravigliosa virtù del signor Ieronimo Ruscelli* (1567), the scholar Girolamo Ruscelli (1518?–66) claimed to be its author, and in the prefatory letter of the book he gave an account of its composition. He maintained that a group of scholars and gentlemen of Naples formed themselves into an "Accademia Segreta," which aimed "to make the most diligent inquiries and, as it were, a true anatomy of the things and operations of nature itself," and was committed "to the benefit of the world in general and in particular, by reducing to certainty and true knowledge so many useful and important secrets of all kinds for all sorts of people, be they rich or poor, learned or ignorant, male or female, young or old." The center of this academy was a "Filosofia"—a workshop or laboratory. The group of scholars and gentlemen of Naples employed eleven specialized artisans as "choremen," or, as Ruscelli calls them, "attendants and servants," to set up and carry out the experiments that produced medicines, dyes, metals, and other useful products. These servants included two apothecaries, two goldsmiths, two perfumers, a painter, and four each of herbalists and gardeners.[89] When he published an account of the academy in *Secreti nuovi* in 1567, Ruscelli claimed that the Accademia had tried every one of the 1,245 recipes three times, and that the elite members of the society "never failed to lend a hand willingly or busy themselves where necessary."[90]

Giambattista della Porta (1538–1615) also founded an Accademia dei Segreti in Naples, whose members were called *Otiosi*, or men of leisure. Della Porta later claimed their experiments were published in his *Magia naturalis* of 1589. Each member of this academy was supposed to bring a new natural fact or mechanical invention to their meetings. The language of utility was strong in the published versions of these collective investigations of practical processes, and the invocation of "use, profit, and pleasure" surfaces alongside the language of reform again and again in these and other such books of secrets. Whether their activities were real or imagined, they show that Ruscelli, della Porta, and their fellow academicians saw a potential in handwork and art for reform of both material life and philosophy. Their new philosophy, organized around the active *Filosophia*, brought scholars and artisans together to work with natural materials and produce useful things. This vision of reform, centered on the manipulation of nature by human art, gathered force through the seventeenth and eighteenth centuries.[91]

To Reduce into Art

Another means of reform touted by humanists and artisan authors was "to reduce" a field of knowledge, or a teachable discipline, into "art." Hélène Vérin and Pascal Dubourg have surveyed a very wide range of practical texts from the sixteenth through eighteenth centuries, and found that, in these texts, "to reduce into art" meant to articulate in writing a method by which a field of knowledge could be organized, and learned by a novice, and by which new knowledge could be integrated or created.[92] The humanist goal of reducing into a codified "art" or a "*methodus*"—an organizing technique or shortcut to knowledge of a subject—became an organizational principle especially for texts by self-described architects and engineers, whose ranks included experts in weaponry, building construction, and fortification. These professions included coordinating a multitude of different activities and realms of expertise, and their texts claimed that their "method" or "theory" gave them oversight and organizational abilities within the complex fields of their actions. Moreover, by writing down technical processes and "reducing" the practical activities into a method, they could develop new skills and new insights into acquiring and employing practical knowledge.

These writers also celebrated "*ingenium*," or inventiveness, and new technical inventions. Invention was connected to "foresight" and "intuition," and held up as the mark of the technician experienced in mechanical effects. As the early dialogue form of engineering books developed into a genre with a more or less standard order of presenting the contents, these works came to be seen as guides to action.[93] For example, one engineer author provided a multitude of circumstances—diverse sites and situations—in which an engineer might find himself. In studying these examples, the reader was supposed to furnish "the storehouse of his memory," by which he would develop judgment, just as the recipe collection provided multiple methods and materials to model the acquisition of experience, and the humanist scholar stored up his experiences as the basis of practical wisdom, or prudence.[94] In the course of the sixteenth century, Vérin argues, the "art" or "method" of engineers came to be seen as an art of decision making, just as Vives had made the development of judgment the goal of the scholar who wished to act decisively in public life and service. Vérin argues that these writings constitute a reform of *method* more than an attempt to communicate and teach technical knowledge.[95]

Another engineer author, the Dutch mathematical practitioner Simon Stevin (1548–1620), sought out a method by which many diverse ways of accomplishing a practical activity could be ordered into simple, often dichotomous, alternatives, sometimes pictured as branching diagrams.[96] These engineer authors propose in their books an ensemble of precepts for making choices, laying out means of proceeding that permit the finding of solutions in a world of particular circumstances. They also attempt to set down rules to guide reasoning, which they often imagined as being capable of mathematical expression.[97] Generally, Vérin argues, these practitioners aimed to write down their art in order to provide a method that would narrow the individual possible cases in order to reduce the field of decision making. Thus, as they struggled to build a professional identity as engineers by formulating a "theory" for their actions, the writers of these treatises aimed to demonstrate both that their sphere of action was liberal, and that it could form a "science of particulars" that could serve as a basis for decision making—judging and deciding when and how to act among the multiple contingent factors in a complex field of action.

The ambitious reformer of knowledge Francis Bacon (1561–1626) also wrote much about the utility of the arts and the need to devise a method of invention. At the same time, he wrote of reducing into art, but observed that we must not be overhasty in the "reduction of knowledge into arts and methods."[98] For Bacon, like the authors and compilers of engineering texts, to reduce into art was to assemble, put into order, to clarify by the aid of precepts or rules, to set out the means of proceeding in a brief and methodical exposition, and to disseminate by writing in a language accessible to all. Bacon and other reformers of knowledge saw the goal of their reform to be the more general diffusion of knowledge and thereby its possible and eventual perfection, built on the cumulative work of generations.[99]

Whether such rules could successfully enable control of the multiplicity of particular cases and the diversity of nature remained an open question throughout the seventeenth and eighteenth centuries.[100]

In their books, then, engineers sought to create an identity as practitioners of a liberal art who possessed a type of knowledge that could be the basis of decision making and, more generally, to reform training in their profession.[101] Like the Spanish attempts in the sixteenth century to systematize the education and licensing of pilots and gunners, such projects of "reducing into art" were integrated into reform and control of training in a variety of trades. Such reform was almost always associated with the central consolidation of power in European states, and it presaged the events of the eighteenth and nineteenth centuries, when the codification of technical practices increasingly became bound up with the aims of administrators wishing to wrest control of production processes from individual tradespeople and from corporate bodies, including trade and artisanal corporations.[102] Interestingly, we already see evidence of this beginning in the sixteenth century, when some Italian towns opposed "engineers" as subversive agents of the centralizing prince's power.[103]

The unprecedented increase and popularity of technical literature was thus due in part to the pedagogical and reforming aims of humanists and the recognition of the utility of the arts to civic authorities, but also to practitioners' recognition of and response to the lively interests of the state in technical processes. In these polities, a new set of relationships among experts, rulers, and artisans emerged.[104] Just as the invention of writing had helped in administering the earliest empires, and written numbers facilitated the levying of taxes, so in early modern Europe, writing down practical techniques and knowledge came to be used as a tool that enabled governments to "act at a distance" to control sites of production.[105]

Conclusion

To answer the question that opened this chapter—who bought, read, and used the many books of art written from the fifteenth century on?—it is safe to say that while artisans owned some such books, they appear more likely to have possessed model books and devotional books, rather than wordy technical books. In any case, the population of literate and book-owning artisans would never have been numerous enough to constitute the main readership for so many editions of these manuals. It was instead a wide-ranging audience that made these little books of art so profitable to printers from the late fifteenth through at least the eighteenth century. This audience ranged from monks to humanist scholars, reading and writing instructors, individuals seeking spiritual guidance, connoisseurs seeking a critical vocabulary, and intellectual and pedagogical reformers, as well as to governments seeking to extend their power and control.

CHAPTER 8

Kunst as Power

Making and Collecting

Things and Processes

In 1456, the sculptor Donatello (1385–1466) gave his physician, Giovanni Chellini, a bronze roundel with a relief of the Madonna and Child. This object had a unique and telling feature: the reverse side of it formed a mold into which molten glass could be poured in order to make replica relief roundels in glass (figs. 8.1 and 8.2). Chellini's account book records that on August 27, 1456,

while I was treating Donato called Donatello, the singular and principal master in making figures of bronze, of wood and terracotta […] he of his kindness and in consideration of the medical treatment which I had given and was giving for his illness gave me a roundel the size of a trencher in which was sculpted the Virgin Mary with the Child at her neck and two angels on each side, all of bronze, and on the outer side hollowed out so that melted glass could be cast on to it and would make the same figures as those on the other side.[1]

As we saw in the preceding chapter, the work of the human hand, or *Kunst*, and *Kunstbücher* were of deep interest to a variety of people in early modern Europe, including patrons who conferred with artists over commissions, humanist scholars alert to the lessons that the practice of the arts could have for self-governance and the making of knowledge about new fields, and court officials tasked with managing projects. As Chellini's account of Donatello's gift intimates, ingenious and unusual objects and the process of making them were equally intriguing to these admirers of art.

In a well-known passage, Albrecht Dürer also celebrated the ingenuity of the artisans in what he referred to as "the new land of gold." He saw the goods brought back in 1520 from New Spain displayed in Brussels, including "a sun all of gold, and a moon, too, of silver […] all manner of wondrous weapons, harness, darts, wonderful shields, extraordinary clothing, beds, and all kinds of wonderful things for human use, much finer to look at than prodigies." Along with noting that these goods were "valued at 100,000 gulden,"

8.1. Donatello, roundel with the Virgin and Child with four angels, Florence, c. 1450. Gilded bronze, diameter 28.5 cm, depth 2.7 cm, weight 4.26 kg. Victoria and Albert Museum, A.1–1976. Purchased with the aid of public subscription, with donations from Art Fund and the Pilgrim Trust, in memory of David, Earl of Crawford and Balcarres. © Victoria and Albert Museum, London.

8.2. Donatello, roundel, obverse. As Chellini noted, this side can be used as a mold. © Victoria and Albert Museum, London.

Dürer claimed that in "all the days of my life I have seen nothing that reaches my heart so much as these, for among them I have seen wonderfully artful [*künstlich*] things and have admired the subtle ingenuity of men in foreign lands; indeed, I don't know how to express what I there found."[2] Dürer's moving reaction to the human ingenuity of the new world exemplifies attitudes to objects and ingenuity found before European expansion, but which developed in tandem with the exploration and exploitation of that new world. He combined in an unselfconscious way an interest in ingenuity with an immediate impulse to mention the monetary worth of objects; these things were valuable not just as testaments to virtuoso ingenuity—they also could stand in for specie.[3]

In *Allegory of Fire* (1608; fig. 8.3), Jan Brueghel the Elder (1568–1625) depicted a copious assortment of artful and ingenious objects in minute detail. Steel armor, porcelain, silver vessels, and gold jewelry all draw a viewer's eye in the foreground, while the deeper ground invites the observer to take in the making of metal objects, from the molten metal to the hammering out of a metal sheet, and the shaping of armor by means of the large, water-driven trip hammer in the right middle ground of the painting. Armor is being polished at polishing wheels, which had come into use in the previous century, and metal rims are being mounted on cannon carriage wheels in the further distance of the workshop. Such paintings were created by Brueghel for a powerful group of learned clergy in Milan and Rome, who appreciated them as objects of pleasure and spiritual exercise.[4] Brueghel's paintings fostered close looking at both objects and the processes by which they were made.

The passion for observing practice and collecting objects reached far and wide in society. It is not possible to determine what impelled the surgeon and scholar of Clare Hall, John Seward, to collect the "hamper full of goldsmythes instruments"[5] that he left at his death in 1552 along with his many medical books and instruments relating to his surgical practice. His motivation may have been similar to that of Basel jurist and city official Basilius Amerbach (1533–91), who also collected goldsmiths' tools and patterns. Interested in humanist practice, Amerbach was fascinated by evidence of Roman antiquity, taking part in the first excavations of the Roman settlement near Basel and proudly claiming to be the first to recognize the Roman amphitheater there as an ancient building. Like other humanists, he collected large numbers of coins, seeing them as valuable historical sources

8.3. Jan Brueghel the Elder, *Allegory of Fire*, 1608–10. Oil on copper, 46 × 66 cm. Pinacoteca Ambrosiana, Milan, inv. no. 68. © Veneranda Biblioteca Ambrosiana / Paolo Manusardi / Mondadori Portfolio / Bridgeman Images.

for Roman life (fig. 8.4). From early in his life, he also had a profound interest in the working processes of goldsmiths. In 1560, when he was training at the Imperial Law Court in Speyer, his father (also a lawyer) desired that he lodge with a fellow jurist, but Amerbach angered his father by choosing instead to lodge with a goldsmith, Jacob zur Glocke.

In the 1570s and 1580s, Amerbach collected the entire contents of at least two goldsmiths' workshops (fig. 8.5), including preparatory drawings, molds, patterns, and tools, all listed carefully in the inventory that he drew up in his own hand.[6] His interest was not just in the materials and tools, but in all the steps of the creative process, as can be seen in his collection—today in the Basel Historisches Museum—such as the stages by which goldsmiths transformed lead plates into ornamental gilt foliage for use on sculpture (fig. 8.6). He collected the unworked templates and each of the progressively more finished metal pieces that documented every stage of the metalsmith's work of hammering these templates into ornamental foliage. Although particularly taken with goldsmithing, Amerbach also collected the contents of a woodcarver's workshop.[7] Amerbach's fascination with documenting the creative and technical process by which artisans produced objects would spread widely among elite collectors in the course of the following centuries.[8]

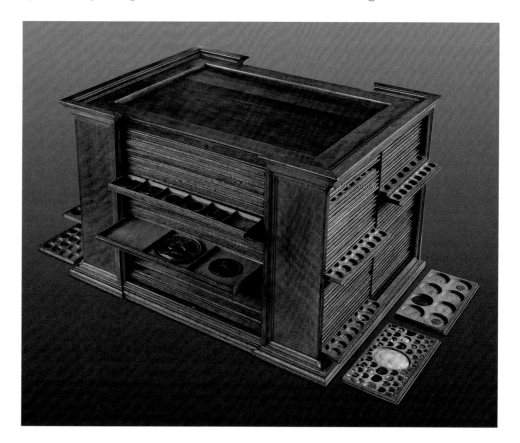

8.4. Coin and medal cabinet, made for Basilius Amerbach by Mathis I. Giger, Basel, c. 1578. Walnut, oak, maple, pear, linden wood, solid and veneered. H 45.9 cm, W 73.3 cm, D 55.6 cm. Historisches Museum, Basel, Inv. 1908.16. Historisches Museum Basel, Peter Portner. CC BY-SA 4.0.

8.5. Goldsmith's workbench, with components and tools, some collected by Basilius Amerbach. Historisches Museum, Basel. Author photo.

8.6. Goldsmiths' models, fifteenth to early sixteenth century. Lead, silver, wood. Historisches Museum, Basel. Among other objects that testified to the creative process, Basilius Amerbach collected goldsmiths' models that evidenced their techniques. Objects in the first and second row at the top right show stages in the process of hammering flat lead plates into ornamental gilt foliage to be used on sculpture.

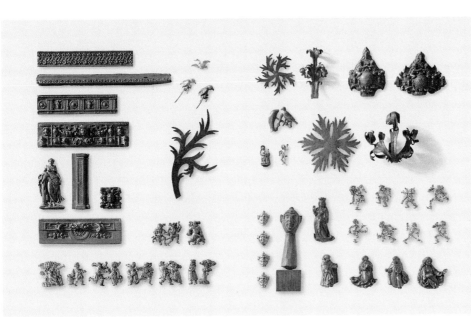

The interest in the process of making art—to be found, as we saw in the preceding chapter, in the books of art, the *Kunstbücher*—also figures largely in the amassing of objects in collections known in northern Europe as *Kunstkammern*, or "chambers of art." The objects brought together in these chambers possessed a host of meanings for their owners, including the representation of status and power, self-definition, specie, spiritual consolation, intellectual stimulation, and aesthetic pleasure.[9] This chapter considers what those collections can tell us about the interest in *Kunst* as knowledge. It focuses in particular on the librarian Samuel Quiccheberg (1529–67), who first helped build the collection of the Fugger family, the long-distance merchants based in Augsburg, before entering the employ of Duke Albrecht V of Bavaria (1528–79). In 1563–65, Quiccheberg wrote a how-to manual of sorts for the collector of objects. It is clear that Quiccheberg's ideal collection, while working to glorify and represent the mastery and power of the collector, focused primarily on the artifice of nature and of the human hand: on craftspeople, their tools, materials, and products. Scholars have viewed Quiccheberg's treatise as a mode of displaying the power of a prince or of representing the plenitude of the cosmos, but we will focus here on how his instructions for collecting encode an imaginary of *Kunst* that tallies with the epistemic dimensions of *Kunst* laid out in chapter 5. This imaginary comes into sharper relief when considered alongside the work of his contemporary, the goldsmith Wenzel Jamnitzer, whom we first encountered in chapter 3. As we saw there, Jamnitzer was known for the figures of plants and animals that he cast from life, and Quiccheberg regarded these life-casts as possessing outsized significance, for, as this chapter will show, these works of the human hand revealed the generative and productive powers of both art and nature.[10]

Chambers for *Kunst*

Keeping pace with the mounting number of *Kunstbücher* written from the fifteenth century on, vast numbers of objects of art were also collected during this time by nobles, scholars, prominent citizens, merchants, and even city councils. Antiquities, decorative and sacred objects, and rare works of nature had been collected in church treasuries for centuries, but in the fifteenth century, nobles in the duchy of Burgundy began to set a European trend for collecting.[11] These collections, or *Kunstkammern*, gathered together natural objects and works of art, both spectacular and mundane.[12] They were filled with (as Holy Roman Emperor Rudolf II's *Kunstkammer* inventory titled them) *naturalia* (natural things), *artificialia* (things made by the human hand), and *scientifica* or *instrumenta* (measuring and calculating instruments).[13] From early on, they included objects from New Spain and the East Indies, measuring instruments, works that combined the artifice of nature and of the human hand, and—demonstrating the interest in process—even artisans' tools. These collections, housed in chambers or within impressive cabinets, were often commissioned by noble patrons in the sixteenth and early seventeenth centuries. The Augsburg merchant Philipp Hainhofer (1578–1647) specialized in stimulating demand among status-hungry

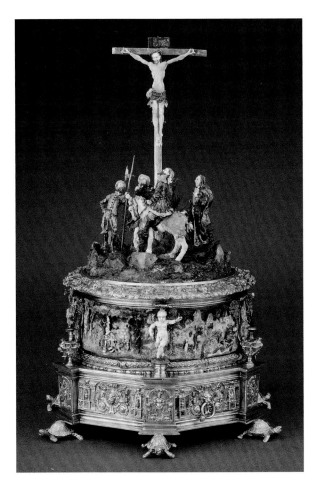

northern German princes, by organizing groups of specialist artisans to construct cabinets made to order for these nobles. Hainhofer's correspondence with these nobles constituted a sustained course of education for them on the skill and industry of artisans, the art embodied in their products, and the necessary categories of discernment for a budding connoisseur.[14]

Objects made for noble patrons often combined mechanical and technical display with typical noble pursuits, such as banqueting, hunting, and weaponry. Noble banquets in the sixteenth and seventeenth centuries often featured complex automata. A very popular one was the pagan goddess Diana mounted on a stag, which moved by itself down the banqueting table, coming to rest in front of a guest and shooting a tiny arrow. The banqueter could then use the stag's head, which doubled as a drinking cup, to toast the table. Another such object, fit for a princely *studiolo*, was a combination *Handstein*, salt cellar, inkwell, and object of devotion made by Christoph Ritter in 1551—the Rolex watch (or perhaps, by now, the smartphone) of its day (figs. 8.7 and 8.8). Even Jamnitzer's stunning meditation on

8.7. Christoph Ritter I (d. before 1573), multifunctional object incorporating a salt cellar, inkwell, and Crucifixion scene, 1550. Silver, gilded silver, enamel, H 23.4 cm. Schroder Collection, London.

8.8. Christoph Ritter I, detail of ore and stones at the base of the object. Schroder Collection, London.

8.9. Wenzel Jamnitzer, *Daphne*, c. 1570–75. Silver, gilded silver, coral, and semiprecious stones. H 66.5 cm. Musée national de la Renaissance, Château d'Écouen, France (E.Cl. 20750). © RMN-Grand Palais / Art Resource NY. Photo: Mathieu Rabeau.

8.10. Jamnitzer, *Daphne*, dismantled at the waist, formed a vessel similar to other banqueting table ornaments that served as drinking accoutrements.

metamorphosis in the guise of Daphne in the process of being turned into a laurel tree could be dismantled to do double duty as a sumptuous drinking vessel (figs. 8.9 and 8.10)!

Although the imperial collections of the Habsburgs were divided into *naturalia*, *artificialia*, and *scientifica*, other noble collections contained objects that accorded with the varied interests and regional concerns of their patron. For example, the collection of Archduke Ferdinand II at Schloss Ambras, begun in 1560, focused on armor, *naturalia* (especially the wonders, or artifice, of nature), and works of art that incorporated natural objects.[15] Duke August I of Saxony was particularly interested in tools and objects associated with mining and metalworking, an important source of his territorial wealth, and he accumulated a large collection of luxury metalworking tools in the seven rooms of his *Kunstkammer*, including a full-size wire-drawing bench (fig. 8.11).[16] He also collected measuring, calculating, and drawing implements and a remarkable body of practical and geometrical manuals, which he bought at the rate of about a hundred per year throughout

8.11. Leonhard Danner (1497/1507?–1586), wire-drawing bench, Nuremberg, 1565. Various woods, wood marquetry, iron (gilded, etched). H 79 × W 440 × D 21 cm. Made for Duke August I of Saxony, this lavish bench produced wire of various gauges. Musée national de la Renaissance, Château d'Écouen (E.Cl.16880). © RMN-Grand Palais / Art Resource NY. Photo: Stéphane Maréchalle.

8.12. François Houard (active seventeenth century), ivory-turning lathe of Maximilian Emanuel, Elector of Bavaria. Bronze, iron, brass, H 2.7 m. Bayerisches Nationalmuseum, Munich, Inv.-Nr. E 1236.1, Foto Nr. D116457. © Bayerisches Nationalmuseum München. Photo: Bastian Krack.

his entire thirty-three-year reign.[17] Some nobles even engaged in handwork themselves. The Wittelsbach dukes of Bavaria learned to work on lathes (fig. 8.12), and, when young, some even performed life-casting. Francesco I Medici, Grand Duke of Tuscany (1541–87), who, as we saw in chapter 4, brought many artists to his court, had himself pictured laboring in an alchemical workshop.

The Most Ample Theater

Quiccheberg's manual for collectors indicates that wealthy merchants set a trend in building collections that nobles eagerly followed.[18] Quiccheberg praised the Fugger collections,[19] which he had helped to create when he entered their service as librarian in 1555, and remarked that one important effect of their collections was to stimulate new and ingenious inventions.[20] Even before these merchant collections developed, it appears that artisans had held extensive collections, for Quiccheberg commented that his own collection was surpassed by the collections of tools and precious items possessed by the artisans of the free imperial cities: "It happened, to tell the truth, that I was energetically surpassed by goldsmiths, painters, sculptors, and others who are more or less unlettered."[21] His treatise made bold claims for the significance of collecting, which he extolled as "a first philosophy." This new activity, Quiccheberg argued, "has brought about the certainty of all scientific areas as well as the most complete methods; like the opening of the doors of wisdom, it has produced the greatest use and godly clarity in the sciences."[22] Quiccheberg titled his treatise, "Inscriptions or Titles of the Most Ample Theater encompassing particulars of the whole creation and outstanding images, or […] a storehouse of artificial and wonderful things […] which, for those who spend time to examine and engage with them, will bring about a singular knowledge of things and admirable prudence" (*Inscriptiones vel tituli theatri: amplissimi, complectentis rerum universitatis singulas materias et imagines eximias […] Promptuarium artificiosarum miraculosarumque rerum […] ut eorum frequenti inspectione tractationeque, singularis aliqua rerum cognitio et prudentia admiranda*). What he meant by prudence (*prudentia*) becomes clear a little later on, when he notes that "it cannot be expressed by any person's eloquence how much wisdom and utility in administering the state—as much in the civil and military spheres as in the ecclesiastical and cultural—can be gained from the examination and study of the images and objects that we are prescribing."[23] Quiccheberg's claims that collections would bring "a singular [implying also "new"] knowledge of things" and "wisdom in statecraft" may seem hyperbolic, but they can be understood as part of a new conception of philosophy that treated knowledge as active, productive, based on nature, and an integral part of civic life, very much in accord with the humanist pedagogical reformers examined in chapter 7.[24]

Quiccheberg's treatise marks an important moment in the history of collecting, for it represents the first time instructions for a collection were put into writing. Quiccheberg, who had been born in Antwerp, raised in Nuremberg, and educated in Basel and Ingolstadt, entered the service of the Wittelsbach Duke Albrecht V in 1559, after service with the Fugger family. In 1565, the same year that Quiccheberg wrote his manual, the duke ordered seventeen precious objects to be kept in perpetuity by the Wittelsbach family. A princely proclamation of this sort was necessary in order to protect them from being melted down or sold as a need for specie arose. The explosive growth of such collections in the course of the century can be illustrated by the fact that Albrecht's original list of items was increased to 27 a few years later, and then still further by Albrecht's son Duke Wilhelm V, finally reaching 3,407 items by 1598, only thirty-three years after Albrecht's first proclamation.[25]

A Storehouse of Artificial and Wonderful Things

Gathering objects in a collection demonstrated princely mastery—mastery of nature and of the world—and Quiccheberg's 1565 manual, the "Most Ample Theater," begins in this way, making clear that a collection should be deployed in the representation of power and wealth, and to show the noble characteristics of the ruler. Quiccheberg's treatise divided the collection into five "classes" of inscriptions—it must be remembered that Quiccheberg is describing not so much a space as a set of objects and the inscriptions on their labels ("Inscriptions or Titles of the Most Ample Theater …"). Such a conception of a collection of material things has more than a hint of the bookishness of a librarian about it.

In the first class, Quiccheberg placed saints' portraits, as saints were intermediaries between the rulers, representing the divine on earth, and God. Following these came family portraits, then maps that delineated the ruler's territory. Along with these could be included maps of the world, but, in line with the celebratory and representational aspects of the collection, Quiccheberg specified that the map of the patron's territory must be more conspicuous, larger, and more richly ornamented than all the rest.

After making clear the ruler's place in a cosmic hierarchy, Quiccheberg moves rapidly over paintings of the ruler's residence, military campaigns, ritual, and spectacle to the things that appear closer to his heart: paintings of large, rare, or unusual animals, especially those found in the patron's territory, and all kinds of models demonstrating human artifice. These models, which were meant to be scaled up for the use of architects and builders, included buildings of all kinds, constructed of wood, paper, and feathers, and ornamented with colors, as well as models of ships, wagons, stairs, fountains, arches, and bridges. In addition, small-scale models of machines, including water pumps, sawmills, grain mills, stamping mills, and dams, were to occupy this first section of the collection, explicitly created for the purpose of scaling them up to full size and thereby discovering whether they functioned usefully or could be improved.[26] The chamber of art was also intended to be useful to the territory.

This first section is followed by four more that all center on the relationship between nature and human *ars*. The second set of titles again begins with the representation of power, as embodied by statues of emperors, kings, famous men, divine beings, and even animals, then passes to the material out of which the statues are formed, including stone, wood, plaster, marble, and all kinds of metals. Quiccheberg again passes quickly over the subjects of the statues to the arts by which materials are worked, alluding to metalworking, turning, sculpting, woodworking, glassblowing, embroidery, weaving, tools and machines, and vessels of all sorts—foreign, ancient, and religious—as well as weights and measures, and everything connected to agriculture and mining, not forgetting the staple objects of collections in this period: coins and portrait medals. Between a discussion of emblematic medals and copper engraving plates (in which he appears to have had a great interest), Quiccheberg alludes to small figures made by goldsmiths, including little ornamental figures with leaves, flowers, animals, and shells. Quiccheberg is here making reference to life-casts, such as those made by Jamnitzer and the author-practitioner of Ms. Fr. 640.

The third set of titles contains natural materials of all sorts, including preserved animals, raw materials of metalsmithing, seeds and herbs, colors and pigments for metal, resins, wax, sulfur, ivory, textiles, and woods, as well as earthy materials including "juices of the earth," both natural and artificial, particularly medicinal earths, chalks, clays, vitriol, alum, and salt. We also find here fluids from dripping hot springs and grottoes like those Palissy and Jamnitzer imitated in their work. Quiccheberg includes in this section prostheses for human limbs as well as parts of animals, such as horns, snouts, teeth, bones, bezoar, bladder and kidney stones, pelts, feathers, claws, skins, and skeletons. It appears that Quiccheberg's principle of organization in this grouping is materials and structures that are part of organic or transformative processes. This explains the presence in this class of impressive stones, such as marble, jasper, alabaster, and porphyry, for all these striated stones show evidence in their appearance of having grown in the earth, like the many specimens of native silver that also filled *Kunstkammern*. At the same time, Quiccheberg again includes in this section casts of animals made from metal, plaster, or ceramics, and other naturalistic objects. Here, lifelikeness seems to be the goal, and he comments, "By this art, they all seem to be alive, for example, lizards, snakes, fish, frogs, crabs, insects, shells," and they are colored so that "one believes they are real."[27] Life-cast plants and silk flowers are also included in this section.

In the fourth section of the collection, tools and machines of all kinds are featured, including instruments for writing, surveying, hunting, gardening, making war, making music, raising weights, or undertaking surgery and dissection, and indeed all the tools by which "artisans all over the world in our time nourish the world."[28] In this section Quiccheberg also includes foreign clothing, which would be displayed on dolls, and, almost as an afterthought, unusual or rare clothing, especially items belonging to the patron commissioning the collection. The fifth and final section is made up of images in all types of media, such as oil paintings, watercolors, tapestries, engravings, genealogies, portraits of famous men, coats of arms, and inscriptions, which were to be both painted on walls and hung on little panels. In this section, too, are listed the numerous cabinets in which small items could be stored and displayed.

Attached to this ample theater and storehouse, Quiccheberg specified the addition of a library, which was to be organized according to the subject matter of the books; a printing press; a lathe room, which actually already existed at the Munich court before Quiccheberg arrived;[29] and a medicine cabinet, which he noted was especially the province of women because they desired to help the poor and sick. Quiccheberg praised medicine as a constant delight for the mind because it always leads to new experience, and observed that the Wittelsbach court already possessed such a medicine cabinet, established through the work of Duchess Anna, the wife of Duke Albrecht. According to Quiccheberg, the duchess also maintained a large aviary holding many different types of birds, and she allowed scholars access to study the birds in it.[30] Workshops, too, were to be attached to the collection, including a casting and stamping operation with both a smith's forge and an alchemical furnace, in which life-casts of plants and animals were to be produced, which could preserve and stand in for these short-lived items.[31]

Quiccheberg counsels the expansion of the collection with further sections that respond to a patron's special interests, such as musical instruments, chests of clothes, and ornaments for plays and masked balls, or weapons. At the same time, Quiccheberg assured the less flush collector that he did not have to include all sections of the collection—just those parts he could afford or that coincided with his interests. But it was essential for a collector to employ a person he could send into all regions of the earth to seek out wonderful things, as well as individuals knowledgeable about the objects in the collection.[32] Quiccheberg also gave advice on how to arrange the objects, suggesting organizing principles that adhered to the "form" of the thing (*formas rerum*), which seems to correspond to a general appearance based on both material and function.[33]

Kunststücke

Quiccheberg's written manual on collecting can be compared to an object, a "chamber fountain" that Wenzel Jamnitzer began in 1556 for Emperor Maximilian II, but only delivered twenty-two years later to Rudolf II in 1578. Like Quiccheberg's theater, scholars have sometimes viewed this fountain as a representation of the cosmos, and it seems cosmic indeed, a true *Kunststuck* (masterpiece). Ten feet high and five feet across, the fountain was assembled within a room (hence the term *Zimmerbrunnen*, or "chamber fountain"), and it comprised several tiers, each pertaining to elements of the divine, human, and political cosmos. It possessed, according to a description made by a seventeenth-century visitor, "not only physics and metaphysics, but also politics, as well as many wonderful philosophical and poetical secrets displayed and proven to the eyes."[34] By examining this fountain as a collection, as a theater or storehouse analogous to Quiccheberg's theater of art and to the recipe collections known as storehouses, we can grasp more clearly the aims of Quiccheberg's collecting and why he considered it a "first philosophy."

Most of the fountain was melted down in the eighteenth century—a fate shared by most such chamber fountains and the vast majority of medieval and Renaissance objects made from precious metals. Our knowledge of the fountain comes from this written description, which probably drew on the booklet written by Jamnitzer to accompany the object when it was presented to his patron (just as we saw in chapter 4 that he wrote a voluminous description and instruction manual for his *Schreibtisch* of instruments).[35] The only components

8.13. Johann Gregor van der Schardt (c. 1530–81), allegory of the four seasons, 1569–78. Bronze, fire gilding, H 71 cm (each). These four figures supported Jamnitzer's chamber fountain. KK 1118, 1122, 1126, 1130, Kunsthistorisches Museum, Vienna. © KHM-Museumsverband.

of the fountain not melted down were four figures at its base. These represent the four seasons: Flora (Spring) with her bouquet of flowers, Ceres (Summer) holding a cornucopia and crowned with a wreath of ripe wheat, Bacchus (Autumn) holding a bunch of wine grapes, and Vulcan (Winter) with a plow he has just hammered out on his anvil. Together, they show, according to the description, the unalterable band of nature that forms a framework for all of human life (fig. 8.13).[36] Above these figures, the fountain rose in five tiers, in accord with the structure of nature based on the four elements (fig. 8.14).

The lowest tier symbolized earth and was represented by Cybele, goddess of the earth and a daughter of Saturn. She was located in a grotto that featured native silver and gold, and all kinds of mining implements, likely similar to the *Handsteine* discussed earlier. Between the ores and the silver formations grew silver and gold flowers, cast from life, looking as if they were growing naturally along lively little brooks flowing with water between the rocks. Life-cast creatures also crept along the stream banks. A polishing mill, stamp mill, saw mill, and hammer mill operated in miniature—all driven by the water of these streams, likely similar to the later mechanical figures installed in Schloss Hellbrunn, the Archbishop of Salzburg's summer palace (figs. 8.15 and 8.16).

The next tier was a basin symbolizing the element of water, represented by Neptune standing on a shell drawn by hippopotamuses around the basin, and battling strange sea monsters that first moved toward him threateningly and then fled from him. The constant movement and the to-and-fro of the battle signified both the ebb and flow of the sea around the earth and the fact that great lords and potentates must battle constantly with enemies of the common good. This section was surrounded by a crown, reminding rulers that they must carefully keep secret their plans of war.

The third tier, symbolizing air, was represented by Mercury, who swung and swooped off the fountain as if actually in flight. Under him a dark cloud spewed raindrops, and images of the four winds portended a furious storm. Flying into the storm were all sorts of birds, symbolic again of the element of air, and four angels carrying laurel wreaths, signifying the serving spirits that intervene between God and humans, and reminding the viewer that God will protect and maintain his children in even the fiercest storm.

Above the tier of air was that of the element fire, represented by Jupiter. Having reached this height, the fountain took on a complex admonitory political program that constituted a "mirror" (or manual) for the prince's rule. It was devoted to showing the order (*polizei*) preserved by the emperor and the hierarchy of nobles as the representatives of God on earth. Surrounding Jupiter were four angels arranged in such a way that as Jupiter circled among them, each bowed in reverence before they turned outward again to the human race,

8.14. (*facing, right*) Wenzel Jamnitzer (or workshop), design for a chamber fountain, 1550/60. Pen, aquarelle, and gouache on paper, glued to parchment, 154 × 62 cm. Kupferstichkabinett, Schenkung Melchior Berri 1849, Inv. 1849.11, Kunstmuseum, Basel. This apparently full-size presentation design is not for the now-lost Jamnitzer chamber fountain; however, it illustrates multiple levels of a fountain corresponding to the spheres of different elements similar to Jamnitzer's chamber fountain.

8.15. Knife-grinder, Schloss Hellbrunn, Salzburg. These water-driven mechanical figures, inset in small grottoes, were constructed 1613–17 by court architect Santino Solari (1576–1646) and Archbishop of Salzburg Marcus Sitticus (1574–1619) for the archbishop's summer palace. These figures continue to operate today; see https://stadtsalzburg.pageflow .io/hellbrunn-en#248188. Their functioning was probably similar to those described on Jamnitzer's chamber fountain. Author photo.

8.16. Water-driven mechanical figure of a potter, Schloss Hellbrunn, Salzburg. Author photo.

to whom they are destined as serving spirits. The angels make clear how any high potentate should organize his regime: as a Michael, a strong hero who protects the subjects and the whole land; as a Gabriel, a wise, powerful speaker and chancellor; as a Uriel, a righteous judge and good preacher who brings the truth to light through his words; and finally as a Raphael, a learned and true physician and an experienced teacher of all the liberal arts. Four eagles flew between the angels, equipped with scepters, to signify the characteristics of the ruler, who among other traits is merciful and generous to "small birds, or subjects."

Above the angels and eagles stood a single large eagle, strangling a basilisk with one claw and holding a stone in the other. Under one wing it held a scepter and under the other, a small flag that moved up and down to honor God. On this eagle rode Jupiter, lord of lords, holding a lightning bolt in his right hand to punish evildoers, and pouring water from a vessel with his left to bless the righteous and bring fertility to the earth. On his head was a crown with eight points, from which small streams of water welled up, gently sprinkling down in a rain of fertility and healing. Finally, in a reflection of his majesty, a flame of fire shot from Jupiter's head.

Under this heavenly firmament rose four arches, symbolizing the imperial crown, beneath which sat four monarchs, the last of which, the Holy Roman Emperor, held a scepter and globe. Around these monarchs was a complete representation of the political structure of the empire and the lineage of the Austrian house up to Rudolf II. The account of this elaborate fountain ends with a description of the self-playing music for two peasant dances, the "Rolandt" and the "Pickelhäring," set in motion through the action of the flowing water.[37]

Noble viewers, visiting scholars, and select artists might obtain permission to see the *Kunstkammer* and Jamnitzer's fountain. Such visitors saw displayed to their gaze the representation of the House of Habsburg's power and reach, as well as the political structure of the Holy Roman Empire. They could consider the lessons of prudence that the fountain imparted to ruler and subject alike,[38] and would absorb the description of nature as made up of four elements in constant ebb and flow. As the viewers came to the bottom tier of earth, with its arresting gold and silver mines filled with ore and native metals, and lifelike animals and plants arranged around flowing streams, along which tiny mills operated, they could not fail to be impressed by the movements of these mills and the self-playing music. They might ponder, as well, the processes of metamorphosis by which ore becomes gold and silver, or the mechanics of water flow, or the workings of mechanical devices. Below these, the fountain rested on the personifications of the seasons, symbolizing the natural processes by which all things came into being, were transformed, and passed away.

This fountain contained the same range of objects as in Quiccheberg's theater—natural materials, works of the human hand, and objects of political representation—and both the fountain and the *Kunstkammer*-Theater were intended to engage the viewer in active consideration. Jamnitzer's fountain provides an interpretative structure for the objects and

8.17. Wenzel Jamnitzer, table ornament, 1549. Cast, enameled, etched, and gilded silver, 99.8 cm. BK-17040-A, Rijksmuseum, Amsterdam. Encrusted with small animals and plants cast from life, this centerpiece is a paean to Terra Mater and her fertility. Plants and small reptiles cluster at the base. Mother Earth supports a basin, with life-cast snakes and lizards around the rim. The floor of the basin is decorated with cornucopias and moresques, while the egg-shaped vase holds plants and flowers, also cast from life.

materials that were displayed as separate elements in Quiccheberg's theater, but, in both, the viewer was confronted with the generative power of human art. Indeed, Jamnitzer's fountain embodied and displayed, in a single and unified work of art, the powers of nature and of human art to which Quiccheberg's collection also alluded.

Jamnitzer's other works of art, the instruments of his *Schreibtisch*, and his *Daphne*, for example, also epitomized the power of art in its engagement with the themes and things of nature. A banqueting table ornament created in 1549 featured Jamnitzer's specialties: grasses and flowers, cast from life and springing from an egg-like vessel at the top (fig. 8.17), while reptiles and insects, also cast from life, creep forth from a grotto-like

8.18. Jamnitzer, table ornament, grotto at the base of the sculpture, from which creep life-cast insects and lizards. Photo: Robert van Lang.

base (fig. 8.18).[39] The central female figure represents mother earth, emblazoned with the words, "I am the Earth, mother of all things, beladen with the precious burden of the fruits which are produced from myself."[40] The table ornament in its entirety symbolized the fertility and generative powers of nature, but it also demonstrated Jamnitzer's own powers of creation and his ability to imitate nature, as with his extraordinary *Daphne*, metamorphosing before the viewer's gaze.

Life-Casting as First Philosophy

Considerable evidence points to the popularity of life-casts among collectors in the sixteenth and early seventeenth centuries. The inventories of the *Kunstkammern* of the Wittelsbachs in Bavaria and of the Habsburgs were full of hundreds of life-casts in silver, tin, lead, plaster, and other media, and the eagerness with which they were sought out can be detected in the anxious letters between Jamnitzer and Archduke Ferdinand of Austria (1564–1602) between 1556 and 1562 about a commission to build a large fountain with animals cast from life around the base (likely the cosmic fountain).[41] Another example of this interest can be found in a letter from an official of the court of Grand Duke Francesco I de' Medici to Duke Wilhelm V of Bavaria (1548–1626) that details a "Pergkhwerckhstuckh," or

Handstein, that the grand duke was sending to Wilhelm. It included life-cast lead patterns for plants that were to be attached to the work. A postscript to the letter states that the live frogs, snakes, plants, and other animals that the duke desired could not be obtained due to the cold weather.[42]

Life-casts in these collections served several purposes: they could prove rare and odd natural phenomena, such as the crippled and seven-fingered hands of peasants and the misshapen lemons cast in plaster in the Bavarian Wittelsbach *Kunstkammer*.[43] They could stand in for the real objects that soon withered and died, and they could display the talent of the artist in producing fine molds and in understanding the casting properties of metals. But they also possessed a more profound significance, as already noted in chapter 3, demonstrating the human ability to imitate the transformative powers of nature. This imitation of nature constituted a form of natural knowledge, both in the techniques used to produce the objects, as we saw in Ms. Fr. 640, and in the epistemic claims made, for example, by Jamnitzer in the *Daphne*. As we saw in chapter 3, the constant investigation of natural materials necessary to creating this imitation are revealed in Ms. Fr. 640, which contains detailed instructions for the processes of preparing the sand and plaster for the molds; then for catching the animals alive, keeping them, killing them, and affixing them to the base of the mold; for constructing the mold and the investment material; for burning out the creature from the mold; then for casting the metal and removing the sculpture from the mold. These detailed descriptions of life-casting allow us to recognize that in creating such works of art, Jamnitzer displayed not only his investigations of nature, but also the products—the proofs—of his knowledge of all these processes of material transformation. His naturalistic works of art constituted statements about the powers of both nature and human art.

When viewed against the background of the material processes and objects of life-casting, Quiccheberg's third class of objects—a seemingly indiscriminate array of natural materials, from "juices of the earth," parts of animals, bezoar stones, and striated stones, to life-casts of animals and silk flowers—takes shape as a similar project to Jamnitzer's. The life-casts discussed by Quiccheberg in this section are related to the other items because their making involves the same processes of natural transformation that brought the natural materials into being, processes that involved, among other components, the earthy materials, vitriol, alum, salt, and the "juices of the earth." Like Jamnitzer's life-casts of animals and plants, these lifelike objects constituted a proof that natural processes are imitable and knowable by the human hand. Quiccheberg specified that the life-casts of animals and plants were to be created in the alchemical furnace, no doubt because fine gold and silver would be worked there, but also because alchemy was the science par excellence that imitated nature.[44] Basilius Amerbach, with whom we opened this chapter, also treasured a silver life-cast lizard inherited from his father, singling it out in his inventory and

8.19. Life-cast lizard. Silver, length 5.5 cm. Listed in the inventory of Basilius Amerbach. Inv. no. 1882.117.64, Historisches Museum, Basel. Author photo.

his last will (fig. 8.19). He, too, kept it with silver casts of flowers, a unicorn horn (about the powers of which he expressed skepticism), and various metal ores of silver, gold, and lead.[45]

Active Knowledge

The historian Dirk Jansen has pointed out that the *activity* of collecting was central for Quiccheberg, in contrast to the Italian philosopher Giulio Camillo (1480–1544), to whose text on the art of memory Quiccheberg's treatise has been compared. For Camillo, ordering—a conceptual process—was the focus,[46] whereas for Quiccheberg this new "first philosophy" was active rather than textual, and it necessitated engaging with matter itself in the workshops and laboratories of the *Kunstkammer*. Moreover, Quiccheberg, like many other new philosophers, believed that objects could teach more effectively than books, for tangible things and detailed images impressed the memory more deeply than written words.[47] Quiccheberg's interest in material things as the basis of a new philosophy was not unique, as we saw in chapter 7. In the calls for intellectual reform of the sixteenth century, scholars attempted to found a new philosophy based on natural things rather than words and on the active investigation of nature. Natural materials and natural objects—all things apprehended by the five senses—became central, as did the craftspeople who could manipulate and transform natural materials and produce artifacts.

We saw in chapter 7 the ways that thinkers such as Rabelais and Vives advocated interacting with makers. In Quiccheberg's and Jamnitzer's day, visits by scholars and nobles to artisans' workshops became ever more frequent, where they avidly took in displays of ingenuity (as we see featured in Benvenuto Cellini's texts).[48] Books that gave a view into artisanal workshops and processes, such as Garzoni's *Piazza universale di tutte le professioni* and Amman and Sachs's *Ständebuch*, also found ready audiences. Quiccheberg's plan to locate artisanal activity and process in the *Kunstkammer*'s workshops brought together process and product at a single site, much as nobles, discussed in chapter 4, accomplished by importing artisans and their workshops into their courts. Not just princes, but all manner of individuals, including physicians, scholars, and city governors, came to regard artisans as possessing the key that could unlock productive powers of nature. Philippe de Béthune, for example, the French court official and collector of Ms. Fr. 640 with its detailed instructions on life-casting, regarded artisans as an important piece of the economy for a territory.[49] In the same vein, Quiccheberg represented the benefits of harnessing the productive powers of nature as a fundamental part of the prudence necessary for statecraft, and he believed that the collection of artisanal practice—objects, tools, and workshops—could assist in gaining this wisdom, for a *Kunstkammer* displayed nature as a source of productive knowledge; it both proved the fruitfulness of nature, and could also stimulate new inventions.

A collection constructed according to Quiccheberg's treatise made clear that practical knowledge was suitable and valuable to rulers and that artisans held the key to that knowledge. Such collections raised the social status of manual work and gave intellectual stature to makers of objects. Such cabinets and chambers were sites that demonstrated and celebrated *Kunst*—the artifice of the human hand—in its imitation and competition with the artifice of nature. This celebration increasingly included the creators of these objects. The presentation of one of the art cabinets assembled by Philipp Hainhofer in the German territory of Pomerania in 1614–15 occasioned a commemorative image that imagined a triumphant procession of all the artists passing before the prince, with each called out by name (fig. 8.20). Such an explicit celebration of the collective work of these craftspeople was unprecedented, although works of art often incorporated reference to the maker, sometimes even with a self-portrait. For example, the joiner Leonhard Danner (1497–1585), who created August I's lavish wire-drawing bench, included his initials more than once on the bench, while his collaborator, an anonymous master of marquetry, included his initials ("A. M.") and a self-portrait in the workshop (fig. 8.21).

As merchants and princes sought out artisanal expertise on the materials and processes of nature and the techniques of transforming them into valuable "goods," the chambers of art also transformed artisans in the eyes of their patrons into "knowers." The many collections, such as that described by Quiccheberg, and works of art, like the chamber fountain created by Jamnitzer, helped in a direct way to form ideas about nature as productive and the potential of natural knowledge to harness that productivity. The objects in the chambers of art consolidated and disseminated the power of *Kunst*, and, in the process

8.20. Anton Mozart (1573–1625), presentation of the Pomeranian *Kunstschrank* (art cabinet) to Duke Philipp II of Pomerania, c. 1614–15. Oil on wood, 39.5 × 45.4 cm. The cabinet, made in Augsburg between 1605 and 1616, was destroyed in 1945, but its contents, including this painting, survived. The presentation by the artisans who created the cabinet, depicted in this painting, did not take place, but the recording of each by name amounted to an unprecedented crediting and celebration of the team of artisans Philipp Hainhofer had assembled. The number above each artisan is keyed to a list of their names. Inv. P. 183a, Kunstgewerbemuseum, Staatliche Museen, Berlin. © bpk Bildagentur / Art Resource NY. Photo: Saturia Linke.

8.21. Leonhard Danner, wire-drawing bench, c. 1565. End of the bench, with the self-portrait and initials (A. M.) of an anonymous master of marquetry, Danner's collaborator, in his workshop. Musée national de la Renaissance, Château d'Écouen, ECL16880. © RMN-Grand Palais / Art Resource NY. Photo: Stéphane Maréchalle.

of collecting objects for their *Kunstkammern*, scholars and their patron princes absorbed and disseminated a new epistemology—a mode of gaining knowledge about nature—that originated with artisans and practitioners, and that eventually became central to the "new philosophy" or the new active science of the late sixteenth and seventeenth centuries.

Making and Knowing

To make by imitation of natural processes, alluded to by the life-casts in Quiccheberg's *Kunstkammer*, would come to be synonymous with natural knowledge. This view is expressed in the sixteenth century most clearly by Palissy's "art of the earth," as discussed in chapter 3, a method he claimed gave insights into natural processes of generation taking place in the earth. In the following centuries, this would come to be a widely held view of the new philosophers. Gottfried Wilhelm Leibniz (1646–1716) would explain the process of fossilization on the model of casting from life:

We find something similar in the art of the goldsmith, for I gladly compare the secrets of nature with the visible works of men. They cover a spider or some other animal with suitable material, though leaving a small opening, they drive the animal's ashes out through the hole, and, finally, they pour silver in the same way. When the shell is removed, they uncover a silver animal, with its entire complement of feet, hairs, and fibers, which are wonderfully imitated.[50]

Leibniz's remarks were made in the wake of the many practical manuals written and published from 1400 on. These texts reflected and fostered a deep fascination with how things were made and how they worked. This fascination, evident also in the depictions of mining machines in Agricola's *De re metallica*, in images such as Jan Brueghel the Elder's *Allegory of Fire* (see fig. 8.3), and in the models of Quiccheberg's *Kunstkammer*, gave rise to a new way of thinking about nature. Knowing nature increasingly became synonymous with knowing how to imitate nature, either by analogy, or directly through hands-on practices such as Palissy's imitation of natural things. In the seventeenth century, many self-described "new experimental philosophers" began to revise Aristotle's definition of *knowing* as the knowledge of causes to declare instead that knowing nature was showing how nature worked to produce things and effects, a domain of expertise held by craftspeople.

Quiccheberg, too, evinces this fascination: "I have recognized how pleasurable it is to visit individual craftsmen, to observe their admirable works." His conclusions about the utility of the *Kunstkammer*, however, hark back to philological goals we observed in marginalia on books of practice in chapter 7: he spoke of the pleasure of researching "when German names can be compared and made equivalent to Latin names,"[51] and saw the collections' use for artisans as a way to "practice distinguishing the difference between individual materials that they work."[52] Despite claims that they possessed a new philosophy—

a new science—Quiccheberg and other sixteenth- and seventeenth-century scholars for the most part could never quite articulate what this productive knowledge was, nor could they entirely appropriate it to themselves. Rather, they and their princely and civic patrons continued to rely on the "unlettered" until well into the nineteenth century.[53]

Conclusion

Like the books of art (the *Kunstbücher*), the chambers of art, or *Kunstkammern*, proclaimed the powers of nature and its transformation by the human hand, and, as this chapter has shown, these chambers of art fostered new ideas about the transformative and productive potential of nature and art, and the ability of ingenious makers to capture and harness that potential. *Kunstkammern* made clear that this knowledge was the business of princes, and elites came to support, promote, appropriate, and gain control of the realm of the human hand represented in the collections.[54] The view that the study of nature and the promotion of the mechanical arts could yield productive knowledge came to be fundamental to the exercise of statecraft, an attitude that continues today in national funding of the natural sciences.

PART 4

Making and Knowing

9.1. Wenzel Jamnitzer, nautilus shell ewer, c. 1570. Gilded silver, partially enameled, 32.5 cm. This ornamental ewer plays on the divide between nature and artifice. At the base, a mass of grass snakes, cast from life, emerge as if from a brood nest. The natural nautilus shell that forms the body of the ewer has had its mother-of-pearl exposed by polishing or stripping with acid. Schatzkammer, Kat. Res-MüSch. 567, Residenz München. © Bayerische Schlösserverwaltung, Maria Scherf / Rainer Herrmann, Munich.

CHAPTER 9

Reconstructing Practical Knowledge

Hastening to Experience

For who could be taught the knowledge of experience from paper? since paper has the property to produce lazy and sleepy people, who are haughty and learn to persuade themselves and to fly without wings […] Therefore the most fundamental thing is to hasten to experience.

THEOPHRASTUS VON HOHENHEIM, called Paracelsus[1]

No objects created by the author-practitioner of Ms. Fr. 640 are known to be extant, but, as we saw in chapter 3, his manuscript focuses extensively on catching, feeding, killing, and casting animals, including snakes. Life-cast snakes can be found on many early modern objects, including Jamnitzer's *Nautilus Vessel* (fig. 9.1). This *Kunstkammer* object, combining natural and human artifice, portrays a mass of common grass snakes at its base, all much the same length and age (fig. 9.2). This object can help us make a leap into the unfamiliar world of snake catchers as they searched out specimens for founders and collectors.[2] We get a sense of how the massed snakes replicate the experience of uncovering the brood nest of a snake.[3] The codification of practical observation and experience in this object, as well as in texts such as Ms. Fr. 640, shows the ways in which craft making in its investigative and exploratory methods informed and intersected with emergent modes of scientific thinking and knowing in the sixteenth century. Over the following centuries, as part of what is now called the scientific revolution, the collaboration and experimentation of the craft workshop would go on to be integrated into the practice of the natural sciences. In the course of this integration, the shared origins of art, history, and natural science were obscured. But much can be gained by renewing a dialogue among the arts, history, and natural sciences, and by resuscitating the concept that the investigation of nature and the investigation of the human world are deeply entwined. Sixteenth-century natural historians took this for granted, and modern sociologists of science, environmental historians, and climate scientists have recently begun to emphasize it as well. Like craft processes, this new mode of material and processual research, and its new historical tools, will need to be deeply collaborative.

9.2. Jamnitzer, nautilus shell ewer, detail of base with grass snakes, cast from life. © Bayerische Schlösserverwaltung, Munich. Author photo.

Accessing the Past

So far, this book has sought to reconstruct experiential or practical knowledge from the texts and artifacts of the European past, with the aim of understanding that knowledge both for itself—how does it compare to other types of knowledge? what were its cognitive, theoretical, practical, and material components?—and within the historical context of its written form and the place of practical knowledge in early modern Europe. As has been pointed out multiple times, books cannot transmit practical knowledge to any great effect—whether for inexpert readers or for historians. These final chapters on "making and knowing" consider one more type of evidence that has come to have increasing prominence recently: physical reconstruction of historical techniques and objects, or, as some practitioners call it, reenactment, re-creation, and remaking. While this is familiar above all to archaeologists and conservators as a method for learning about techniques and materials, I believe it can, with caution, be used to gain insight not only into the material world but also into the mental world of the past. It bears stating at the outset that physical reconstruction must always be grounded on historical texts, objects, and contextual information, and it never pretends to be a fully "authentic" experience of the past.

Since Wilhelm Dilthey's (1833–1911) positing of "empathy" and R. G. Collingwood's (1889–1943) "reenactment" as part of a historical method by which the "inner experience" of individuals can be known and understood, theorists of history have debated whether "empathy" and "reenactment" are legitimate heuristics or even true analytic methods for historians.[4] Historians who have called for an empathic approach to the past aim at comprehending an unfamiliar lifeworld. For scholars of material culture, such as Robert Blair St. George, objects can help the historian to "learn to sense and comprehend the world from the point of view of those who created it and felt its living structures of power, desire, and mystery; to be able to see that specific artifacts, customs, and occasions were structurally linked to one another; and to explore the rules that set limits on but still allowed for innovative action."[5] Some historians see this focus on everyday life and material things as particularly useful in evoking in readers feelings of *Betroffenheit* (engaged concern and interest) and *Anteilnahme* (active identification with the lives of others), goals that have been particularly important to those writing about perpetrators and victims of historical violence of the fascist and Nazi era.[6] From my own perspective as a historian of science, I have found a focus on making practices useful in teaching, especially in jolting students out of their sense of what is self-evident and "natural." I have been surprised in thirty years of teaching that, despite the critical analyses of sociologists of knowledge, despite climate change and environmental crises, most students come to class with a fundamental sense that the progress of modernity has created an unbridgeable divide between modern science and historical systems of knowledge they often label "pseudoscience." A jolt is needed to enable students to make an imaginative leap back into the past and take seriously the worldviews of earlier eras in order to understand better the historical formation of the knowledge system we know today as "modern science." An education in the bodily skills and material intelligence needed for the production of sophisticated works of art by craftspeople who worked within a material imaginary that connected red, gold, blood, and lizards can administer such a jolt. I have come to see, however, that material reconstructions can do more than serve pedagogical aims for the historian, as elaborated in this and the next chapter.

In *The Art of the Maker*, Peter Dormer makes the case for learning a craft in order to understand it as "knowledge." He argues that the cognitive knowledge of craft is not recoverable by language, but instead must be demonstrated: "The thinking in the crafts of oil painting, pottery, modelling, lace making and so forth resides not in language, but in the physical processes involving the physical handling of the medium." In order, then, to understand craft, he notes that constant practice in the craft is necessary, which gradually tutors the senses, makes it possible to recognize patterns in the behavior of materials, discriminate between qualities, and develop a fine sensibility to the nuances of difference among them. These abilities are the result of the "extended dialogue" between maker and materials, but also, of course, within a community of makers, who subscribe to a "framework of values" about the craft and its products, and who judge the work of the novice.[7]

This view of the necessity of practice to craft leaves little option *other than* to reconstruct technique and to engage in bodily learning in a workshop. Indeed, anthropologists, in particular Michael Coy, have argued that participant observers of craftspeople should in fact undertake an apprenticeship. This has raised a methodological stumbling block for other thinkers. By studying craft as an apprentice, "the subject of study and the object of study are merged." Hence the claim that such reenactment is simply too subjective.[8] But how else can we gain knowledge that is incomplete in written form? As discussed in previous chapters, experiential knowledge is acquired through observation and imitation, rather than through texts, in large part because written descriptions could never sufficiently take into account the always changing conditions of the workshop, or the unpredictable qualities of the materials, and the constant adjustments to on-the-spot conditions that a craftsperson must make.[9] Written documents simply cannot provide full evidence for practical knowledge. Historian Leora Auslander makes this point powerfully with regard to material objects:

> people do different things with words and with things […] Whether it is in the making of things or in their using, people use them differently from words to create meaning, to store memories (or enable forgetting), to communicate, to experience sensual pleasure (or pain) […] People who make things have repertoires of forms (and the meanings conveyed by those forms) in their heads, hands, and eyes. They do not necessarily have words for them. Even if they have words, those words may be less adequate to the meaning or feeling of the object than the object itself […] That doesn't mean that such attachments or practices are without meaning or that their meanings are inaccessible to historians.[10]

To capture the fullness of past cultures and their systems of knowledge, we must be attentive to such meaningful objects and their making—an attention that can be heightened through reconstruction.

Auslander's point has even more force in the effort to understand the making of material things in the past, and what those practices meant to the practitioners who undertook them. No one claims that reconstruction alone is the answer, or that reconstruction can transport us fully into the material or mental world of the past, or that it can recapture aspects of the past with complete faithfulness or accuracy. No historical method can enable us to step fully into the shoes of past individuals. The foregoing chapters have suggested that we can see practice as a category of knowledge, the active dimension of which historians have not considered in its totality. Like craft practice itself, the texts of history cannot fully convey this active dimension, thus the historical recovery of this active knowledge demands additional modes of research.

Reconstructing Material Knowledge

Object-based inquiry and the techniques of reconstruction are familiar to museum scholars, curators, conservators, and archaeologists, especially those dealing with prehistoric artifacts, but some historians remain unfamiliar with the work on reconstruction of technique, reenactment, and bodily knowledge.[11] In this chapter, I provide a short overview of the types of insights reconstruction has made possible before turning to the reconstructions of the Making and Knowing Project.

At the simplest level, reconstruction and remaking are extremely efficient ways to gain a sense of how and why particular techniques were used in the past; reading an account of a technique is very inadequate in comparison. Two elementary examples from my own experience illustrate this. First, when taking a course on historical techniques of oil and tempera painting, and following the early modern practice of tracing over the lines of an experienced artist, I suddenly realized how powerful such imitation could be. Tracing the work of other artists was often the beginning of drawing instruction, but the historian simply reading such admonitions in early modern texts would not see, as I did, that what today might be viewed as a "rote" act of tracing actually turns out to be a kind of shortcut to seeing the function of, and being able to replicate, the swelling and diminishing of lines, as well as coming to understand the most effective techniques of shading and perspective. Similarly, when I took a silversmithing course, we were taught the two different sounds to listen for when raising silver on a stake—first the dull blow that closes the metal in on itself, and then the sharp sound as the metal is squeezed by the hammer.[12] Only hearing these sounds in person as one performs the action activates them in one's sensory consciousness as meaningful (a useful realization in its own right).

A second, broader advantage of the reconstruction of practical knowledge for historians of the early modern period (and perhaps of all preindustrial periods) is that we scholars are steeped in text-based sources and trained from an early age in reading, writing, and propositional knowledge, with the result that we may fail to grasp the greater part of human experience in the preindustrial world, when most learning and knowledge was experiential and acquired by observation.

Testing Recipe Texts

Archaeologist Eleanor Robson cautions against taking recipe texts at face value without fully understanding the making processes of contemporary material remains. Her comparison of techniques for making seventh-century BCE glass artifacts with contemporaneous glassmaking recipes indicates that the recipes were actually centuries out of date. They had been copied and recopied, not for their instructional value, but for their cultural significance and ancient authority. The recipe texts, long read by archaeologists for information on techniques, could easily lead the researcher astray without corresponding examination of material remains.[13]

Another example of testing recipes comes from archaeometallurgists Thilo Rehren and Marcos Martinón-Torres, who have led the way in investigating the tools and artifacts of metalworking, including crucibles, assay remains, and many other pieces of material evidence from metal production deep into the past and around the world.[14] Martinón-Torres has argued that reconstruction is essential for illuminating historical techniques, as well as for testing specific hypotheses posed in the absence of other kinds of evidence. In their experiments on purifying silver by cupellation, in which the crucible is fabricated from a material that absorbs the lead and copper impurities, they tested a multitude of early modern crucible recipes and the very specific material ingredients to which they refer, by reconstructing the crucibles and then subjecting both the historical objects and their reconstructions to scientific analysis.

The analysis of material remains is key to reconstructing technique. This is especially clear when no texts exist. Scholars in the field of archaeometallurgy have undertaken many laboratory and field experiments to understand the extraordinary diversity and apparent effectiveness of past production processes. Experiments have investigated, for example, why there is so little evidence of tin smelting in Bronze Age Britain—no remains of hearths, venting pipes, or crucibles. The results of the reconstructions showed that clay-lined smelting hearths of various types that were dug into the ground smelted the ore efficiently and left little trace. The simple, fired clay pipes and crucibles likely decomposed fairly quickly, and thus left no identifiable remains.[15] Iron smelting in Roman and medieval-style bloomeries has also been undertaken in the field to try to reproduce the technology of smelting ores from these periods. Such reconstructions rely upon the findings from archaeological excavations of iron smelting sites, as well as historical accounts: for example, in the case of iron smelting in the Bristol Channel ore fields, a detailed 1531 written account of bloomery iron smelting of Glamorgan hematite ores indicates that, at that time, a bloomery could produce two 50 kg blooms a day, using three men pumping the bellows at a time.[16]

The Goals of Reconstruction

Martinón-Torres has noted the variety of aims to which reconstruction experiments can be directed, from "experiential," in which sensory interactions with the material and entire systems of knowledge can be investigated and illuminated, to "experimental," which seek to rigorously test individual variables and are usually carried out in laboratories with modern equipment that can control temperature and other variables. An example of metallurgical reconstructions by conservation scientist David Bourgarit and his collaborators demonstrates this. They carried out both laboratory and field experiments on the complex metallurgical processes of producing brass by cementation, processes that are undocumented except by the brass vessels themselves. In this study, they highlight the different aims of laboratory tests, which can test precise temperatures needed for the diffusion of gaseous zinc into solid and/or liquid copper, and fieldwork experiments, through which they could begin to tease out and understand the extremely complex interaction of the many factors involved in carrying through the process successfully. Their fieldwork was conducted by building furnaces based on those excavated in the thirteenth- to fifteenth-century workshops at Dinant and Bouvignes in the Meuse Valley.[17] The complexity and variability of factors—chemical composition of the metals (which can be affected in some cases by weathering), temperature, oxygen content of the furnace, weather, fuel, time, and so on—means that all such work will form only one pathway along which historical insight and understanding can be accumulated. Like all historical understanding, it will rely upon a complex causal argument made on the basis of empirical investigation (whether of documents, objects, or materials-in-process) that will always be open to—perhaps call out for—reconsideration of one or more of its multiple causal variables.

Reconstructing Culture from Materials

Reconstruction of techniques also results in insights about culture. For example, reconstructing the tools of reading and writing can transform our interpretation of texts and the underlying human experience they encode. Literary scholar Peter Stallybrass and book historian Roger Chartier worked with book curators and conservators J. Franklin Mowery and Heather Wolfe to give a new interpretation to Shakespeare's phrase "erasable tablets" in *Hamlet* and in various sonnets. Most scholars have understood Shakespeare's use of this phrase to refer to printed books or, metaphorically, to ideas "written" in the mind, but through reconstructing inks and gessoes, the group found that Shakespeare meant actual erasable tablets—paper coated with a gesso ground and bound into very small codices or notebooks that could be carried in a pocket. Depending on how the note taking was done, the writing could be erased by wiping with liquid (even spittle), rubbing with bread, or sanding off. This very material referent adds another layer of complexity to Hamlet's utterances about memory and forgetting, and completely transforms Shakespeare's meanings in some of his sonnets.[18] This kind of excavation of meaning by understanding processes of making has clarified other historical reading and writing techniques, including a green-tinted reading frame for preventing eyestrain, which was reconstructed by art historian Marjolijn Bol.[19]

Historians of science have a long tradition of reconstructing crucial scientific experiments, and some such reconstructions provide insights into the character of scientific activity and its social dimensions. In a well-known study, Otto Sibum reenacted James Joule's 1840s experiments to determine the mechanical equivalent of heat, which were crucial to the development of thermodynamics.[20] By re-creating Joule's machines and processes, Sibum discovered several important points. First, he came to appreciate the high level of skill and expertise Joule had to possess in order to take the heat measurements, an embodied skill Joule had learned as the son of a Manchester brewer. Brewers valued accuracy of heat measurements (which they carried out by touch, smell, and sight) and termed this exactness "nicety of attention and performance."[21] Sibum terms this embodied, mostly tacit knowledge "gestural knowledge," or "*im Verlauf der Werk verbundenen*

Wissens" (knowledge bound up in and obtained in the course of work—perhaps "processual knowledge" is also apt).[22] Second, although Joule had never mentioned it in his accounts of the experiments, he must have had a workmate to help him in taking the measurements, as he could not have moved fast enough by himself. Sibum found that managing the machines and instruments, and taking the measurements, required quite strenuous physical labor, but in his published work Joule never mentioned the brewer's mate who must have worked alongside him.[23] Third, precision instruments were necessary to Joule's work, and Joule collaborated with the instrument maker John Benjamin Dancer to have them made, but in his published work Joule portrayed this collaboration as one in which Dancer occupied more or less the position of a servant. Finally, no other experimentalist could corroborate Joule's measurements by witnessing his work because the presence of the necessary additional person in the room raised the temperature of the room too much.[24] His experiments thus could not easily be replicated outside of the particular space in which he had worked because of the need for a cool room at a constant temperature.[25] Indeed, American physicist Henry Rowland had trouble trying to replicate Joule's experiments as late as 1875.[26] Sibum concludes that Joule left out the work and the workers in order to portray himself as a disembodied observer of nature, reading off precision measurements from a machine, making his work more "objective" than the embodied brewers' judgment. In doing so, he was acting in accord with the emerging standards of "science," as articulated by William Whewell, who had coined the term "scientist" in the 1820s and who privileged sight and hearing (not the touch and smell of the brewers) in obtaining natural knowledge.

Historian of science Peter Heering reconstructed an eighteenth-century solar microscope and found that historians could understand the importance of solar microscopes to Enlightenment writers only by seeing them in action. Such instruments had fallen out of favor in the nineteenth century when they were reported as producing fuzzy images unfit for scientific activity. Heering's reconstructions show that these instruments actually produced strikingly clear magnifications. He concludes that solar microscopes were excluded from scientific research in large part because they were associated with nonprofessional social gatherings of amateurs, which did not mesh with new ideas about the professionalized and exclusivist nature of science.[27] Without his reconstructions, however, historians would have gone on believing the nineteenth-century researchers who criticized the microscopes' out-of-focus images.

Other reconstructions by historians of science, such as Lawrence Principe's adventurous reconstruction of the "Bologna stone,"[28] have revealed the multitude of factors that can account for the difficulties in replicating past processes. In other cases, reconstruction can function as a philological tool, like that of Shakespeare's tablets, to provide a material concordance of terms that have been understood by historians as rhetorical. William Newman and Lawrence Principe have reconstructed alchemical practices such as the "tree of Diana," a tree-like silver structure described as being grown in a laboratory vessel, as well as an alloy of metallic antimony and copper called Diana's Net—both phrases in alchemical texts that historians have previously taken to be allegorical or metaphorical, but which these historians have demonstrated to have material correlates that can be reproduced in the laboratory.[29] Similarly, in her work reconstructing the meaning of intentionally obscured substances, such as the famed "Green Lion" of alchemical texts, Jenny Rampling has made the important point that the reading of such texts, and the substances themselves, changed over time, and their different compositions can be illuminated by reconstruction.[30]

Reconstructing Materials of Art

Reconstruction has also served an essential role in the conservation of art, sometimes now called "technical art history." Among many insights emerging from this field, one has been the recognition of how much reconstructions of technique reveal about the material itself as part of the cognitive process. Materials can have properties that constrain the maker in particular ways, producing results that lead to new techniques and new knowledge. In general, only repeated experience with a material itself can provide these kinds of insights, as inexpert reconstructors may not recognize whether the constraints come from their own inexperience or from the materials. For example, a conservator at the National Gallery of London, Jill Dunkerton, reconstructed fifteenth-century painting techniques for a BBC program, *Making Masterpieces*, and she found that "stand oil" (oil that has been prepolymerized by heating gently, or standing in the sun, and was used for making oil paints dry faster) also suppresses brushstrokes. It thus contributed to the "glaze effect" of Northern European glazes in oil painting, which became such an important part of early Netherlandish painting. She also discovered that the use of diluents, usually assumed to be added to difficult-to-handle pigments such as azurite and lapis lazuli, actually makes paint thin rather than giving it better handling properties. From her many reconstructions and long experience, Dunkerton thus concluded that the accepted understanding of these diluents based on modern practices was faulty.[31] Scholars in this field have noted that the skill of the experimenter in the historical techniques can be the most important variable of all, and one that can only be "controlled" by much experience with the materials and the processes.

In the same vein, sculptor and scholar Andrew Lacey's reconstruction of the mold making and casting of a Renaissance bronze depended on his own expertise as a sculptor and founder, as well as very close study of period objects and sources.[32] Firsthand experience with the material, like that possessed by Dunkerton or Lacey, is essential, and it often leads to an upending of long-held assumptions about those materials, as well as questioning of art historical categories and interpretative frameworks that have become so ingrained as to become second nature.[33] As will be treated at more length in the next chapter, this questioning of established frames of reference—the nature of a material, for example, or the conception of the boundaries and aims of art—is one of the most important outcomes of reconstruction, even when this is not the explicit goal. The material world of the past and its apperception and understanding by the historical actors are very effectively masked by our own taken-for-granted experiences of moving through and perceiving the material world (trained by schooling, discipline, and scholarly analysis), and thus questions that challenge our own perceptions and training often can be the most difficult to formulate.

Historical tools and materials are masked by our unexamined conceptualizations of what materials are and how they are used, and these can only be unmasked by essaying them according to historical processes. For example, in a replication of two recipes for mosaic gold (an imitation gold pigment), conservators re-created a gold pigment made by combining mercury and tin. They showed that the process for making the pigment actually brought previously unknown materials into being; materials that possessed unfamiliar crystal structures and interesting decay phenomena. These materials are produced by phase transitions, which can cause great variation in material structures.[34] This allowed the conservators to understand what constituted the "authentic" pigment on the illuminated manuscripts and what were actually symptoms of decay, and thus to achieve a better understanding of the artist's original intention.[35]

Imitating gold pigments has also led to insight into the broader understanding of nature on the part of the artisan. Spike Bucklow has demonstrated that replicating recipes for pigments can indicate which ingredients are "active"—that is, which ingredients influence the outcome of the product—and can help distinguish those from ingredients that are added because they fit conceptually within the artisan's view of the world, but that play no active part in the pigment.[36] As noted in chapter 3, red components, such as the pigment vermilion (composed of sulfur and mercury), were often added to processes involving gold, even when they seem not to have had any practical effect on the chemical process.[37] Sulfur and mercury were regarded by scholars and practitioners as the basic elements of all metals. Similarly, some recipes for vermilion add sal ammoniac (ammonium chloride). Although Bucklow's reconstruction shows that no practical advantage comes from addition of this salt, he believes that by adding a salt, the practitioners viewed the final material as more perfect because it comprised three principles of mercury, sulfur, and salt, which could represent the three principles of spirit, soul, and body, or the three elements of water, fire, and earth.[38] In another recipe, sulfur, mercury, and salt are all specified in a process for the transformation of gold leaf into powder, although only salt is actually needed to effect this transformation because, in the actual procedure, sulfur and mercury vaporize upon heating. Again, this recipe seems to require these additional ingredients because sulfur, mercury, and salt were regarded as components of all metals and thus were seen as crucial in any metallic transformation.[39] None of this could be known through study of these texts alone. Instead, by following the recipes contained in such writings, the function and meaning of these processes can become clearer.[40]

Close study of objects and experimental reconstruction of techniques employed to produce the earliest European small bronzes in the fifteenth century allowed Metropolitan Museum conservator Richard Stone to demonstrate that artists such as Pier Jacopo Alari Bonacolsi, called Antico (c. 1460–1528) and Severo Calzetta da Ravenna (active c. 1496–c. 1543) used innovative improvisations that allowed them to produce multiple copies of their highly sought-after sculptures.[41] Because no writings allude to these practices, this information can be gained only through the investigation of multiple objects and proof-of-technique through reconstruction. This finding questions historians' assumptions about the significance of "originality" and "mechanical reproduction" in the Renaissance.

Reconstructing the Viewer's Experience

Even when practical writings exist, they do not tell the whole story. Remaking early modern clothing has revealed that many steps are left out of tailors' manuals. The texts must be compared with extant historical objects as the first step in reconstruction. This working between text, object, and active remaking provides missing information about the construction of dress—for example, elements that were counterintuitive in making, but when fitted, turn out to increase the range of movement possible in a shaped bodice stiffened with thin reeds.[42] Ulinka Rublack's research and collaborative remaking of the costumes of sixteenth-century Augsburg burgher Matthäus Schwarz (1497–c. 1574), underscores this point. She notes that the rarity and fragility of early modern dress, including its color change with time and the necessity of viewing it in low lighting, has prevented historians from appreciating the "complexity and sensuous vibrancy of early modern dress." In an expert physical reconstruction of Schwarz's outfit for the Imperial Diet of Augsburg in 1530, undertaken by the director of the London School of Historical Dress, Jenny Tiramani, on the basis of a textual and pictorial description, Rublack shows how the vibrant colors of clothing signaled alliances and contributed to the political unfolding of the Diet. She concludes that the make of clothing (in some cases only known with certainty through remaking) "can help us decode the meanings of dress as ingenious visual acts and the aesthetic as well as emotional arguments in a particular milieu."[43]

Rublack notes the degraded state of most early modern pieces of clothing; this is another important reason for reconstruction, as it gives insight into the original appearance of works of art. In many works, colors have faded or otherwise degraded so much that we no longer know what an artwork originally looked like. This is especially true for sculpture, which has also been the victim of modernist aesthetics that has taken the essence of sculpture to be form, rather than color. Roberta Panzanelli notes that the "history of art has tended to dismiss polychrome sculpture as quirky and not quite true to the essence of sculpture."[44] But color, which has been used on statuary since the most ancient times, always formed an essential part of the viewers' experience of the work of art. Without having a realistic idea of the appearance of an object, we cannot hope to understand the artistic aim of the maker or the sensory and affective experience of a work of art in the past. Indeed, reconstructions of polychromed wood statuary and oil paintings have revealed that one important component of the artistic aim in the Middle Ages was to achieve a variety of surface textures that reflected light in different ways—for example, the sparkling glazes playing off each other and contrasting with the areas of burnished and matte leaf gilding.[45] None of this can be fully appreciated solely from texts or even from the surviving objects, given their various states of disrepair or the results of nineteenth-century cleaning campaigns.[46] In addition, reconstructions of polychromy on statuary, as was undertaken in an exhibition directed by Panzanelli at the Getty Museum, can tell us much about period conceptions of the body and of gender, and about notions of vitality and life.[47] Without a realistic idea of the appearance of an object, we can understand neither the intentions of the maker nor the viewer's experience.

Early music, most of which relied only partly on written forms, poses the same problem of reception and experience. Music of the early modern period was improvised within certain formats, much like an Indian raga today. When musicologists now attempt to reconstruct early music, they must reconstruct the instruments as well as appreciate the tacit knowledge of improvisation. The reconstruction of historical music provides a useful point of comparison for other kinds of practical knowledge because no scholar would recommend the reconstruction of early music from texts alone; it must be played and performed, always with the underpinning of period descriptions of reception, of course, in order to begin to understand the sensory experience of music in an earlier time. Naturally, this raises the question of whether a twenty-first-century listener could ever authentically approximate the aural context of a fifteenth-century hearer, but, surely, all would agree that for such ephemeral, performative acts of creation scholars need all the sources of insight they can muster.

Reconstructing Collaborative Practice

Finally, reconstruction can show us how collaborative practice operated. Garden historian Mark Laird has been instrumental in the reconstruction of Painshill Park, an eighteenth-century garden southwest of London in Surrey, which included some of the first plantings of North American species (fig. 9.3). His reconstructions brought into focus two important points: first, the aesthetic aims of the garden to which we have no access now (without verbal or visual descriptions of these early gardens); and second, the recognition that the gardens were really a result of a collaborative process among designer, gardeners, plants, *and* the environment. If one worked only from written plans and designs for the garden, this latter fact would never emerge.[48] This insight is much more significant than it first appears, because, like other sources from the crafts, it gives evidence that distributed cognition and collaborative working methods were the norm, in contrast to current models

9.3. Reconstruction of "shrub-
bery" plantings, directed by Mark
Laird. Painshill Park, Surrey, UK.
Author photo.

of cultural production. As a result, the historian's hunt must move from a search for the "theorist" or "designer" who created such a garden to focus instead on the dynamics of collaborative practice by which the garden came into being.[49] By laying bare the dynamics of collective knowledge-making, reconstruction can thus help us text-centered moderns to "think outside an author" as the agent of cultural production.

Hastening to Experience—the Making and Knowing Project

As I was writing this book, I was also studying the manuscript at the core of research in the Making and Knowing Project, Bibliothèque nationale de France, Ms. Fr. 640. This late sixteenth-century manuscript has appeared repeatedly in this book because it is full of recipes for and observations on art objects, in some cases containing the most detailed accounts of techniques that we have in writing from any period. It is a unique record of the techniques, materials, and practice of art, broadly conceived, from early modern Europe. What caught my eye first in this manuscript were the sketches for casting from life (see figs. 3.5–3.12).[50] While the sketches were fascinating, they were also incomprehensible to me at that time, and trying to follow the text accompanying these sketches led me into a chaos of repetitious entries in the text, and a sense of considerable intimidation.

At that point, I realized I needed help, and I began an extraordinarily fruitful and stimulating collaboration with practitioners and museum scholars (principally Tonny Beentjes, now head of the Metal Conservation Program at the University of Amsterdam) in which we attempted to reconstruct the life-casting techniques detailed in Ms. Fr. 640. Tonny and I began by examining all the life-cast objects we could locate in museums, then we turned to reverse-engineering the work of the author practitioner of Ms. Fr. 640. Where he engaged in a laborious translation of making and doing into words and writing, we sought to turn his words into processes and products (fig. 9.4). This involved a back-and-forth movement: we combed through the manuscript for all relevant parts of the process, which in itself was not simple, then we tried the techniques, then returned to the text. It was an intense—sometimes tense—exchange between a text-centered scholar and a skilled practitioner. In the process, we found that the minute traces of facture on the museum objects matched the techniques detailed in Ms. Fr. 640, such as a thickening agent like butter on delicate wings, as shown in figure 3.13, details of posing and attaching the animal or plant on the clay base, and methods of molding that were not intuitive to a modern caster, such as pouring from the tail of the lizard, the narrowest part of the mold.[51]

In reconstructing these processes, we pored over the manuscript, coming to see that "reading" was not a simple matter of drawing out passages that dealt with mold materials or metal alloys. Rather, every recipe for making a mold was embedded in a mix of other techniques, sometimes specific to the animal or object being molded. One had to return

9.4. Reading and reconstructing. Workspace for making life-casting molds. Tonny Beentjes and Pamela Smith, Ateliergebouw, Amsterdam. Author photo.

9.5. Ms. Fr. 640, fol. 125r. Like many pages in this manuscript, this one shows a complex process of iterative composition. For a transcription and translation, see *Making and Knowing Project et al. 2020*, https://edition640 .makingandknowing.org/#/folios /125r/f/125r/tc. Bibliothèque nationale de France, Paris. Source: gallica.bnf.fr.

again and again to different sections of the manuscript. It was not a set of instructions, and we realized that reading the recipes could not be separated from trying the methods. The manuscript's author-practitioner set down his text at various intervals in a fair copy, followed by more experimentation, which gave rise to failures and new ideas, followed by further observations and trials, all recorded in increasingly cramped script in the margins, as on fol. 125r in his trials on molding and casting (fig. 9.5). From the evidence of the manuscript, it would appear that the composition of the text could not be divorced from the lived experience of actually performing the actions. The text enacts the trying again and again of the author-practitioner, and his essential need to proceed in gaining an understanding of materials and processes by what we would call the method of "experimentation."

9.6. Jamnitzer, table ornament,
life-cast snakes and leaves around
the lip of the basin of the sculpture
painted with oil-based paint.
Photo: Joosje van Bennekom.

Moving from text to experimentation and back again has given me a hint of what it must have meant for those not brought up and steeped in reading and writing from an early age to cross the divide from lived experience to the written word. It resonates with Paracelsus's words that serve as an epigraph to this chapter, as well as with Bernard Palissy's refusal to put his working methods down on paper:

Even if I used a thousand reams of paper to write down all the accidents that have happened to me in learning this art, you must be assured that, however good a brain you may have, you will still make a thousand mistakes, which cannot be learned from writings, and even if you had them in writing, you would not believe them until practice has given you a thousand afflictions.[52]

One important lesson of reconstruction for us was that the process of reading and doing when reconstructing involves constant back-and-forth movement between text and handwork. We could not understand this book of art without testing its processes. Reconstruction forms a close and engaged mode of reading texts of practice.

We learned much about the techniques and tricks of life-casting: we realized that "metal casting" is a misnomer—that it would better be termed "mold making," because identifying and producing the correct materials and techniques of making the matrix of the mold is the essential key to casting metal objects. Indeed, all early modern works of metal casting attest to the constant experimentation on mold and investment materials.[53]

Reconstructing recipes in the manuscript also clarified techniques, such as the painting of metalwork (fig. 9.6), to choose one among many. Ms. Fr. 640 admonishes the metalworker to keep the live specimen (whether rose, crayfish, or green leaves) on hand: "As in this & all other things, have always the natural one in front of you to imitate it."[54]

While such insight into technical details is invaluable, the view into the experiential and intellectual world of artisans revealed by the intensive textual and laboratory study was revelatory. One of the most important outcomes of the process of reconstruction was a new attention fostered by the active mode of reading that allowed us to see things we would never otherwise have noticed in the sources or on the objects.

The Digital Manuscript

The more deeply I read and researched Ms. Fr. 640, the more I came to see how unusual its detailed descriptions of artisanal processes were, and I realized that it merited a critical edition to make it more accessible. The magnitude of the project seemed overwhelming, because producing a definitive transcription and translation of the entire manuscript would require reconstructing a huge number of diverse processes simply in order to understand the text. How would that be possible? I assumed the edition would be made available as a published print book, but, as I considered the publication of such a tome, I came to see that words alone would be inadequate to convey the manuscript's contents; we needed an abundance of photographs and videos. In my almost complete ignorance in 2014 about the creation of a digital book, but thinking there was no better way to learn than by immersion, I decided to produce a *digital* critical edition. With input from the Digital Humanities Center at Columbia, especially Terry Catapano (now Making and Knowing digital lead) and Alex Gil, and many serendipitous meetings, such as that with Marc Smith, who became the Making and Knowing paleography lead, I set about establishing the Making and Knowing Project (www.makingandknowing.org/). Drawing from my experiences working with Tonny Beentjes and his MA students, and teaching a V&A/RCA MA course in history of design, "Matter, Practice, Form, and Meaning in Early Modern European Sculpture" (2012), as well as co-organizing various conferences that combined study of material culture and history of science,[55] I planned a "laboratory seminar" on "Craft and Science," in which students would undertake hands-on reconstructions of "recipes" from Ms. Fr. 640. After much negotiation, I was immensely fortunate, just weeks before the start of the first class in fall 2014, to be given a 1950s-era chemistry laboratory, and to hire the first of three cohorts of immensely creative and intelligent postdoctoral scholars. These individuals, working with the project from 2014 to 2020, were central in developing the project and creating the edition. The Making and Knowing Project resonated with new directions in the history of science and technical art history, and we were very gratified over the course of six years to receive multiple grants to support the project. It turns out that the creation of a digital edition is very expensive![56]

Hands-On History

Through a process of research-driven pedagogy, with the intriguing Ms. Fr. 640 at its center, the project worked collaboratively to create the edition.[57] The project began in June 2014 with the first of several French text workshops, and by February 2020, it released the edition, *Secrets of Craft and Nature in Renaissance France: A Digital Critical Edition and English Translation of BnF Ms. Fr. 640*, https://edition640.makingandknowing.org.[58] The creation of the edition owes much to the nature of Ms. Fr. 640 as an anonymous text full of technical detail, which aroused the interest of all those who strayed into its orbit, as well as to the many wonderfully creative, thoughtful, and supremely dedicated participants in courses, meetings, and brainstorming sessions. The Making and Knowing Project fused pedagogy with a strong focus on this fascinating research object, which partially adapted the the lab-based scientific research group to the practice of history. This proved a powerful research model. The project created the edition through a series of "expert crowd-sourcing" workshops and regularly scheduled university courses that involved students, practitioners (including sculptors, painters, taxidermists, and technical art historians), scholars of the humanities and social sciences (history, art history, anthropology, and museum scholars), natural scientists (chemists and conservation scientists), and specialists from the digital humanities and computer science (data scientists and librarians). The research process employed novel methodologies for history, such as large-scale collaboration in cross-disciplinary research groups, historical reconstructions of past techniques, and prototyping of the edition in digital courses. The project's creation of *Secrets of Craft and Nature* consisted of four interrelated and iterative components: (1) transcription, translation, and encoding of the manuscript; (2) critical commentary, including in-depth, multifaceted research of the manuscript's "recipes," notably by hands-on laboratory reconstructions; (3) working group meetings for critical review and oversight; and (4) digital development of the final online environment of the edition.

The first stage, transcription and translation of the manuscript, was carried out in a series of three-week summer paleography workshops that brought together both experts and graduate students. Each summer in 2014, 2015, and 2016, approximately fifteen graduate students from European and North American universities gained skills in Middle French script and textual analysis by transcribing, translating, and encoding the manuscript. These workshops resulted in three versions of the text: diplomatic and normalized transcriptions, and an English translation. All three versions were comprehensively marked up in TEI (Text Encoding Initiative) encoding, which developed from year to year, as we engaged with the students in user stories and much brainstorming about what to tag. The centrality of metadata to interpretative analysis—an obvious point for digital humanists—came as an epiphany to the project.[59] The text workshops continued through 2018, focusing on translation and markup, then the Making and Knowing team (myself, three postdocs, and a project manager) continued to intensively refine and revise translation and encoding right up to the first release in 2020, and, indeed, beyond.[60] We all learned enormously from each other in this crucible of intense collaboration, and I feel supremely privileged and fortunate to have worked alongside so many wonderfully talented individuals.

The collaborative editing of the manuscript text took place via Google Drive (which in 2014 was unfamiliar to all participants). This enabled collective work on the manuscript text, as paleographers worked simultaneously on the same part of the text and saw edits in real time. Google Drive also crucially permitted all participants (including working group members and visiting experts) to write and view comments on any part of the shared documents. These comments proliferated from year to year, facilitating the collective transcription, translation, and encoding work that took place over five years, and informed the critical apparatus as participants left questions, citations, external research, and, most important, notes about their decisions during all part of the research and editing process.

The paleographers' transcription and translation formed the basis for the laboratory seminar, "Craft and Science: Making Objects in the Early Modern World," offered each fall and spring semester from 2014 to 2018. Students carried out research by studying historical texts, museum objects, and hands-on laboratory reconstruction of the recipes. Laboratory research focused on understanding materials and processes by means of experimental reconstructions of selected recipes from the manuscript, in which the students comprehensively investigated historical materials, ingredients, processes, tools, and their associated terminology, availability, origin, and scientific significance. The students kept field and laboratory notes in which they recorded their reconstruction experiments as well as their experience as humanities and social science students doing hands-on work in the lab; this gave them the opportunity to reflect on methodology and the use of reconstruction as a historical source. Reconstructing the manuscript's technical recipes has played a crucial role in deciphering a complex and significant text, and in understanding the changing practices of creating, codifying, and transmitting knowledge about nature in early modern Europe. With oversight from course instructors and visiting expert practitioners, the students brought together this research in multimedia essays that now form the historical and material commentary for *Secrets of Craft and Nature*.

The focused research in paleography and laboratory activity culminated each year in the third component of the project, annual working group meetings. Each meeting brought together about twenty expert scholars and practitioners with approximately twenty students from the year's two offerings of the lab seminar to discuss and critique the student-authored annotations. The meetings provided the necessary expert oversight on the critical edition and also brought rich new insights from the scholars' varied disciplines to inform the project's research. In the same way, the year's laboratory research cycle informed the ongoing transcription and translation activities of the paleography workshops from year to year. The interpretation of the manuscript evolved continually in light of the material reconstructions of the lab seminar and knowledge exchange of the working group meetings. This iterative approach formed a core methodology of the Making and Knowing Project.

The final component of the project was the transformation of the four versions (facsimile, diplomatic and normalized transcriptions, and translation) of the manuscript and the voluminous multimedia research and critical commentary into a public-facing digital environment. Like the other three components of the project, digital development followed the principles of collaborative research, interdisciplinary knowledge exchange, iterative design, and the integration of research and pedagogy. For this phase, the project worked with new collaborators to offer courses in the digital humanities. In spring 2017 and 2019, these courses introduced students to the concepts and tools relevant to the creation of a digital edition, and they helped to prototype the edition as they trained the students and the project in digital skills. The first course experimented with selected passages to produce a minimal edition, using Ed software, developed at Columbia's Digital Humanities Center (https://cu-mkp.github.io/GR8975-edition/).

These four interrelated research and pedagogical components highlight the project's collective and iterative design: through each cycle from paleography workshop through lab seminar to working group meeting to digital seminar and prototyping, new insights were gained, information was accumulated, and questions were generated for the next phase in the cycle. The strength of the project's collaborative research also derived from the fact that the participants not only came together from different disciplinary backgrounds but also possessed varying degrees of expertise. Teaching and researching through collective workshops, in which more experienced participants who are overseen by discipline experts work closely with novices, fruitfully facilitated both training of the novices and the consolidation of knowledge by the more experienced participants.

The Historian's Laboratory

Our most sustained exploration of both the material practices and the conceptual world of Ms. Fr. 640 came in the seminars undertaken in the Making and Knowing Project's laboratory (fig. 9.7). These laboratory seminars began with skill-building exercises, including reconstructing historical culinary recipes (generally not from the manuscript, as it contains very few food recipes), which the class then discussed and (literally) digested.

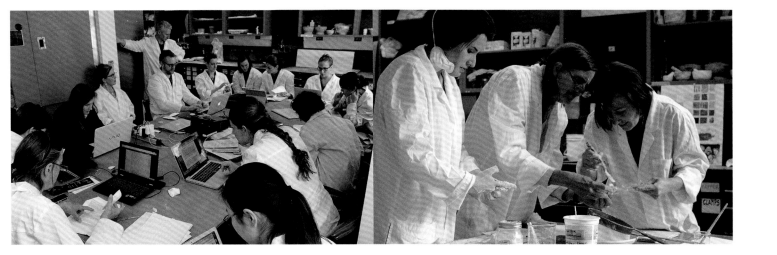

This exercise aimed to spur the students to consider the methodological challenges of reconstruction, as well as to formulate a template for a rigorous reconstruction procedure, which included principles and steps for reconstruction. From culinary recipes, we moved to related entries in Ms. Fr. 640, such as using bread as a mold to quickly try out a sculpted pattern, into which various mixtures of wax, tallow, and sulfur were cast. In this exercise, students had to learn how to make sixteenth-century bread. Through the process of making molds from their freshly baked breads, the students became familiar with the properties not only of the live cultures of the sourdough starter used for making the bread, but also of various materials—bread, beeswax, tallow, and sulfur—in their different states. They were introduced to the molding and casting process, which was continued with cuttlefish bone molding, and finally box molding with "sands" and binders, a progressively more complex set of materials and techniques. From there, we moved on to verdigris growing, red lake pigment making, and in one year, we reconstructed sixteenth- and seventeenth-century painters' techniques.

Every semester, a visiting expert practitioner helped lead the students for two weeks in skill building relevant to the year's focus on a subgroup of recipes.[61] One year, for example, Dr. Marjolijn Bol led the class in creating a test panel for their pigment, varnish, and gilding work. This involved preparing and applying nine layers of gesso, allowing each layer to dry, then giving it many hours of sanding. We then experimented with pigments in a variety of binders, and learned to gild on the gesso (fig. 9.8). Throughout this skill building, Marjolijn encouraged the students to think about the optical properties of pigments and the ways in which they changed according to substrate and binder. In the first four weeks, students began to consider topics and recipes they might select for their annotations.

9.7. The Making and Knowing Laboratory, Columbia University, New York. © Making and Knowing Project. CC BY-NC-SA.

9.8. Marjolijn Bol led the laboratory seminar students in creating a panel with a historically authentic gesso ground. The panels then served the students in testing and practicing painting and gilding techniques. © Making and Knowing Project. CC BY-NC-SA.

By the seventh week, the students had chosen their recipes for annotation, and were ready to begin work.

To reconstruct the recipes, the students had to engage in quite extensive text-, document-, and object-based historical research. They needed to understand terminology and process, gain a sense of the object that was the goal of the recipe, and try to construct a "genealogy" that would clarify whether the author-practitioner was drawing upon a written source or at least a practice common to the period. An important aspect of the initial stages of the students' work was to prepare a protocol for safe practice.[62] For the remaining eight weeks of the semester, students worked on their annotations in groups of two or three. Aiming for much more than simple accounts of hands-on reconstructions, the students tried to identify a historical question that their hands-on reconstructions could help answer. The annotation—really a concentrated research essay—formed an integrated text that wove together conventional historical research, based on documents, texts, and objects, and hands-on historical experiments, to make an argument about the students' research into the topic and what their reconstructions revealed.

Insights from the Making and Knowing Laboratory

Our reconstructions—which will be treated at more length in the next chapter—have given many kinds of insights into the manuscript. Two essential but elementary ones: (1) we are now certain that the manuscript does indeed accurately describe the techniques and materials of the sixteenth century, and that the manuscript itself is an account of firsthand practice; (2) we have clarified a whole series of techniques, and demonstrated that various (and sometimes implausible) materials could be employed to produce the intended products of the recipes. The insights that I as a text-centered scholar gained into the nature of experience and experiential knowledge, although perhaps familiar to craftspeople and artists even today, were invaluable and would not have been possible simply by reading the manuscript. Indeed, I came to realize that understanding a practical text cannot be complete without appreciating experientially the practice it contains.[63]

The Making and Knowing reconstructions have, of course, yielded much information about the manuscript, including deciphering marginalia; gaining clues to the author-practitioner's identity, as well as his level of education, knowledge and practice; revealing the author-practitioner's familiarity with techniques outside of his French-speaking domain; and finding evidence of his learning new techniques or experiencing failures in his practice.[64] They have also provided insight into obscure techniques and materials used in the sixteenth century, such as the material referred to as "spalt," "laspalt," and "spat" by the author, as well as about the nature of experiential or practical knowledge, as revealed, for example, in his use of specific terminology, such as "impalpable." The reconstructions have also given insight into daily life, in the form of medical remedies; a recipe that mentions the use of noisy mortars by apothecaries to attract business; a method to train a dog (involving cheese held in the armpit); millet consumption in Languedoc; and the surprising *chaine opératoire* of sand for hourglasses.

Basic elements of everyday workshop practices taught us much. The manuscript, for example, contains very few precise quantitative measurements. It most often uses ratios (for example, one part lead to two parts tin), as seems to be common in early modern technical writings. Thinking in ratios also seems to be common in other kinds of literature, such as merchant manuals, and for authors such as Michael of Rhodes,[65] which seems to indicate a common conceptual field worthy of further investigation. In our reconstructions, we soon learned to employ other kinds of indicators that appear in the manuscript in place of quantities—namely, descriptions of the state of the material in its final form, such as, "Make it clear like a pureed broth, or like starch water that women use to make their starch" (fol. 113v), or comparisons of the desired consistency (fols. 89v, 113v, 121v). This practice involves a kind of material imitation in which materials are modified in order to take on the appearance and the properties of another substance, against which they are then assessed.

9.9. Measuring heat intensity with a "paper test" to determine if the molten metal is too hot to be cast into a cuttlefish bone mold, as noted in Ms. Fr. 640, fol. 145r: "And to know when it is at a good heat, dip a little piece of twisted paper into it. If it blackens it without lighting, it is at a good heat. But if it burns it & catches fire there, it is too hot." The paper employed here is handmade to reproduce the properties of early modern paper, composed of 50/50 hemp and cotton, heavy weight, gelatin sized, and of third quality. It was supplied by Timothy Barrett, University of Iowa Center for the Book. © Making and Knowing Project. CC BY-NC-SA.

Similarly, important parts of the process, such as temperature, could be measured by the combustion point of other materials (fig. 9.9): "to know when it is at a good heat, dip a little piece of twisted paper into it. If it blackens it without lighting, it is at a good heat. But if it burns it & catches fire there, it is too hot."[66] We came to see that, in an early modern workshop where materials were not standard, heat was provided by wood that burned with different intensities, and environmental conditions might change with the weather, quantitative measures would actually have been *less* useful than qualitative descriptions of the aimed-for consistency, or the appearance of the material at certain points of the process. In casting, for example, metals had to be heated to a high enough temperature to penetrate the entire mold, but not high enough to ruin the metal, thus visual signs were important, especially the color of the metal, such as "when gold reaches its perfect heat, it is green like an emerald," or, "when [lead] is very hot, it becomes blue, let it then pass this color & rest a little before casting."[67] Once again, we saw how necessary was the combination of repeated experience with the behavior and properties of materials, close observation, and repeated trials of materials. Certainly, Ms. Fr. 640's author-practitioner was always thinking about the properties of materials and how they manifested themselves.

The more research we did on the manuscript, the more we realized that the author-practitioner of Ms. Fr. 640 was interested in imitation in multiple guises. His recipes for life-casting are by far the most detailed written accounts of this technique, and include, for example, casting the multiple petals of a rose (fig. 9.10), and clever "secrets" for a method by which an exact imitation in concave can be produced of a portrait relief medal in convex—called "incuse reverse casting," or as the author terms it, "Molding hollow on one side, and in relief on the other [*Mouler cave d'un costé et de relief de laultre*]." Columbia students were able to achieve proof of concept (fig. 9.11), and then collaborators in Amsterdam were able to make a successful medal. He was interested in sleight of hand, such as to make red

9.10. Life-cast rose. Tin-lead alloy, c. 20 cm. Created by Giulia Chiostrini and Jef Palframan, 2015, following processes described in BnF Ms. Fr. 640, fols. 155r and 155v. © Making and Knowing Project. CC BY-NC-SA.

9.11. Collaboration between the Making and Knowing Project and the University of Amsterdam Metalwork Conservation Program on Ms. Fr. 640, fol. 92r, "Molding hollow on one side and in relief on the other." Left to right: Making and Knowing team members Rozemarijn Landsman and Jonah Rowen, "proof of concept," cast tin medal: concave side, convex side. University of Amsterdam team members Michaela Groeneveld and Marianne Nuij, cast tin medal: pattern, convex side; pattern, concave side. © Making and Knowing Project. CC BY-NC-SA.

wine made with brazilwood turn to "white wine." The spectrum from deceptive trickery to secrets and tricks of the trade to the honorable practice of a profession has always formed part of craft self-representation, and the power of art to deceive has resulted in sometimes vehement critiques of craftiness and artifice. But imitation is more than representation (or deception), as we shall see in the next chapter, and it pervades the manuscript's recipes in much more material and significant ways.

The Craft of Learning

The combination of text-based, object-based, and laboratory reconstruction research has allowed us to decipher obscure materials and techniques, and to learn more about the repeated experimentation of workshop practice. Interestingly, we were able to see the author-practitioner of Ms. Fr. 640 himself learning in the course of trying the recipes. He appears to have learned and recorded his experiences in some of the processes, while other recipes he seems not to have tried. Our experiments have taught us about the author-practitioner's bodily intimacy with some materials he used, and how this sensory engagement shaped his classification of materials into taxonomies. We have been able to begin to delimit some parts of his taxonomies and imaginaries of materials—as treated in the next chapter—to see the ways in which he reasoned and experimented across types of materials in order to hypothesize about useful materials and probable results of his trials. We learned that he manipulated his materials to take on the often contrary properties of other materials in order to create useful substances. Over the years of the project, we thought much about why the author-practitioner might compile this record of practice, and we were always fretting as well about the potentials and pitfalls of using our reconstructions as historical evidence.

Ideally, this material and collaborative work of reconstruction develops skills of observing, both of one's fellow workers at the bench, and of changes in the material. Students realized they needed to learn the skill of self-consciously attending. This kind of training not only inculcates some basic hand skills, but also can enhance cognitive abilities of attention. Moreover, the work in the lab is long and hard, and takes discipline and the cultivation of habit. In all, we have come to appreciate the constant experimentation of the workshop, and we have come to see that failure, repetition, and "extension" are regular and very important components of learning through experience in the workshop, a useful lesson in today's high-stakes testing educational regime.[68]

In this process, the students also come to see the power of attending, imitating, and modeling as tools of learning, practices that have been demonstrated as effective by the collectivity of artisans for centuries.[69] Ideas of inborn individual talent and "genius," and the counterproductive personal insecurities that arise from such ideas, can be tempered by such training. Also useful in that regard is the experience of collaborative research for humanities students. I, too, had to learn the remarkable efficacy of collaborative work, something that, as a historian of science, I should have realized earlier from being familiar with the model of modern natural scientific research. We are so much more intelligent when working together on a problem. Finally, I sense (although cannot yet demonstrate long term results) that the hands-on exploration makes the students bolder scholars, more willing to take intellectual risks, to make themselves vulnerable, and to challenge themselves and their materials, whether in the lab or on paper.

Even in the short time the project has been active, we have come to see how odd it is that historians whose objects of study are historical materials and techniques (which include historians of material culture, of art, and of science), have generally not considered engagement with the materials of their historical topics as an essential part of their training and research. In this way, we may have missed an essential part of our intellectual toolbox, namely, a literacy in materials and techniques. The only efficient—perhaps the sole—way of beginning to acquire this literacy is through hands-on work with materials and techniques. The reconstruction of historical techniques and objects provided to students by the Making and Knowing Project cannot bring any of the participants to true proficiency in this literacy, but it can *begin* to provide such a training. It can make the students aware of what they *do not* know, even what they do not yet know how to ask. While students in a semester-long course will never reach the stage of true literacy or skilled and expert practice, they will begin to appreciate the rigor and time needed to attain it, and they will ask new questions of texts and objects.

CHAPTER 10

A Lexicon for Mind-Body Knowing

The previous chapter surveyed various historical insights gained by reconstruction; this final chapter will build upon those insights to think about how we might best characterize the kind of embodied natural knowledge produced in an artisanal workshop in early modern Europe. This chapter strives to characterize more fully the relationship between "making" and "knowing," and considers possible terminology and concepts for the processes, principles, and taxonomies that emerged out of an artisan's engagement with the world of materials through skilled practice, experience, and the work of mind-body. Or, put another way, this chapter seeks to further flesh out the terms we have used in previous chapters: "skill," "*Kunst*," and "practical knowledge." Despite all the developments in twentieth- and twenty-first-century philosophy, history, and social sciences,[1] there is no consistent methodology and terminology to analyze or historicize experiential knowledge, in part due to the inherent difficulties of putting such knowledge into words, and, in part due to the pervasive and enduring hierarchy between mind and hand. The desire of artists to efface their own labor in a work of art, as well as centuries of art historical training and of viewing objects in museum display cases with little access to the sense of how it feels to handle objects, among other developments, have narrowed the focus of historians' attention to the *products* of crafts and their meanings, rather than the *processes* of manipulating materials in order to produce these objects.

Mind and Body in Cognition

Today, scholars from all disciplines, whether the natural sciences or the humanities, seem to agree that cognition involves both mind and body, but in scholarly articles and in the public realm these continue to be discussed in the vocabulary and conceptual framework of separate realms of mind/thought and body/action, rather than in terms that show the entirely intertwined phenomenon of what is essentially "mind-body thought-action." Scholars writing about embodiment often find language an awkward collaborator in their efforts. For example, in his inspiring work on the making of embodied knowledge, Trevor March-and employs analogy to understand the "dynamic syntax" of motor cognition in craft that relies upon "motor representations," a kinetic analogue to propositional representations formed in the mind.[2] One of the most important theorists of *gestes* and gestural knowledge of the twentieth century, François Sigaut, asserted that "knowledge" and "skill" are entirely different realms: the realm of knowledge is that of design and thought, the realm of skill that of effect and action.[3] Another scholar pits abstraction and embodiment against each other in his claim that historians have missed the essential story in ignoring "skill" to concentrate on the "abstractions of an ill-defined activity called 'science.'"[4] To escape this dichotomy, science studies scholars have tried to level the playing field between "practice" and "knowledge." A philosopher of science sees the dichotomous vocabulary as a matter of definition; "knowledge" needs to be carefully defined because it is only the judgments by epistemic communities that render a mode of cognition and/or a corpus of externally codified standards and norms as constituting "knowledge." Thus, we have to pay attention to how participants in an epistemic community "know they know."[5] Materials scientist and anthropologist Heather Lechtman posits a unified theory of knowledge in her demonstration that the categories by which communities order their intellectual and social world can be rendered through "technological behavior just as they are rendered linguistically,"[6] thereby providing a way to group both action and language-based cognition into a larger concept of "knowledge." As quoted in previous chapters, Tim Ingold, too, has provided an analysis of mind-body thought-action. Recently, Maikel Kuijpers has suggested a method to assess the traces of skill in ancient artifacts in an empirical manner, thereby providing components of a thing- and material-based vocabulary for discussing skill.[7] These scholars emphasize the primacy of practice as the source of knowledge, thereby challenging a long philosophical tradition that assumes mind- and language-based cognition to be the sole generator of abstract or conceptual knowledge about the world.

Despite this scholarship, one still frequently reads that craft is "knowledge how," while science is "knowledge why." In previous work, I used the term "artisanal epistemology" to challenge this facile dichotomy, and to recognize that the ability of craftspeople to produce material things rests upon experientially derived systems of knowledge that are simultaneously practical and conceptual and can be employed rigorously and methodically to generate new knowledge. In this book, I have provided historical case studies to suggest that we use the historical understanding of *Kunst* to describe craft knowledge. Recognizing, however, that this term, too, is not sufficiently differentiated from "art" and all the baggage it carries in the present day, I offer in this chapter further approximating concepts and terms by which language can describe practical knowing. In doing so, I strive, like some of the artisan authors I cite in this book, not only to describe practice in words but also to erase the long-standing division between "mind" and "hand" that has informed philosophical analysis and helped foster social stratification for millennia. Overturning this hierarchy of knowledge will be a long road with many branching paths of scholarship and action; in what follows, I offer the testimony of history.

Working Knowledge

The practices of craft knowledge to produce things like life-cast lizards involved a sensory apperception of materials—how do they look, feel, smell, sometimes sound, and, often, taste? How does the craftsperson know the materials are good, authentic, and useful to their purpose? How do they behave when mixed with other materials, when heated, when cooled, when burned to an ash? How are the necessary materials transformed into their required state? Making involves techniques of interacting with materials through the bodily senses. This sensory knowing of materials and the tools of their transformation are the fundaments of craft knowledge. For example, how is a mold made for a marsh marigold (see fig. 3.7) that will successfully pick up its delicate detail? The answer is not simple, because we generally do not have historical examples of such intermediary materials and objects as molds (and, if they are extant, they generally are not valued). This knowledge of materials and techniques—the effective interaction with and manipulation of materials— was transmitted through the course of apprenticeship in which a novice observed, imitated, and was directed by a master, but it could also be transmitted in part by the contents of the workshop (see figs. 3.1 and 3.2)—the materials and the casting flasks, the tools, and other implements. Even the physical space of the workshop itself could contain a form of embedded knowledge. Knowledge was also transmitted in the techniques. A regularized workshop practice could form a template that made the transmission of knowledge more rapid. Sometimes, as we have seen, such "secrets" were codified in writing, but that was only one (and probably the least effective) of several ways they could be codified and transmitted.

The objects produced by this material and gestural knowledge were proof of the craftsperson's material and technical knowledge—they constituted a proof of the craftperson's *working knowledge*, and, more than that, they indicated a body of *knowledge that worked*. A vivid illustration of the process of acquiring knowledge that worked, and one worth quoting at length, is Cellini's account in the *Two Treatises of Goldsmithing* (1568) of learning from one of his Paris goldsmiths:

Whilst in Paris I used to work on the largest kind of silver work that the craft admits of, and the most difficult to boot. I had in my employ many workmen, and inasmuch as they very gladly learnt from me, so I was not above learning from them; the plates I planed with such diligence gave them cause for such marvelling; but, nonetheless, one charming youth, on whom I set great store, said to me, with the utmost modesty, that in Paris it was not customary to plane the plate in the way we did it, and albeit our method seemed very clever, he would undertake to produce the same result without all this planing, and so gain much time.

To this I replied that I should only be too delighted to save any time; so I gave him a pair of vases to do, weighing 20 lbs a piece, and my models for them. Before my very eyes the youth melted his silver in the way I told above, and cast it between his iron plates. Then he cleaned some of the edges off and set to right away to hammer it into shape and give it its rotundity […] without paring it in any way. Both vases he turned out in this way with great care and admirable technique. It is just because in Paris more work of this kind is done than in any other city of the world that the craftsmen, from constant practice, acquire such marvelous technical skill. I should never have believed it had I not seen it for myself. Then, at first, I thought that it was the quality of silver that gave them a vantage, because they work here with a finer quality of silver than anywhere else; but my workman said no, and that silver of baser alloy would serve his purpose equally well. I tried him and found that it was so. From which I conclude, therefore, that a man can start straight away with shaping what he wants out of his silver without wasting needless time in planing it up first […] I do not go so far as to say that it is bad to plane the metal first: nay, I have found either way good.[8]

This passage points to the variety of different techniques to achieve the same result that typifies craft knowledge. Cellini's observations on his workshop also make clear the social dimensions of craft knowledge; the proof of knowledge that works is made within a community of practitioners.

Material Imaginary, Materialized Theory

The terms "working knowledge" and "knowledge that works" do not, however, go far enough, for as part 1 made clear, works of art were not just "products" that proved "knowledge that works," but instead often incorporated conceptual components and philosophical claims, and investigated causal questions, such as generation. As treated in previous chapters, one such example is life-casting, which, like so many other preoccupations of craft processes, began to overlap with the scientific objects of the new experimental philosophers in the seventeenth century. For example, studying the text of Ms. Fr. 640 reveals the material imaginary (a term we first explored in chapter 2) of the author-practitioner. He strives for deep red in his molds (fig. 10.1), likening it to vermilion or cinnabar (fol. 161v) by adding rusted red iron filings in molds. He aims to "make the molds strong," and it seems he referred to the "strength" of the blood-red molds produced with this technique. Biringuccio also instructed his readers to mix sifted iron scale in the composition used to make crucibles, which needed to withstand high temperatures and thermal shock.[9] The editor of Biringuccio's *Pirotechnia*, Cyril Stanley Smith, notes that such a practice would be useless in creating a stronger composition. In our reconstructions, however, our preliminary findings suggested that using the full amount of iron oxide created an extremely durable mold that stood up well to casting.

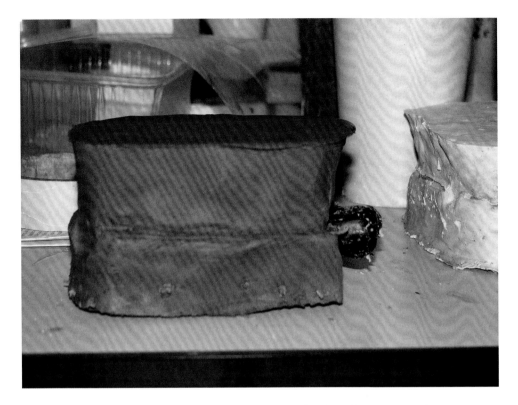

10.1. A deep red, two-piece mold following recipes in Ms. Fr. 640 for combining plaster of Paris, ground and previously fired molds, ground bricks, water, iron oxide, ammonium chloride, and eggs. Created by Pamela H. Smith and Tonny Beentjes. Photo: Pamela H. Smith and Tonny Beentjes.

The quality of the mold may explain the value of red in this instance, but as we saw in chapter 3, blood and red were implicated metaphorically in metalworking, and red substances were used as ingredients in many recipes. As we saw, red and blood were also connected to gold, which was seen to promote health and healing; this linked it to the most balanced and healthy temperament, the sanguine, itself associated with the humor of deep red blood in medical theory. Red is connected to gold in Ms. Fr. 640's recipe for making imitation rubies, in which one ingredient is gold leaf.[10] As in other metalworking recipes, lizards, too, crop up in Ms. Fr. 640, as a reported method for turning metals a gold color.[11] As we saw in chapter 3, red, blood, gold, and lizards were part of a knowledge system that underpinned metalworking practices and techniques, a material imaginary that is articulated by metalworkers in their works of art, their practices, and their texts, including Ms. Fr. 640.

Indeed, this material imaginary appears on the very first leaf of Ms. Fr. 640 in a recipe for "*coral contrefaict* [counterfeit, or imitation, coral]," where the author-practitioner has headed the page with a suggestive cross (fig. 10.2). In placing this cross on what is apparently the first full leaf of the manuscript text as a whole, he might have been following the practice of commonplace and account-book writers who began their text with a cross: "at the beginninge of their writing is to put fyrst the name of God, makinge the signe of

10.2. Ms. Fr. 640, fol. 3r, with a
cross just under the heading of
the recipe for *coral contrefaict*
(counterfeit coral). The cross could
indicate that this is the first page
of the text block; however, the
cross was also used by illuminators
to indicate the color vermilion, a
use that draws a close connection
between vermilion and the blood of
Jesus shed on the Cross. Vermilion
is the colorant used in the counterfeit coral recipe on this page.
Bibliothèque nationale de France,
Paris. Source: gallica.bnf.fr.

the crosse the wiche is most commonly usid amongest all Christen men."[12] It is, however,
suggestive that this leaf of the manuscript contains a recipe for imitation coral (fig. 10.3),
a *red* substance, created by the addition of vermilion, a pigment, as we learned in chapter 3,
used by the painters and scribes of illuminated manuscripts, who placed a cross in the
text to signify where the red pigment vermilion should be applied, thereby connecting the
pigment to Jesus's spilling of blood on the Cross.[13]

   As Palissy had understood the making of his naturalistic ceramics to be an imitation—
and through that imitation, an investigation—of natural processes, it would appear that
making imitation coral could also provide insight into generative natural materials. The
coral recipe on fol. 3r of Ms. Fr. 640 is made by swirling buckthorn branches in a red molten resin mixture, which coats the branches, then hardens on cooling.[14] This produced
an object that was viewed by at least one sixteenth-century observer as both replicating

10.3. *Coral contrefaict* (counter-feit coral), created by the Making and Knowing Project, following the process described in Ms. Fr. 640, fol. 3r, with assistance from Elizabeth Berry Drago, 2014. © Making and Knowing Project. CC BY-NC-SA.

and explicating the growth of coral: the Neapolitan apothecary Ferrante Imperato (1525?–1615?) wrote that coral could grow as a plant both on rocks and on other shrubs by covering their branches "like a dressing [*in modo di veste*]." This is why, he reports, when coral accidentally snaps, we can occasionally see an inner core of wood (*si scopre l'interno le-gnoso*).[15] It seems likely that he was looking at a branch of imitation coral, taking it for the real thing, and theorizing about the process of its generation. Such an epistemic process and the knowledge that it generates could be termed a "materialized theory."

While the author-practitioner of Ms. Fr. 640 did not proclaim, as Palissy had, his pro-cesses of imitation to be a bodily investigation that gave insight into processes of nature, Ms. Fr. 640 allows us to see that such practices were common to early modern workshops. In his recipe for *"jaspe contrefaict* [counterfeit, or imitation, jasper]" on fol. 10r, the author-practitioner provides brief instructions for producing imitation jasper stone—

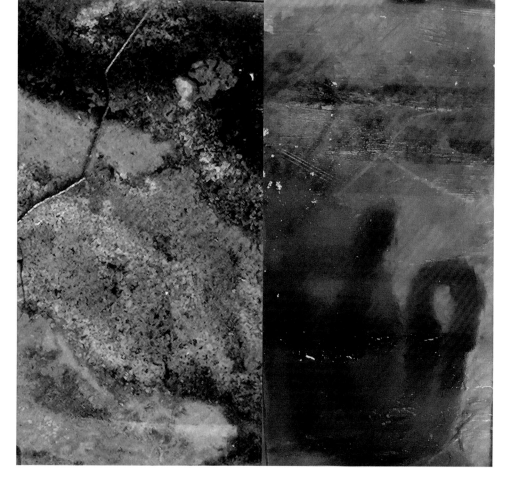

10.4. Left: Sample of natural red jasper. Right: Counterfeit jasper made from bovine horn, red paint (madder lake, spike lavender varnish), and gold leaf, following instructions from Ms. Fr. 640, fol. 10r. Created by Isabella Lores-Chavez, 2016. © Making and Knowing Project. CC BY-NC-SA.

one of the striated stones that, as we saw in chapter 8, Samuel Quiccheberg viewed as formed by the generative juices in the earth. By an elaborate and confusing process of planing animal horn into thin strips, then painting, gilding, scratching, and creating striations on it with dyed and painted wool, a translucent imitation stone, shining with what the author-practitioner views as the characteristic "luster and fatty polish" of jasper, is formed (fig. 10.4). In making imitation jasper, with which he intended to ornament bed frames, the author-practitioner was not aiming to deceive, but rather was carrying out an everyday practice of the workshop in which materials were transformed to imitate other substances. Although making something as quotidian as inlay for beds, he nevertheless participated in preoccupations about the relationship between human art and nature's artifice that also underlay the collection in the *Kunstkammer* with its displays of natural and *contrefaict* materials.[16]

Material Metaphors

A material imaginary makes use of "material metaphors"—materials used to think and experiment with. In his work on embodied cognition and the material enactment theory, Lambros Malafouris considers how metaphors work in human communication: they draw from a familiar domain to direct understanding toward an unfamiliar domain.[17] He sees objects working as material signs that do not so much "represent" something else, as "enact" and "objectify" that meaning from another domain.[18] As many historians have pointed out, for example, everyday religious objects can channel and partake in the spiritual, immaterial realm.[19]

The Making and Knowing Project has demonstrated that this metaphorical enactment happens not just at the level of the finished object, but also during its making process, and even in the materials themselves. One example of how this works in a making process can be found in a burn salve recipe in Ms. Fr. 640, fol. 103r, entitled simply, "Against burns, excellent," through which spirit is materialized in an ointment. Oil-based salves and ointments were widely employed in early modern Europe to cure wounds and relieve swellings and joint pain. Linseed oil and "the newest wax" are the basis of this salve, both ingredients common in cooking, lighting, and many other types of household and craft processes. Versatile and useful materials, they appear in multiple recipes in Ms. Fr. 640—mostly for artistic rather than medical purposes. Linseed oil was not usual in medicinal preparations, whereas wax had many healing properties, often deriving from anagogical power, being compared to the light and humanity of Christ, and its malleability used as a metaphor for the Creator's handiwork. Grades of wax were judged by purity, the best "not very yellow," and "not too fatty." To purify wax for expensive white candles, wax was crushed in a mortar, then simmered with seawater, alum, and saltpeter. When it was left to settle and cool, the remaining impurities rose to the top to be skimmed off. Repeated several times, this process could produce wax "as white as God's own beard."[20] Such purified white wax was most likely not the "new wax" specified in the burn salve recipe, as the "newest" wax had probably not undergone this purification and was still yellow and "fatty."

For the salve, one part wax was melted in three parts hot linseed oil, then the mixture was washed several times in holy water, the process being timed, as was common in craft procedures, by reciting the paternoster (the Christian "Our Father," or Lord's Prayer). Like wax candles and other material objects—known as sacramentals—that played a role in devotional practices, holy water was a bridge between material and spiritual realms, sometimes being used as a cure-all or to expel evil spirits.[21] The burn salve requires nine iterations of mixing holy water with the salve while reciting the paternoster, nine times at first, then decreasing incrementally by one, forty-five times in all. Nine was a significant, sometimes magical, number, connected to the Trinity, conveying perfection, and defining the norm for human gestation. The white ointment that results is to be applied to a burn

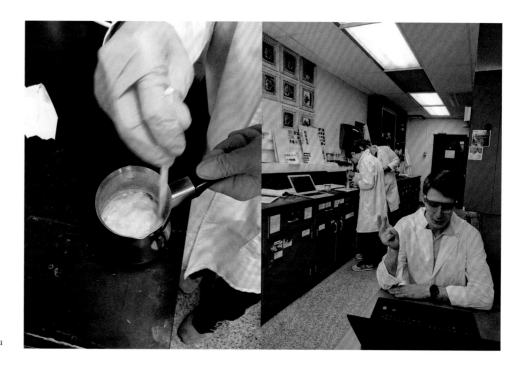

10.5. The burn salve mixture after the first cycle of stirring and washing, as described in Ms. Fr. 640, fol. 103r, prepared by Xiaomeng Liu and Sasha Grafit. Their stirring is being timed by paternosters, read by Ludovic Touzé Peiffer (right), 2017. © Making and Knowing Project. CC BY-NC-SA.

twice a day for nine days. At each application, the burn should first be washed with a linen cloth moistened with tepid water and wine, then the salve applied, and, finally, ivy leaves laid over it to make the beard grow back and avoid scars. Ivy remained green through the winter, and could symbolize eternal life, although in this case, its "hairy" stems perhaps were intended to engender hair growth.

The symbolic actions and ritualized process are clear, but what did this salve look like? Only re-creating the recipe can make this clear. Our reconstruction in the Making and Knowing lab revealed that bringing the burn salve into physical being was a surprising process of spiritual materialization.[22] We began by measuring the one part new yellow beeswax to four parts linseed oil by volume, and heated the mix somewhat higher than the melting point of beeswax to about 85°C. On cooling, the mixture began to "curdle," as noted in the recipe. Distilled water sufficed to replace holy water. With some skepticism, we added the water to the curdling mixture, while one student intoned the prayer in Latin, appropriate to the author-practitioner's time and place.[23] To our surprise, the mixture immediately transformed, becoming whiter, creamier, and lighter with each washing, increasing in volume about five times by the end of forty-five paternosters (figs. 10.5 and 10.6).

The anonymous author-practitioner of Fr. 640 claimed he learned the recipe from a *Pouldrier*, a gunpowder maker. Gunpowder making was dangerous work, and this maker apparently sustained disfiguring burns of which the author-practitioner saw no trace. Such a salve recipe is perhaps unremarkable—in modern chemical terms, it is an emulsion, commonly employed today to make lotions and ointments. The reconstructing of this

10.6. The burn salve after nine cycles (forty-five paternosters) of stirring and washing. The salve has increased in volume and been rendered a bright white. © Making and Knowing Project. CC BY-NC-SA.

process (in contrast to simply reading it) revealed that the process of mixing [holy] water while repeating paternosters caused a tangible "purification" and "inspiriting" of the raw materials. Like all reconstructions, ours could neither achieve, nor strive for, full authenticity—we could not capture the "subjective," period-specific spiritual significance of the process (we did not even use holy water!)—but, even so, the material transformation of its making converged with and made tangible the spiritual process.

In the case of the burn salve, it was not an object that worked as a "material sign," in Malafouris's term, but, instead, the making process transported us from one realm to another. In other sections of Ms. Fr. 640, we can see how materials themselves work as "material metaphors" through their enactment of another domain, as in vermilion "representing"—or, in Cennino's recipes, "enacting"—the blood of wounds, or standing in for the passion of the Cross, perhaps even echoing the process in the crucible through which vermilion comes into being. Materials could also work as metaphors through their activation of human engagement in the material world, as is demonstrated in the role of sulfur in Ms. Fr. 640. Sulfur is ubiquitous in the manuscript. In alchemical writings, sulfur is one of the fundamental principles (along with mercury) of all metals, but in Ms. Fr. 640, sulfur is quite practically and widely used, most often as a casting material due to the very fine detail it takes. When its native properties—its bright color, lack of luster, or brittleness—

10.7. Spectrum from pure sulfur to pure beeswax, experimenting with combinations described in Ms. Fr. 640. From left to right: pure sulfur; sulfur with more wax; sulfur with less wax; and pure wax. The brightness of the pure sulfur and the translucency of the wax make details hard to see. The mixtures of wax and sulfur make fine detail visible and avoid the brittleness of sulfur. Created by Rozemarijn Landsman and Jonah Rowen, 2014. © Making and Knowing Project. CC BY-NC-SA.

make it inappropriate for certain tasks, the author-practitioner attempts to transform its properties by combining it with other materials. He works ceaselessly to transform sulfur into more workable form, and the students followed his lead (fig. 10.7). The many uses of sulfur in Ms. Fr. 640, including as an ingredient in pigments, for casting, and for "boiling an egg in cold water without fire" (fol. 35r), attest to the material's availability and versatility, and such properties seem to have invited the author-practitioner to enact ever more trials. The students who reproduced his intensive experimentation concluded that sulfur, known for its versatility, both enacts and represents the process of trying and assaying, thus becoming a material metaphor for "experimenting" itself,[24] the essence of craft work.

Fundamental Structuring Frameworks

Materials are good to think with. The author-practitioner of Ms. Fr. 640 thought through and across materials and their properties as he made decisions about what to "try" next.[25] He hypothesized about the behavior of materials both on the basis of the properties they exhibit, and on the processes through which he puts them, such as calcining (burning to a powdery ash), to produce very fine, white "impalpable sand" made from oyster shells, or bovine bones, wheat flour, or alabaster (fig. 10.8), or trying to find a suitable and effective substrate for a yellow pigment by trying in turn lead white, chalk, and finely ground eggshells. He made generalizations about the properties of materials, based on their appearance, sensory look and feel (as well as smell, taste, and sometimes sound), but most often on their usefulness for his practice. In the process, he formed a taxonomy that cut across and organized materials and processes, performing what all craftspeople did (and do) as they scrounged their local environs for materials they hypothesized might be useful.

As the author-practitioner expresses it on fol. 90r:

10.8. Experiments in calcining hard materials to create impalpable sands for molds, following instructions in Ms. Fr. 640. From left to right: calcined alabaster; calcined bovine bone; calcined oyster shells. Created by Sofia Gans, Celia Durkin, Yijun Wang, Michelle Goun Lee, Diana Mellon, Emogene Cataldo, and Julianna van Visco. © Making and Knowing Project. CC BY-NC-SA.

Artisans who work in large works & who need to further their profit by seeking things already prepared in nature, because she does not sell her wares to her children, and to also save the time they would use for grinding finely & for artificially preparing sands, seek the [earth] of the mine, which is not too fatty, the one that is a kin of earth, not too lean & consequently without bond, but rather that which is pulled from the depths of the sand-bed in bricks & clods that show its natural compaction, which is quite difficult to break & which has a very small & delicate grain, & which is found soft when handling it between the fingers.

There is much in this short passage, but I quote it here because it shows the search in his surroundings of Toulouse for a kind of working material that matches the categories he has in mind, that is, a sand/earth that is neither too fat nor too lean, neither too dense nor too brittle. The passage reveals the categories of a taxonomy formed through the author-practitioner's sensory experiences in pursuit of useful materials. This taxonomy informs his hypothesizing about, and repeated testing of, the kinds of sands that might be useful as molding materials. This kind of testing, categorizing, and hypothesizing is far from simple "trial and error."

TAXONOMIES OF USE

Early modern things and substances were often organized according to their uses, rather than to their morphology, structure, or appearance. The contrast to today's largely morphological taxonomic classification system was brought home to the Making and Knowing Project in spring 2015, when we sought elm roots for a "magistra"[26] which was used as the liquid binder for the dry powdery "sand" for making molds. Ms. Fr. 640 specifies on fol. 87v, "Founders take the roots of a young elm when it is in sap & boil it in wine, or better yet vinegar, and keep it all year long in a barrel." We tried this binder recipe in the fall with the readily available slippery elm bark, well known for its mucilaginous properties, and it worked beautifully. We were aware that slippery elm was native to North America, and would not have been available to the author-practitioner in late sixteenth-century Toulouse. When the ground thawed in New York City and trees were "in sap" again, we therefore hunted for an authentic sixteenth-century French elm. We began a cross-country email exchange with curators of historical gardens, during which we learned about the nostalgia for the European elm, killed off by elm disease in the nineteenth and twentieth centuries. Finally, we were persuaded to use a hybridized elm species that, according to our informants, was developed to resemble this sixteenth-century European elm.[27] When I inquired whether this species, considered taxonomically similar today, would necessarily share the properties sought out by a sixteenth-century practitioner, or if taxonomic systems today instead operate by a different set of categories, I got the answer that in today's hybridization, the property of "gooey sap from the roots would be one of the last things they were looking for."[28] We came to realize that our taxonomies for classifying species based on morphological properties—determined by the visual appearance of leaf, bark, and flower—differed profoundly from the taxonomy of our author-practitioner. His system of classification was based on the properties of the material as they were expressed or manifested in the process of human manipulation—a system also found in early modern herbals, which organized plants according to their "virtues," or properties useful to humans for healing and other processes. Such a taxonomy of use formed an overarching structuring mechanism for materials, processes, and knowledge in craft procedures.

CHANGES OF STATE

We discovered another such working framework when we investigated the author-practitioner's intensive experiments with different types of sand (*"sable,"* by which he meant the dry components of a mold) and binders (the binding medium, sometimes called a "magistra"). In the metalworking recipes, he employed sands and binders to produce box molding "sand," or powder, which will take a fine impression, is cohesive enough to allow the opening of the mold, but releases the pattern easily, and is strong enough to withstand the entry of the metal into the mold. In order to create this ideal molding matrix, dry and wet materials of various kinds are mixed carefully together to form a sand that gives "a nice hold," but still comes "apart easily" (e.g., fol. 118v), instructions that the lab came to refer to as the "squeeze test" (fig. 10.9).[29] In creating both the sand and the binding medium, the author-practitioner is preoccupied with transforming materials from one state to another: hard, brittle oyster shells, for example, into fine powdery "impalpable" sand.[30] Indeed, throughout the text, he is preoccupied with the native properties of materials— liquid, congealed, solid, vaporous, hard, soft, malleable, brittle, among others—and their

transformation. The aim of transforming the states and properties of a material provided another means of organizing working procedures and classifying the knowledge obtained through these procedures.

This focus on changes of state is in line with many recipes in early modern practical writings that attempt to "fix" volatile materials to make them endure the fire, or, conversely, that strive to increase the plasticity of solid materials. Petrus Kertzenmacher, who reorganized and republished the metalworking text *Rechter Gebrauch der Alchimei*, noted that the entire secret of alchemy was contained in "*corpora* and *spiritus*," that is, making volatile *spirit* into a fixed *body*, able to remain in a solid state even when heated. To fix the essence of a substance was to isolate and hold its particular power. But, conversely, many of Kertzenmacher's recipes also had the goal of dissolving fixed bodies.

Such recipes sometimes aimed at changing the properties of materials by putting them through phase transitions—typically by slow, constant, and long-term heating, such as that of an egg sitting under a broody hen, or by the heat that is given out by the slow fermentation of manure, caused by thermophilic bacteria that maintain a relatively constant temperature. Recipes for softening hard substances, such as liquefying stone or horn in order to be able to cast them, and the related process of making brittle materials malleable, as in the transformation of iron into steel, are also to be found in large numbers in such recipe collections.[31] Learned and lay alchemy had in common with craft practices this attempt to transform the states of substances. For "alchemy," to many practitioners, meant not primarily the transmutation of base metals into noble ones, but rather the knowledge of how to effect transformations of state in natural materials.

10.9. Finding the right consistency of mold material by the "squeeze test," which in Ms. Fr. 640, fol. 118v, is specified as the sand giving "a good hold, nevertheless coming apart easily." © Making and Knowing Project. CC BY-NC-SA.

A Spectrum of Material Properties

Recipes in Ms. Fr. 640 and in other recipe collections aimed at varying the qualities and properties of one material to resemble or imitate the properties of another. For example, in Ms. Fr. 640, the author-practitioner sought to transform sulfur into a material with new properties—he mixed melted beeswax with tallow, which produced a material more capable of being carved; he combined melted sulfur with different pigments in order to make casting details more visible in the surface of the sulfur. In another, rather complicated process, he "passed" solid sulfur through melted wax to bring into being a material that, as students discovered in their reconstructions, produced very smooth surfaces and took fine detail when cast, as sulfur does, but which was less brittle, and remained flexible and malleable at a lower temperature, as wax can. This "sulfur-passed wax" thus had the workability of wax but the opacity and ability to take fine detail of sulfur, thus overcoming the translucency and softer lines of wax, which make it difficult to see details in the carved or cast wax.[32]

The author-practitioner's preoccupation with the transformation of material properties became especially clear in reconstructing a series of recipes for making molding material that seem implausible when read as text—implausible because, instead of mixing a dry sand and a wet binder to produce a suitable casting matrix, the recipe calls for two dry materials, ox bone and rock salt, ground together to form a dry powder. This dry, crumbly mixture appears entirely useless as a mold material, until it is, as the recipe directs, left moistened "in paper, folded in a wet napkin, which is previously made in the damp air [*serain*] of the night or in the moisture of the cellar." In their reconstruction, the students jerry-rigged a humidifier to provide this moisture, which caused long crystals to form in the salt by which the unlikely powder was transformed into a very effective mold material that took detail well and was extremely durable.[33]

Our reconstruction of these "implausible" mold recipes revealed that the author-practitioner was exploring materials along a spectrum of their possible states and inherent properties. This spectrum of possible states or properties, defined by opposing end points, such as brittle vs. malleable or solid vs. liquid, formed an underlying interpretative grid within which the author-practitioner could design his trials and interpret their results, as we have seen in his work with wax, sulfur, and molding sands. He viewed materials as possessing attributes that included the Aristotelian qualities—hot, dry, wet, and cold—but, like other contemporaneous recipe collections, Ms. Fr. 640 made reference frequently to a ubiquitous set of binary qualities: hard vs. soft, brittle vs. malleable, sour (*aigre*) vs. sweet/soft (*doux*), and fat (*gras*) vs. lean (*maigre*). Similarly, although Biringuccio disavowed what he referred to as the alchemical idea that sulfur and mercury formed the underlying principles of metals, believing instead with Aristotle that metals were composed of water, he discussed metals in terms of qualities of hot and dry or cold and wet (which, incidentally, corresponded to the qualities of sulfur and mercury, respectively). Furthermore, Biringuccio viewed metals, earths, and minerals as lean (implying brittle and more arid) or fat (implying unctuous and more liquid).[34] Like other metalworkers, he believed all metals could be rectified of their "corruptions" through tempering—that is, by the addition of other substances, by treatment with another metal with the opposite qualities, or by heating or cooling. Tin could alter the "tractable sweetness" of other metals, corrupting them, which in some cases was desirable, such as when tin was added to copper in bell metal to make it more sonorous, "just as if it puts the spirit there and vitalizes the substances."[35] For Biringuccio, transforming a metal to make it harder was a corruption of the metal, for hardness or brittleness reduced the essential metallic quality of metals, which he regarded as their malleability. When a metal was lean, it caused porosity in casting, and the metal needed to be tempered with an unctuous substance. Conversely, a mold made of a lean substance received the unctuousness of metal particularly well.[36] This fundamental structuring framework within which the properties and behavior of materials were manipulated is visible also in Biringuccio's instructions for making a strong, lean clay for crucibles that will be resistant to fire, by mixing it with talc, iron scale, and young ram's-horn ashes, or with crushed peperino, flintstone, or some other stone that "seems to be arid and resistant."[37]

The fat-lean binary can also be found in discussions of agricultural soils, and in foundries where, for example, investment and mold materials were made from a buttery, unctuous earth produced by long, generative fermentation of clay and wool (as Benvenuto Cellini tells us in the *Two Treatises*).[38] The fat-lean spectrum also appears in mining texts, where fat earths form a central goal of mineral prospectors because they were believed to be generative of rich ores. In the seventeenth century, a chymical practitioner, Johan Rudolf Glauber, and the Leiden professor of practical medicine, Herman Boerhaave, identified fatty earth in their writings as the substance out of which metals could be generated, and it remained an object of scientific investigation into the nineteenth century. For seventeenth-century (al)chemical theorists, "fatty earth" (*terra pinguis*) would come to be seen as a foundational and transformative element in many material processes, of interest to Isaac Newton and important in developing first the concept of "phlogiston," then, in the eighteenth century, Lavoisier's "oxygen."[39]

This framework of binary qualities is part of an ancient, overarching paradigm of balance that underpinned the health worldview, discussed in chapter 2, and other systems of knowledge. Gunners, for example, believed that gunpowder exploded because contraries came into contact.[40] The properties of fat and lean, at opposite ends of a spectrum, provided a framework and a vocabulary within which early modern craftspeople articulated and understood their strivings.

This spectrum of material properties was brought home to the Making and Knowing lab when students making the implausible sands struggled with making salt "fatty," as called for in a recipe for molding sand in Ms. Fr. 640. It is hard to know what that means within a modern understanding of materials. After grinding rock salt, they found the texture of the salt had become "sticky," and, as they recorded, "we *felt* the 'fattiness' of the rock salt, which also afforded a new workability for our molds." They concluded that such terms as "fat" and "lean" rely very directly upon sensory interaction and an intimacy with matter, resulting from the proximity of the craftsperson's sensory testing of materials and the ingestion of foods and medicines. Humans, materials, foods, and medicines all involved the same types of qualities, including cold, hot, wet, dry, fat, and lean, and this contributed to the intimate connection between practitioner and materials.[41]

Our modern, abstracted knowing of material through chemical composition differs from the early modern practitioner's more intimate bodily and sensory knowing by means of consuming, digesting, smelling, tasting, observing, and acting and thinking within bodily processes. As noted in chapter 5, the body functioned as a tool in the workshop. Bodily fluids—including urine, blood, saliva, phlegm—were an integral part of the author-practitioner's practice,[42] as were everyday foods, such as butter, eggs, garlic, and bread. He measured in proportion to the human body: "the height of a man,"[43] "as much as you can take with 4 fingers and the thumb, or a small double handful."[44] As revealed in examining the consumption of butter, the craftsperson's body and the materials being worked interpenetrated each other, as when the metalworker inhaled the cold vapors of the metals. Both humans and metals had to be tempered and brought into balance by ingesting substances that counteracted undesirable states, such as excess coldness caused by the evil smoke of the metals, for which, as we saw, the author-practitioner of Ms. Fr. 640 provided a remedy with properties at the opposite end of the spectrum: bread and butter.

Imitation

Ms. Fr. 640 is full of recipes that instruct in how to imitate materials, including, besides *coral contrefaict* and counterfeit jasper, among many others, imitating marble by painting on wood, mimicking vitreous rouge clair (a red glass) made with oil paint or with red resin from the dragon tree, and making imitation gemstones, such as emeralds, rubies, and topaz, from red lead and glass. A recipe common to painters' workshops, "Color of gold without gold on silver," imitated not the material of gold, but its visual properties, as students found when they reconstructed a mixture of turmeric extraction and turpentine varnish applied over silver foil.[45] All these quotidian practices of art making strive to counterfeit another—often more costly—material. In *The Nature and Art of Workmanship*, David Pye regards this imitation as part of an aesthetic system: "perhaps the most constant and delightful aesthetic phenomenon throughout the history of sculpture has been this very expression in hard stone of the properties of soft materials like flesh, hair, and drapery. The stone remains recognizably stone yet the hair is recognizable as hair and the cloth as cloth." Plastic arts, such as casting, ceramics, and glassworks, especially lend themselves to this system because metals, clay, and glass changed state as they were melted, worked, hardened, and shaped to form the artwork. According to Pye, this effort to imitate the representation of another material constitutes the essence of skill, or, as he terms it, "workmanship," or "craft."[46] To Pye's insight, we can add that the craftsperson often also strives to imitate and even compete with the artifice of nature in such a representation. Such imitation and rivalry possessed a long and important genealogy in European craft and art, of course, embodied above all by the enduring employment of Ovid's *Metamorphoses* in representing and philosophizing about natural and artistic processes of transformation, such as we saw in Jamnitzer's *Daphne*.[47]

The effort of imitation in early modern art had a much more profound significance, however, than simply the mimicking of another material by depicting or representing it. As we have seen, sulfur was transformed by combining it with pigments and with wax to endow it with the properties of other substances. Lean substances were pushed along a spectrum toward fat ones; sour and brittle ones manipulated into sweet and soft, or workable, ones. The type of imitation by which material properties were transformed by manipulating them into a similitude of another material's properties was pervasive in the workshop, and much more than an aesthetic system. The framework of binary qualities that gave shape to a spectrum between two poles, along which materials could be manipulated, was a means of planning, organizing, and conceptualizing work with materials. Such working and thinking through material properties provided a way to meet the challenges that natural substances presented to a craftsperson. It provided a means for the everyday exploring, hypothesizing, and testing by which a practitioner came to understand the properties and behavior of materials and the processes and products they made possible. Imitation by transforming properties simply—yet significantly—articulated and codified a *mode of work*. This mode of work seems to be an essential component of the earliest human engagement with natural materials, for what is bronze but manipulating the properties of tin and copper between the poles of hard brittleness and soft malleability? Sixteenth-century practitioners expressed in their texts and their objects how necessary were repeated trials of materials, challenging the materials to the limit, testing various means of transformation, and attempting to push materials to take on the properties of their polar opposites. This process was a fundamental framework within which skilled practice could go forward.

Conclusion

In his many trials, the author-practitioner worked to challenge his materials, transforming them to take on the often contrary properties of another material: the coarse into the "impalpable," the lean into the fat and workable, the hard or brittle into the malleable, the sour into the sweet, and the elements of earth and stones into the simulacra of gemstones.

When studied and reconstructed, the processes recorded in Ms. Fr. 640 reveal systems of belief and knowledge, and testify to the material, physical, and philosophical engagement of craft with the generative and transformative powers of nature. As we saw above, imitation jasper constitutes an attempt to mimic a precious material, but through this imitation, a craftsperson like Palissy also investigated the genesis of the stone. This investigative form of imitation allows the practitioner not only to explore the properties of these raw materials, but also to theorize about the formation and composition of the precious material in nature. Ms. Fr. 640 demonstrates that its recipes (and craft knowledge more generally) were not just productive, but also investigative and "theorizing"—making was knowing.

The search for material knowledge that is effective is carried out through action and practice, and must be extracted from the materials of nature by the practitioner using the senses and the body. This process of bodily engagement with the material world often takes the name of "experience," and is learned and explicated through imitation and replication of other craftspeople and of the processes of nature. This imitation of nature produces an effect—an intervention in the material world, such as a quotidian object or a work of art—which displays, proves, and constitutes knowledge. This cognitive practice was simultaneously an aesthetic striving, an articulation of an epistemology (sometimes in textual, but most often in material form), and an investigation into the properties of materials. We can call this epistemic complex "knowledge that works," a "material imaginary," "materialized theory," a set of "fundamental structuring categories," a "taxonomy," a "framework for action," or an "artisanal epistemology"—but at its most basic and all-encompassing, it was simply and straightforwardly a "mode of work" in the human engagement with the materials of the environment.

Global Routes of Practical Knowledge

This book is about producing, codifying, transmitting, theorizing, and studying practical knowledge. Vernacular systems of craft knowledge and their codifications in early modern Europe formed the subject of part 1, "Vernacular Theorizing in Craft," and in it, we examined the material imaginary that underlay practices and techniques of European craftspeople, especially metalworkers. Part 2, "Writing Down Experience," provided an overview of the codification and transmission of writing about experiential knowledge, techniques, and knowledge systems from the ancient world through the era of the *Encyclopédie*. It considered, as well, the content, import, and ambitions of the artisan authors and the types of knowledge provided in their little books of art. Part 3, "Reading and Collecting," examined the reception of artisanal texts—*Kunstbücher*—and the perception in the sixteenth century that these texts, and the practices of making and objects they described—collected together in *Kunstkammern*—held potent knowledge, of utility to rulers. Part 4, "Making and Knowing," detailed how historians can engage in the study of artisanal texts and objects through hands-on investigation, and provided an overview of some insights into practical knowledge that such study offers.

Global Perspectives

The "early modern" period covered in this book encompasses long-term intellectual and cultural transformations based in Europe that have been known by names such as the "Renaissance," "Reformation," and "Scientific Revolution," as if these were paradigm-shifting ruptures of short duration. In a related vein, the teleological label of "early modern" suggests that this tumultuous period ushered in "modernity" (implying cause for celebration or lament). Over the last three decades, historians have begun to rethink these perspectives and have turned their attention to the greater integration of various parts of the globe, including Europe as only one among other regions, in the period from 1400 to 1800.

From this perspective, the subjects and period covered by this book must be understood within the increasing integration of many parts of the globe, especially with regard to the trade in natural materials, precious metals, and luxury commodities across long-distance commercial networks. Trade had flowed across Eurasia, around the Indian Ocean, and over the Mediterranean for thousands of years, but in the period treated here, larger parts of the globe became connected by the establishment of more or less regularized trading routes. Global integration occurred in this period in several ways, including by the spread of pandemics across the entirety of Eurasia and into the Americas, the recurring "plagues" representing only one type. The lands of Eurasia and the Americas also shared the bloody advent of new weaponry and gunpowder. The new gunpowder warfare contributed to a greater consolidation of central power in the hands of fewer rulers, whose growing numbers of bureaucratic functionaries began to implement and administer systems of expansionist, imperial, and even "universalist" rule.

As commodities and tribute bounced and jostled over trade routes, knowledge flowed along with them. Embodied knowledge moved in individuals as they migrated or were forcibly transported to new regions, and it moved along with sailors, soldiers, and merchants as they pursued trade and waged war. Knowledge also traveled in objects, instruments, manuscripts, and printed books, as elites avidly sought out rare and beautiful things to stock their collections and their *Kunstkammern*. And it moved as factors sent back specimens, new things, and information to the metropolis. Polities at both ends of Eurasia sponsored information-gathering projects in pursuit of trade and conquest. Knowledge moved within and among these fields not just *geographically* but also *epistemically*, as knowledge systems of different social and cultural groups intersected. Knowledge formed and transformed, was filtered and sometimes disappeared, as goods and people traveled from local settings and vernacular modes of expression—ranging from the ceramics manufactories in Jingdezhen and Puebla; shipbuilding arsenals in Calcutta, Venice, Ragusa, and Istanbul; artisanal workshops, collectors' cabinets, and gardens in all parts of the globe, to the knowledge and written forms of *Bencao* (materia medica) and *pulu* texts (catalogues of various things), or into the codifications of astronomy and astrology produced, for example, by the Mongols as they absorbed Chinese scholarship. These itineraries helped to shape the new epistemologies of *kaozheng* (evidential studies) and the "experimental philosophy" in Europe.[1]

The history of Europe in this period has come to look quite different as a result of these new global perspectives. Europe has begun to be decentered as a driver of cultural and economic change in the first half of this period, as many sources of novel techniques, objects, materials, and ideas that entered Europe via diverse routes have been documented and studied. In the second half of this period, European expansion into global commerce fostered the destructive hegemony of European imperialism and colonialism, which, among many other processes, saw the emerging new sciences and the concept of "modern science" come to be employed as tools of empire and "civilization."[2] What used to be told as a history of the European roots of a universe-changing intellectual transformation from the Renaissance forward is now more often being recounted against the background of routes of commerce, extraction, exploitation, and appropriation.

The new focus on movement and routes instead of originary moments, or "roots," has transformed perspectives on the dynamics of knowledge making in this period. Knowledge is no longer viewed as a discrete corpus of ideas, texts, and practices that can be disseminated or transmitted from one point to another, to arrive as a whole or in parts. Instead, knowledge is understood to be constituted in part by its motion across time, space, and intellectual and social distance. Motion causes the production of knowledge through the intersecting relations among communities and within intellectual fields that it calls into being, the disruptions it triggers, and the new perspectives it opens up. It has come to be fairly self-evident now among historians of science that the routes that materials, practices, and knowledge trace out over time and space are often more important to their formation than their roots or origins.

Material Imaginaries across Time and Space

VERMILION

In the light of this global perspective, this book concludes by returning to the writing down of practical knowledge by Michael of Rhodes and Cennino Cennini, with which it began. In chapter 1, I took their books to represent the beginning of a great expansion in writing down experiential knowledge. In this epilogue, I partly upend that view by setting it within the expanded perspective of global history. In previous chapters, I examined vermilion as one of the nodes and material metaphors in the material imaginary of European metalworkers' practice. What I left unsaid in those chapters was that vermilion followed a complicated material itinerary before the sixteenth century, which illustrates how knowledge moved across epistemic distance, especially between anonymous practitioners and book-writing scholars. The techniques of manufacturing artificial vermilion, which replaced naturally occurring cinnabar, spread westward in the early medieval period, probably from China to the Islamic world and thence to Western Europe.[3] The first written recipe for making vermilion appeared in Europe in the eighth century, although there is evidence that it was being produced in Europe before then.

Vermilion making interested medieval European scholars, such as Albertus Magnus. Albertus was part of the generation of European scholars intensely involved in assimilating the ancient Greek and Latin corpus of texts entering Europe as a result of expansionism and conquest across the Mediterranean and in Islamic Spain. Albertus had witnessed vermilion making[4] on the "long journeys to mining districts," that he made as a clerical administrator.[5] Albertus found vermilion compelling because he had just come into contact with Arabic alchemical theory, which taught that the "principles" of sulfur and mercury—which in material form were the components of vermilion manufacture—formed the underlying substratum of all metals.

The theory that sulfur and mercury formed the principles of all metals had only just arrived in Europe in Albertus's lifetime via the books of alchemy translated from Arabic into Latin. Before the eleventh and twelfth centuries, European scholars knew of recipe collections such as the *Mappae clavicula* (A Small Key of Handiwork) for operations using gold, precious stones, and gems, as discussed in chapter 4. Although these recipes contained fragments from earlier Greek alchemical writings, these mysterious interpolations had been detached from any sort of conceptual framework. In the twelfth century, however, books of alchemy began to be translated from Arabic into Latin. Although alchemical theory stretched back to a conglomeration of Greek matter theory and Gnostic spiritual practices, such as those found in Hellenistic Egypt, the sulfur-mercury theory of metals was an innovation of Arabic alchemical writers, worked into the older Hellenistic texts. It appeared for the first time in the Latin West as an entirely new field of knowledge, and one translator began his text, "as your Latin world does not yet know what alchemy is and what its composition is, I will clarify it in the present text."[6]

Albertus had discovered the sulfur-mercury theory in the texts of the Arabic (al)chemical scholars, including the compilations of the composite ninth-century authority, Jābir ibn Ḥayyān, and Abū Bakr Muhammad Zakariyyā Rāzī (854–925), whose works were translated from Arabic into Latin in the twelfth century.[7] The philosopher's stone, the substance believed to be capable of instantly transforming a mass of base metal into shining gold, was often described as a red powder, like vermilion. Vermilion making and alchemical theory thus had at their core the active, generative agents of sulfur and mercury. While alchemical theory regarded these as two principles that composed all metals and had to be manipulated to make the philosophers' stone, vermilion production employed both as material ingredients. Creation of the philosopher's stone was said to involve spectacular transformation that ended in a red powder, just as the process of making vermilion also caused a series of impressive color changes from black, to blue-silver, to end in a red powder.

When alchemical theory arrived in Europe in the twelfth century in textual form, then, craftspeople had already been combining mercury and sulfur to produce a red powder for at least four centuries.[8] Indeed, the *practice* of vermilion production seems always to have predated the articulation in texts of a theory of metals. It is almost impossible to separate alchemical "theory" and metalworking practices: as Albertus said in the thirteenth century, "[W]hat artisans have learned by experience is also the practice of alchemists […] if they work with nature."[9] It appears likely that the sulfur-mercury theory of metals actually emerged from the practice of making vermilion, in other words, from the work of craftspeople and their techniques. Thus, one of the most pervasive and enduring metallurgical theories of matter and its transformation—the sulfur-mercury theory—probably emerged from the making of a valuable trade good that resulted from the worldwide circulation of craft techniques and texts.

Vermilion is thus a story of knowledge moving across both geographic and "epistemic" distance. The alchemical books traveled over long distances and through multiple translations, but an equally important knowledge-making process was that in which craft practices shaped and informed the theories of text-oriented scholars. As we saw in chapter 3, European metalworkers oriented their practices through the material imaginary of red, vermilion, and blood, and they viewed lizards as agents of transformation. In the same way, Albertus Magnus, relying upon the Arabic alchemical works that had come to him along a complex path of translation and compilation, saw the making of red vermilion he observed in a workshop as confirming his core theory that mercury and sulfur were central components of metals and of metallic transformation. In both, blood red was the heart of the transformation of matter.

TRANSFORMATIVE AGENTS: LIZARDS AND SILKWORMS

Ideas about lizards as generative agents can be traced even farther over distance and time: In the early twentieth century, anthropologists working on the oral culture of unlettered South Asian villagers recorded oral recipes that used reptiles to produce light, like that of the lizard-tail recipe from the medieval book of secrets, taken up by the Latin West from Arabic books of magic and technology, treated in chapter 3.[10] Similarly, in China, lizards were regarded as agents of transformation; indeed, ideographs for "transformation" and "lizard" are closely related. Moreover, lizards and red pigment are explicitly connected in in the third-century CE text, *Bo wu zhi* (Comprehensive Record of Things) by Zhang Hua, which includes a recipe for keeping lizards in a vessel, force-feeding them cinnabar until they have turned a deep red, then grinding them and dabbing the resulting pigment on a woman's body where it was supposed to glow without extinguishing until she had sexual intercourse (apparently a method for keeping the emperor's concubines chaste).[11] The material imaginary of European metalworkers in which lizards, vermilion, and generation were connected to metallic generation appears to have been informed by components that journeyed across Eurasia. One connecting thread between these distant locales may have been "books of secrets," recipe collections, and books of wonders and marvels, popular at both ends of Eurasia.

While I have only hints of this connection in the case of vermilion and the material imaginary of lizards, red, and generation, more concrete evidence of the travels of materials is provided by the late sixteenth-century French practitioner's manuscript, Ms. Fr. 640. It records the movement of materials within Europe—the blue dyestuff woad traded between southern France and Spain, dyes and pigments from Italy, amber from the Baltic, and metals from Germany—as well as the global movements of materials, such as turmeric and stick lac from South Asia, cochineal from Central America for dye and pigment making, dragon's blood resin from trees growing in the Canary Islands and North Africa, a method for damascening armor from the Near East, and a recipe for "damasking" cloth by resist dyeing it with "Moresque" templates, a design likely derived from the Ottoman empire.

Among the evidence of the movement of materials across Eurasia, the manuscript includes a recipe entitled "Work done in Algiers" (fol. 52r–v) that appears to have followed a similar itinerary to vermilion.[12] This recipe represents almost the sole "alchemical" entry in the manuscript, but unlike some alchemical texts, it contains quite practical considerations and ordinary ingredients. In many ways, apart from its alchemical aim of making a powder that transforms antimony into gold, this recipe is of a piece with the rest of the manuscript in its attention to practical details, first-person perspective, and materials common to a workshop. It is worth quoting this entry in full, as it is reminiscent of the recipes that enclose lizards in a flask, which are then fed with unlikely materials. The "work of Algiers" begins with the instructions for feeding the horse that will supply the manure for heating the substance at a constant low temperature:

Take a colt of three or 4 years & feed it on ~~rye~~ barley & straw cut in the manner one feeds horses in Spain, and water it with good fountain or river water. I do not know if it would be good to water it occasionally with water of sulfurous baths, & to sometimes give it fenugreek or other hot foods, for the intention of the worker is ~~to it~~ to use the heat of its dung, & the climate here is cooler than that of Algiers. Keep it in a warm & close place & so that none of its dung & urine should be lost, of which you will make a heap or two in order that while one cools, the other will be at the appropriate heat to continue.

It continues with instructions for hatching the silkworm eggs:

Also take a large glass mattras [long-necked vessel], as thick as you can, & one finger thick if it can be done, & of the capacity of one pitcher or earthen jug. Around the feast of St. John, put into it a dozen & a half chicken eggs, that is to say, the yolk without the glair & the (germ see) [unfinished sentence]. Others say lx yolks. And with this dozen & a half egg yolks put in half an ounce (others say for lx eggs half a lb) of female silkworm seeds. And after having luted the mattras well (I do not know if it is at all necessary for the generation that there be air), put it & bury it in the heat of the dung up to the neck, and leave it there until several ~~[illegible]~~ worms are engendered.

From the mass of silkworms, one will emerge that must be carefully nurtured:

And then remove the mattras & do not bury it in the dung any longer, but only keep it placed on the hot layer of the dung until all the worms will have eaten & consumed one another, bustling & stirring, and only one remains. When this is the case you need to feed it at regular intervals, day & night, with the aid of two men, who by intervals will take care of it, and you will feed it with an egg yolk covered with a gold leaf, or with a liquid egg yolk with the gold leaf incorporated; & take good care that it does not want for such food (some say one egg yolk per hour, others three, but the thing itself will demonstrate the practice).

Through such treatment, the worm metamorphoses into a snake from which wings, like those of a basilisk, begin to form, at which point it must be put to death. Careful safety precautions must accompany this process:

Nourished in this way it will achieve its growth in two months or seven weeks & will become like a snake, one *empan* & 4 fingers long, & one lb in weight, and as its wings begin to grow, one will need to put it to death by making a ring of charcoal fire around the bottle one *empan* distant from it, and then stopper & lute the bottle well in order that it does not exhale. Or to be safer, retire from there until the fire has died down & all is cooled, for its exhalation would be dangerous. And for the occasion when you feed it with pincers, wash your mouth with good vinegar & take some preservative & plug yourself up well.

The dead creature is then dried and ground in a mortar, producing a powder that can be used to create gold:

Once it is dead, put it in a linen cloth or a canvas of silk & fold it & hang it from the ceiling, where the air & sun dry it. Once it is quite dry, pulverize it in a mortar. And keep this powder carefully, because one ℥ [ounce] of this thrown on iii lb of molten ♁ [antimony] reduces it to finer ☉ [gold] than the other one. But it does not have as much weight. For this work you also need to choose the oldest ♁ that you can, which has often been melted & finely hammered into sheets or other works, & purify it before by melting & throwing it into honey & vinegar. The term of the work is nine months from the feast of St. John until the 25th of April.

    One folio further on in the manuscript, there appear purely practical instructions for raising silkworms (the Spanish silkworm eggs, it is noted, producing one and half times as much as the ones from Languedoc). The two entries have no obvious connection, although both involve growing silkworms from eggs. From the "Silkworms" entry it is clear that the author-practitioner is familiar with early modern sericulture practices, and his detailed account of raising silkworms seems to be the result of firsthand experience. Compared to other contemporary works on sericulture, however, the author-practitioner's account is lacking in one crucial particular—it exhibits no interest in deriving silk through the process, but instead only eggs. This is a marked difference from other literature on silkworms. Of course, it may be a coincidence that a recipe for using those silkworm eggs to create a gold-producing powder, "as they do in Algiers," occurs one folio earlier, but there are other similarities that could indicate that the purpose of "Silkworms" is not to extract silk threads from cocoons, but rather to prepare freshly laid silkworm eggs for the "Work done in Algiers." The dates mentioned in "Algiers" for starting the process of growing silkworms in the flask are aligned with the date that ends one complete life cycle of silkworms as recorded in "Silkworms"—from worms, to pupation, moths, mating, and finally laying of eggs to begin the cycle once again.

The two processes of growing silkworm eggs and making a gold-producing powder both connect to a cycle of growth and rebirth in nature and in the Christian liturgical year. They both begin on John the Baptist's Day, occurring close to the summer solstice, the longest day in the year, followed by the significant period of nine months—the length of human gestation—to the feast of the Resurrection, occurring near the spring equinox, the turning point of the solar year and a time of rebirth in nature. Moreover, the author-practitioner is clearly thinking of resurrection when he writes about the silkworms, on fol. 53v, that they "remain 4 or five days resting without eating, as if they were dying to be reborn again." Other writers at this time described the silkworm explicitly in terms of resurrection and rebirth, and also capable of bringing wealth. He and his contemporaries clearly prized silkworms not just for the substance they produced, but for their symbolic uses as metaphors of regeneration and resurrection.

The "Algiers" recipe also involved a transformation through death and rebirth: the silkworm eggs are supposed to hatch in the enclosed flask as they are fed on egg yolks. When the sole surviving silkworm grows to resemble a serpent, and then, as its wings begin to sprout, metamorphoses into a basilisk, it is burned, producing a toxic vapor, and its ash is to be used to transmute metals into gold. The toxic fumes generated by burning the surviving silkworm in the flask are significant in the light of another recipe in Ms. Fr. 640 that also produces a potent agent, a "horrible poison," by enclosing lowly creatures—in this case, snails—within a flask. This recipe, which spans folio 55r–v, calls for allowing snails to putrefy into a poison that will kill a person when "spread on a board over which the person walks, or on stirrups." Although snails rather than silkworms are involved in this particular recipe, it contains several parallels to "Algiers": the snails are sealed in a glass bottle, fed boiled egg yolks, and kept in the bottle buried in horse manure until the substance turns into a powerfully toxic substance. Like "Algiers," the preparation of this poison also happens around midsummer, "in the month of June and July."[13]

These recipes for enclosing silkworms and snails in flasks to transform them into powerful agents of transformation appear to be related to those in the twelfth-century metalworking account by the pseudonymous Theophilus in which basilisk powder is the key ingredient for making "Spanish Gold," as discussed in chapter 3. In that recipe, basilisks are raised in sealed brass vessels. They develop from chicks hatched from eggs laid by cocks, growing serpent tails in the process, before being burned into ash, from which "red gold" can be produced. Similar accounts in the *Rechter Gebrauch der Alchimei* (1531) also lay out a process of sealing lizards in a vessel to create gold. There are many similarities between the recipes in Ms. Fr. 640 for the work of Algiers, the "horrible poison," and the processes of creating gold and gold-producing agents in the texts of Theophilus and the *Rechter Gebrauch*: enclosing low creatures in a flask, feeding them foodstuffs not normally eaten by them (eggs for silkworms and snails; brass, dirt, and milk for lizards), and transforming the animals into highly valuable or inordinately poisonous substances through a process of putrefaction and regeneration. On fol. 98r, the author-practitioner explicitly references a relationship between metals, lizards, and gold: "Leadsmiths say that making a lizard die in the melted tin it makes the tinning become very golden," adding more prosaically, "Or else putting in sal ammoniac."

No earlier European metalworking texts make use of silkworms as a powerful alchemical agent to produce gold. Instead, the author-practitioner's use of silkworms has an intriguing correspondence to a body of knowledge from China that dates back to 1400 BCE, the concept of *Gu* (or *Ku* in the Wade-Giles system).[14] *Gu* refers to procedures associated with certain—usually minority—cultures in the southwestern regions of China. These procedures had the aim of producing wealth or poison, and worked by sealing creatures such as poisonous snakes and insects in a vessel, where they consume each other until there is but one survivor, called the *gu*.[15] The traditional Chinese character for *Gu* is succinctly captured in the strokes of its written form. *Gu*, in fact, is formed from two radicals, *Chong* and *Min*: 蟲 (*chong*) + 皿 (*min*) = 蠱 (*gu*). *Chong* refers to small, short-lived creatures between worms and insects,[16] and *Min* means vessel or receptacle. The *Gu* character is formed by the character for *Chong* in its upper register, while *Min* constitutes the lower register. Thus, the character itself embodies the placing of worm-bugs in a vessel for the process of *Gu*.[17]

Around the twelfth century, *Gu* began to be associated with another word: *Jin can* (金蚕), which translates to Golden (金) Silkworm (蚕). The concept of the Golden Silkworm predates its associations with *Gu* as "the silkworm that generates silk thread, literally spitting out money." Xu Xuan's (916–91) *Jishen lu* (Records of the Investigation of Spirits) "tells of a man finding (and discarding) a gold-colored silkworm larva inside of a perfectly round stone."[18] Twelfth-century Chinese scholars classified several varieties of *Gu* based on the surviving creature from the vessel, of which the Golden Silkworm (*Jincan*) was the most powerful.[19] The twelfth-century author Cai Tao, a scholar and official in the Northern Song Dynasty, was the first to record the practice, noting that the Golden Silkworm originated from Sichuan and spread to provinces including Hunan, Hubei, Fujian, and Canton.[20] From that point on, a variety of texts began to identify and classify the *Jincan*, including medical texts that provided antidotes for its harmful properties. One text from the Southern Song Dynasty, most likely from the first half of the thirteenth century, contains a concise and informative account the golden silkworm:

The silkworm has a golden color and feeds on silk textiles from Sichuan. Its excrement is used to poison people […] Whoever eats this will die. The silkworm can bring fortune to its owner, and make him rich. It cannot be killed by water and blade. [If one wants to get rid of it, one must] place the silkworm into a gold or silver vessel and leave it on the road. Whoever picks it up will become its new owner. This is called "marrying out the golden silkworm."[21]

These accounts, like the silkworm and poison recipes in Ms. Fr. 640, make clear the dual nature of these powerful agents of transformation—they are capable of bringing good fortune or catastrophe, and can act as both poison and a source of increase.

Another similarity of the "Algiers" process and *Gu* practice can be found in the timing of the processes: recall that in "Algiers," the silkworm eggs are to be sprinkled on top of chicken egg yolks in a flask and buried under manure around the feast of Saint John, the date associated with the summer solstice in the Gregorian calendar. The traditional Chinese *Gu* preparations occurred on the "fifth day of the fifth month,"[22] which is the Duanwu festival, the date associated with summer solstice in the lunar calendar.[23]

The intriguing recipe, "The Work of Algiers," seems to indicate yet another trace of the material imaginary that associated red, gold, vermilion, lizards, and the transformation of metals—this time apparently traveling across Eurasia from China to Europe along with silkworm cultivation. Part of a knowledge system that sought to explore and explain material transformation, this material imaginary seems to have been created by the amalgamation of significant materials and practices of cultivation and transformation as it moved from China, through the Islamicate world of North Africa and Spain, to Toulouse in southern France, where it crystallized in writing in a recipe compilation of a writing artisan.[24] In this imaginary, the silkworm, a thing variously valued and meaningful in different spheres of human interaction—agrarian, economic, social, and religious—also appears to have functioned as a material metaphor that enacted metamorphosis and ennoblement. Such amalgams contain written and tacit knowledge, theories and practices. They involve materials that are meaningful and productive in nature and culture (or, perhaps better, nature-culture).

Conclusion

The global account of the metalworkers' material imaginary I have provided here suggests that the lived experience of metalworkers in sixteenth-century Europe emerged from a long-term itinerary of materials and trade goods, such as pigments and silk. Over the long history of these goods and materials being cultivated and worked within different communities, a "material complex" of written texts and material practices—of making and knowing—formed around them and journeyed with them as they moved. What we see written down in the sixteenth-century European metalworking texts is just the tip of an iceberg in a process of knowledge formation. Following the processes of amalgamation and agglomeration by which such material complexes and their material imaginaries emerged can provide a new understanding of how human beings produce knowledge. By focusing on the material dimensions of the human engagement with matter over the deep human past, and by following the flows of material objects and techniques, we can delineate the formation of knowledge systems as they emerged from material, social, and cultural fields. A historical analysis that begins with natural materials, then follows them through their reciprocal interactions with human bodily practices into objects that are given meaning, used, consumed, desired, and studied by human communities, can be illuminating. Indeed, each of these stages—the materials, the human-material interactions via skilled practice, and the objects and their meanings in production, use, consumption, and in their afterlives—can form whole, self-contained sites of study and analysis. Today, researchers in different disciplines share the view that mind and hand are not separate in human cognition and action; however, we do not have a concept or vocabulary for the amalgam formed by the actions of brain, mind, and body. If we agree that making and knowing are an inseparable whole, then new accounts of where mind and hand intersect in the interface with the material world—material histories—may be able to bring them together to provide a foundation for thinking and writing in non-dichotomous ways about mind-hand knowledge and action.

As we have seen throughout this book, writing down craft and skill has been employed almost as long as writing itself has existed in order to make arguments about the relationship of mind and hand, as well as to argue ethical, political, intellectual, aesthetic, and economic positions. Besides being an account of the writing down of craft and skill, this book is also a recognition of the use of craft in making arguments of many kinds for the reform of knowledge and society. As such, I recognize that my call to question the binaries of mind and hand joins a long tradition. At the same time, I hope that the one of the goals of this book—that of recognizing the amalgam of mind and hand in human actions and capacities, and working toward a fuller understanding of "thinking with the hands"—will be implemented, not just in university training, as the Making and Knowing Project has done, but also more generally in valuing, supporting, and celebrating the trades and training by apprenticeship as alternatives to the text- and test-based learning at present ascendant in our culture. *Longue-durée* material histories that use new sources of evidence and modes of analysis, in tandem with new pedagogical practices, can bring hand and mind together to attain a richer, more unified understanding of human experience and action today and in the past.

Acknowledgments

The remark of brilliant sculptor-scholar Elizabeth King that "only process saves us from the poverty of our intentions" aptly describes for me the past decade of researching and writing this book. Its composition in fits and starts was enmeshed with the prodigious adventure of the Making and Knowing Project (www.makingandknowing.org). The project grew out of research for this book, but soon took on a life of its own. The collective creation by the Making and Knowing Project of a critical edition of the remarkable manuscript compilation of more than nine hundred sixteenth-century artisanal techniques, BnF Ms. Fr. 640 (published in 2020 as *Secrets of Craft and Nature in Renaissance France. A Digital Critical Edition and English Translation of BnF Ms. Fr. 640*, https://edition640 .makingandknowing.org), ended up informing two chapters and many parts of the present book.

The extraordinarily rich and all-consuming six years of bringing the digital edition into being, and the much longer period of writing this book, demonstrated Elizabeth's wisdom in myriad ways. My intentions had indeed been limited by the conventional format of a printed book, by my own training as a historian normally working solo on research, by the time and energy that collaboration across disciplinary divides consumes, and by the familiar and always daunting sense (which every researcher feels) that the more I researched, the less I knew. I now recognize that this book has only scratched the surface of a largely unsounded reservoir of writings about artisanal process that needs infinitely more practical and digital research to explore fully. As the interim product of that process, this book aims to provide a survey of and short immersion in that reservoir. I hope it will serve to spark the interest of, and be of use to, my many dear young colleagues and students whose intellectual lives I have been privileged to share over the last decades. May all these wonderful collaborators transmit the promise and excitement of this material to their own students, apprentices, and audiences! I hope, too, that the Making and Knowing Project and its digital critical edition of Ms. Fr. 640 will successfully convey the sheer adventure of process and of improvisation that hands-on work involves, and provide a template of sorts for how to pursue research of this kind through pedagogy. There is nothing better in scholarship than researching alongside students and young scholars.

We all worked together intensively, mostly joyously, and often hilariously on *Secrets of Craft and Nature* for six years of paleography and text workshops, laboratory seminars, working group meetings, digital humanities course, and any number of spin-off projects (including augmented reality annotation tools, high school chemistry experiments, and an extremely intense but ultimately futile three-day pursuit of historical ruby-red glass). I was so privileged to have been bound up in the all-consuming communion around this unique manuscript. I worked and learned in daily contact at different times with Ann-Sophie Barwich, Tonny Beentjes, Donna Bilak, Jenny Boulboullé, Terry Catapano, Philip Cherian, Colin Debuiche, Soersha Dyon, Hannah Elmer, Steven Feiner, Alex Gil, Clément Godbarge, Alexis Hagedorn, Pascal Julien, Jo Kirby, Joel Klein, Andrew Lacey, Siân Lewis, Lan Li, Margot Lyautey, Sara Muñoz, Jef Palframan, Miriam Pensack, Sophie Pitman, Naomi Rosenkranz, Claire Sabel, Marc Smith, Ad Stijnman, Caroline Surman, Tillmann Taape, Dennis Tenen, Tianna Helena Uchacz, and Heather Wacha. The students who took part in various aspects of Making and Knowing activities were a constant source of stimulation and inspiration. It was just plain wonderful to work with you all—you know who you are!

A multitude of other collaborators contributed to the project, and all can be found in the credits page of the digital edition: https://edition640.makingandknowing.org/#/content/about/credits. The depth of my gratitude to all is profound. I acknowledge with heartfelt thanks the supporters of the Making and Knowing Project, including Columbia University, in particular, David Madigan; the National Science Foundation (grants nos. 1430843, 1734596, and 1656227); the National Endowment for the Humanities (grant no. RQ-249842–16); the Science History Institute; the Henry Luce Foundation; the Gerda Henkel Foundation; the Gladys Krieble Delmas Foundation; the Florence Gould Foundation; the Maurice I. Parisier Foundation; and Howard and Natalie Shawn. For support of text editing and illustrations for this book, I gratefully acknowledge the Schoff Fund of the University Seminars at Columbia University. Material in this book was presented to the University Seminar on the Renaissance.

Long before the Making and Knowing Project, I realized how fruitful it was to spend time with museum curators, artists, and practitioners. For the year I spent as a Fellow at the Getty Research Institute, I owe a long-time debt to Jane Bassett and Malcolm Baker. During that time, Amy Meyers and I were also conspiring to bring together scholars and practitioners from different disciplines to converse and think around historical objects, knowing how much light could be shed on the relationship between making and knowing by doing so. Together with Hal Cook, we concocted an unforgettable conference held in several different institutions in London in 2005, *Ways of Making and Knowing: The*

Material Culture of Empirical Knowledge, subsequently published as an edited volume. I continued such thinking with objects during the year I spent at the V&A Museum, generously supported by a New Directions Fellowship from the Andrew Mellon Foundation, with Marta Ajmar, Malcolm Baker (to whom I owe special gratitude for helping set up my residency at the V&A), Peta Motture, Giorgio Riello, Carolyn Sargentson, and John Styles. In 2007, Tonny Beentjes and I spent an intensive summer examining as many extant life-cast objects in European museums as we could locate, and then worked on and off together for the following four years puzzling through Ms. Fr. 640 to try to understand if the objects we examined could be reconstructed from the instructions in the manuscript. I am grateful to Elizabeth Cropper, Peter Lukehart, Therese O'Malley, and the Center for Advanced Study in the Visual Arts for supporting our research with a Samuel H. Kress Paired Research Fellowship. I remain deeply indebted to Tonny for all his knowledge, observations, and good collaboration.

My wonderful colleagues have helped me think through materials, including, at Columbia, Dianne Bodart, Michael Cole, Martha Howell, Matt Jones, Joel Kaye, Dorothy Ko, Alessandra Russo, Avinoam Shalem, and Kavita Sivaramakrishnan; and, at Bard Graduate Center, Ivan Gaskell, Deborah Krohn, Peter Miller, Andrew Morrall, and Ittai Weinryb. My co-editors and co-authors of the volume *The Matter of Art: Materials, Practices, Cultural Logics, c. 1250–1750* (2015), Christy Anderson and Anne Dunlop, and the contributors to that volume, as well as the panelists in the series of panels at the memorable 2010 Renaissance Society of America meeting in Venice, also stimulated me to think more deeply about the properties of materials and the reciprocal dynamic between makers and materials.

A year of research at the Davis Center, Princeton University, in 2009–10 gave me the opportunity to begin thinking about the movement of materials and artisanal practices across long temporal and geographic distances, which Lorraine Daston and the Max Planck Institute for the History of Science, Berlin, further generously supported as a 2015–17 working group, which resulted in an edited volume, *Entangled Itineraries of Materials, Practices, and Knowledges Across Eurasia* (2019). I am deeply indebted to Raine not just for support of the working group, but for being such a rigorous, stimulating, and constructive interlocutor for my entire career. In the same realm of rigorous, stimulating, and constructive interlocutors, as well as just plain good friends, I am immensely fortunate to count Christine Göttler, Pamela O. Long, and Christia Mercer, who read versions of this entire book. I have spent wonderful hours with Ann-Sophie Barwich (a font of useful bibliography for chapter 5), Rachel Berwick, Francesca Bewer, Marjolijn Bol, Karen Brudney, Caroline Bynum, Carol Cassidy, Sven Dupré, Paula Findlen, Erma Hermens, Vera Keller, Elizabeth King, Ann-Sophie Lehmann, Annapurna Mamidipudi, Alexander Marr, Christina Nielson, Ulinka Rublack, Linda Seidel, Nussara Tiengate, and Jane Wildgoose.

For giving me a sense that this book was worth writing, and for much feedback, I have many audiences of lectures and workshops to thank. I want to single out Cris Miller, Mary

Terrall, Helena Wall, and Peggy Waller, who, without knowing it, were present at the first formulation of this book for the Ena Thompson Lectures at Pomona College. Among the numerous other institutions and audiences for the material of this book, I also want to note the short residency I spent at the Toronto Center for Medieval and Renaissance Studies, which gave me the good fortune of meeting subsequent Making and Knowing Postdoctoral Scholar Tianna Helena Uchacz.

The last ten years have been a blur of "process," and of learning how much more consequential it often is than "product." I hope, however, that the publication of this book as product will allow time for more relaxed exchange with all these good friends and brilliant scholars.

I would be remiss if I did not give heartfelt thanks to the assistants who toiled over the editing of many different drafts of this book, including especially Alexander Lash. At different times, Shabnam Shirazi Aslam, Josue Lecodet, Christian Macahilig, and Kelsey Troth also worked on various tasks. At the University of Chicago Press, I am thrilled to have worked with Susan Bielstein, James Toftness, Caterina MacLean, and Rebecca Brutus, and I am very grateful to Lys Weiss of Post Hoc Academic Publishing Services for her careful copyediting.

Finally, I acknowledge with pleasure my great debt to my loved ones, near and far, who were with me through all these years, including the world-changing pandemic. My scholarship was, as ever, my solace during this strange time. Another form of solace came from my parents, Ron and Nancy Smith; my siblings, Jenny Rowe and Scott Smith; my beloved children and grandchildren, Muki and Ady Barkan and Rachael, Willow, and Carl King; and my steadfast partner, Zori. I dedicate this book to my family.

Notes

INTRODUCTION

1. Quoted in Schneider 2009, 124.
2. Plato 1953, 520 (274c–75b).
3. Cellini 1967, 19.
4. Moxon 1970, Preface.
5. Diderot 2003; Diderot [attrib.] 1751. For more on these social dynamics, see Hilaire-Pérez, Simon, and Thébaud-Sorger 2016; Belhoste 2012; De Munck 2007; Jones 2016; and Bertucci 2017.
6. Ong 1986.
7. A point also made by Keller 2001, 35; and Hilaire-Pérez, Nègre, Spicq, and Vermeir 2017b.
8. For overviews, see Zilsel 2000; Halleux 2009; and Long 2015.
9. For overviews of the literature, see Pappano and Rice 2013; and De Munck 2019.
10. An exception is Polanyi 1998; and Polanyi 2009.
11. For an overview, see Ingold 1994; Sigaut 1994; Clark 1997; Ribard 2010; Malafouris 2013; Schatzberg 2018. An outstanding contribution by historians of science came out as I was completing this book: Leong, Creager, and Grote 2020.
12. Sutton 2007; Tribble and Sutton 2012; Wolfe and Gal 2010; Glatigny 2017; Collins and Evans 2006; and Myers 2008.
13. Bouillon, Guillerme, Mille, and Piernas 2017; and brilliantly illustrated by Hilaire-Pérez 2013.
14. Ingold 2000. Interestingly, in a 1931 essay, George Sarton, founder of the History of Science Society and its journal, *Isis*, called for viewing science as a part of human cultural evolution, to write a history of science that was the history of, as he termed it "human civilization" itself. See Sarton 1956. Sarton's call was neglected by subsequent generations of historians.
15. See Coward 2016; Iliopoulos and Garofoli 2016; Tallis 2003; and Ingold 2000.
16. Tarule 2004, chap. 1.
17. I treat this subject at more length in Smith 2011. See also Lecain 2017.
18. Merton 1961; reprinted in Merton 1973, 343–70. More recently, scholarly and popular accounts describe the collective nature of invention and innovation, see Pacey 1992; Conner 2005; and Sawyer 2007.
19. Chaplin 2008.
20. Mukerji 2006.
21. Watson-Verran and Turnbull 1995.
22. Bray 2008.

23. Tim Ingold (2000, 312) writes that to use the word "science" is not to denote a thing but to make a claim—a claim that has to do with the supremacy of the human mind and abstract reason.

24. Bray 2008, 320–21. Thomas P. Hughes introduced the view of technology as a "system," especially in Hughes 1983. On the definition of technology, see Edgerton 1999.

25. BNF, Ms. Fr. 640, fol. 79r.

26. Hermann Weinsberg (1518–97), quoted in Jütte 1988, 260.

27. Debuiche 2020.

28. See, among others, Sennett 2008; Schon 1995; Davidson 2012; Korn 2015; Crawford 2010; and Kuijpers 2019.

29. For example, Epstein and Prak 2008.

30. For an overview, see Adamson 2010.

CHAPTER 1

1. Eccles. 38:24–25 (King James Version). Parts of this chapter are drawn from Smith 2010.

2. Quoted in Lloyd and Sivin 2002, 16. Although prejudice against handwork was pervasive in Europe and China, anthropologists argue it is not universal: Howes 2005, about bodily intelligences valued by the Peruvian Cashinahua; and Geurts 2005.

3. Barney 2009.

4. Tell 2011.

5. Lis and Soly 2012, 43.

6. Löhr 2008, 160.

7. Emma Capron informed me that the signature on the underside of the wing would have been very difficult for a viewer to see, thus the signature and the object may have possessed a votive dimension.

8. Averlino 1965. For more on the self-consciousness of artists, see Ames-Lewis 2000.

9. Aristotle 1933, 1.980a–81b.

10. Cuomo 2007, 13.

11. Aristotle 1933, 1.981b.

12. Aristotle 1999, VI.1–8.

13. Aristotle 1944, 1.1258b.

14. Aristotle 1944, 1.1260a. For the social and political context of this attitude, see Lis 2009.

15. Parry (2014) sees this emerge fully in Aristotle's *Posterior Analytics*.

16. Parry 2014. For the practice of *technē* in the Athenian republic, see Long 2001, chap. 1.

17. Cuomo 2007, 29.

18. Cuomo 2007.

19. Quoted in Lis and Soly 2012, 78.

20. Daiber 1990, 102–4.

21. Kheirandish 2007, 948–49.

22. Ovitt 1983, 97.

23. Druart 2020.

24. Ovitt 1983, 99.

25. Truitt 2015, 47.

26. Hackett 1997, chaps. 8 and 12; Ingham 2015, 71.

27. On this reevaluation of handwork: Geoghegan 1945; Le Goff 1980; Whitney 1990; and Ovitt 1987.

28. For these developments, see Ovitt 1987, Whitney 1990, and Summers 1987; Long 2001; and Long 2000.

29. Michael of Rhodes 2009.

30. For these assessments, see Michael of Rhodes 2009, vol. 3.

31. Amelang 1998, 48.

32. Amelang 1998, 124–25. Amelang points out that artisanal works before Bräker's had no model of *Bildung*, that is, a goal of interior transformation. The notion of writing as a private textual space is a modern conception (246).

33. Amelang 1998, 169.

34. Chartier 2011. Chartier remarks on journeyman glazier Jacques-Louis Ménétra's journal (composed c. 1764–1803), which allowed Ménétra to recompose his own life through the prisms of the books he read (139).

35. On that power, see Goody 2000; and Goody 1987.

36. Franci 2009, 133.

37. Stahl 2009, 87–91.

38. McGee 2009, 211–41, this discussion 238–41. See also Shelby 1997.

39. Bondioli 2009, 273–74.

40. Reith 2005.

41. Franci 2009.

42. Wallis 2009, 313.

43. Goody (1977) argues that writing develops abstract thought and perhaps even gives rise to cognitive change through fostering an internal monologue.

44. Long 2009, 12–20.

45. Franci 2009, 140–42.

46. Wallis 2009, 318.

47. Long, McGee, and Stahl 2009, vol. 2, 321 (fol. 110a).

48. Cifoletti 2006b; and Cifoletti 2006a.

49. Lorenz Lechler, *Unterweisung*, manuscript written in 1518 for his son, in Shelby and Mark 1997, 89.

50. Frankl 1945, 46–47.

51. Prak 2011.

52. Biringuccio 1943, 280.

53. Long 2009, 31–32.

54. This point is made by Piero Falchetta (2009, 197) with reference to the portolans in Michael's book.

55. Biringuccio 1943, 145.

56. Biringuccio 1943, 401.

57. Kastan 2002, 39. See also Chartier and Stallybrass 2013.

58. Chartier 1994, 32–37. Also Carruthers 2008, 262–75.

59. Chartier 1994, 52.

60. In some genres, print prolonged the tradition of the composite collection: Chartier 1994, 56. Crowther-Heyck (2003) notes the compilatory nature of vernacular works of natural history. Printers often compiled practical books and added sections to existing ones in order to create something different from their competitors: see the history of Brunschwygk's *Cirurgia* of 1497 in Sudhoff 1908. See also Sigerist 1941. Amelang (1998, 144–48) notes that artisan autobiographies are often compiled in no particular order, which he regards as the writer's intention not to "author," but rather to "witness," which could also account for their illustrations.

61. Blair 1999, 225.

62. Carruthers 2008, 264.

63. Cole and Watts 1952, 14. The British Society's History of Trades project was unsuccessful; the French version did not come to full fruition until 1761–88.

64. Amelang (1998, 120–21) comments that those few artisans who speak directly of their craft in their books link it to personal identity and self-worth.
65. Cennini 1960, 131.
66. Cennini 1960, 3.
67. Cennini 1960, 60–61.
68. Cennini 1960, 65.
69. Cennini 1960, 113.
70. Michael of Rhodes 2009, fols. 111b, 183a, 204a, and 185a, in vols. 1 (transcription) and 2 (translation). The first examples of double-entry bookkeeping in fourteenth-century Italy began "in the name of God and profit," and Luca Pacioli's (1447–1517) 1494 treatise on bookkeeping advised every merchant to begin his account book in the name of God. Heading each page with a cross could also have apotropaic effects, resembling textual amulets written on paper and carried with a person. Skemer (2006) discusses the use of the cross to bring divine protection (89–90), and the use of crosses in textual amulets (172–93).
71. Cennini 1960, 1. Vernacular works often used the story of Creation as an organizing principle. See Crowther-Heyck 2003, 260–61; and Bennett and Mandelbrote 1998.
72. "*Bitte(n) got für hanssen muoltscheren vo(n) riche(n) hofe(n) burg(er) zu ulm haut daß Werk gemach do ma(n) zalt mcccc xxx vii,*" quoted by Brandl 1986, 58.
73. In a study of funerary monuments designed by three fifteenth- and sixteenth-century artists for their own memorials, Irving Lavin (1977–78) makes the point that each of the monuments was self-consciously innovative and difficult to execute, and involved great bodily labor, with each artist clearly identifying this bodily sacrifice with the passion and body of Christ.
74. Broecke 2015, 190, 192. I also treat this in Smith 2004.
75. Kruse 2000. See also Lehmann 2008.
76. Crowther-Heyck 2010, on the interpenetration of spiritual and bodily realms in early modern Europe.

CHAPTER 2

1. Baxandall 1966. Parts of this chapter are drawn from Smith 2014.
2. Baxandall 1966, 140.
3. Smith 1943, xiv.
4. See Dormer 1994, 8–10.
5. Valérie Nègre (2016) provides a list of the various terms. See Watson-Verran and Turnbull 1995; and Egmond 1999. See also Adamson 2018. Pickstone (1997, 102) called for such studies: "we need to put up-front the logic of pre-industrial craft products, and the knowledges within which they were presented. These logics are definable and persistent." See also Geertz 1983. Diverse efforts to account for artists' knowledge are provided by art historians such as Fehrenbach 1997; Cole 2002; Löhr 2008; Suthor 2010; Leonhard 2013; Felfe 2015; and Neilson 2019. Other notable efforts to identify a philosophy in the work of artisans include Kamil 2005; Prown and Miller 1996; and Spike Bucklow 1999, 2000, and 2001, a series of articles entitled "Paradigms and Pigment Recipes."
6. Cellini 1967, 95, 104.
7. Theophilus 1986, 112.
8. Albertus Magnus 1967, 196. Pliny (1952, 93): "Persons polishing cinnabar in workshops tie on their face loose masks of bladderskin, to prevent their inhaling the dust in breathing, which is very pernicious, and nevertheless to allow them to see over the bladders."
9. Ms. Fr. 640, fol. 79v.

10. Ms. Fr. 640, fol. 123v.
11. Van Schendel 1972, 77–78. See also Barclay 1992, 100.
12. De Avila 1531, chap. XLVIII.
13. Paracelsus 1941, 112.
14. Biringuccio 1943, 81.
15. Kertzenmacher (1538, xxxxviii recto) lists the signs of the zodiac (what he calls *"Planeten"*) under which the seven metals should be worked, for example, gold should be worked under Aries and Leo, not under Aquarius and Libra, while silver should be worked under Taurus and Cancer but not under Scorpio and Capricorn. Paracelsus (1941, 113) saw the influence of the planets operating like the tincture of the dyer, except that it was invisible.
16. Biringuccio 1943, 27.
17. Aristotle wrote briefly about two exhalations that emerged from the earth: a dry smoke that produced stones and a watery vapor that produced metals. Medieval authors including al-Razi, Geber, and Albertus Magnus amplified Aristotle's brief reference in *De meteorologia*. See Albertus Magnus 1967, xxx–xxxi.
18. Dym 2006, 169–72.
19. Dym 2006, 172.
20. Sisco and Smith 1949, 39–40.
21. Sisco and Smith 1949, 34–35.
22. Sisco and Smith 1949, 47.
23. Bucklow 2001b, 451.
24. Theophilus 1986, 81. This is noted by Bucklow 2001b, 452.
25. Cited in Schürer 1994, 64.
26. Biringuccio 1943, 373–74.
27. Biringuccio 1943, 114.
28. Michael of Rhodes 2009, vol. 2, 321, fol. 110a.
29. Michael of Rhodes 2009, vol. 2, 323–24, fol. 111a–b.
30. For an introduction to the view of the body as permeable to celestial influences, see Paster 2004; and Floyd-Wilson 2013. Walker (2015) shows how vernacular texts, such as almanacs, fostered such a view.
31. Biringuccio 1943, 17. See also *Das Schwazer Bergbuch*, a compilation of mining ordinances, mining information, and images from c. 1490–1556; and Bingener, Bartels, and Slotta 2006.
32. Piccolpasso 1980, 109 and 89.
33. Schreiber 1962, 79–80.
34. Daniel der Bergverstendig 1533, fol. 3v.
35. Johannes Mathesius, a pastor in St. Joachimsthal in the 1560s, mentioned that miners had given him all kinds of minerals and rock specimens: Mathesius 1562, Vorrede, n.p. See also Schreiber (1962, 621–22), who gives examples of mineral specimens used as votive offerings and in church ornament in the sixteenth through eighteenth centuries, as well as a miraculous mineral specimen that exhibited an image of Mary with the baby Jesus, found in 1669. For an introduction, see Huber 1995; and Haug 2012.
36. Biringuccio 1943, 13.
37. In an inventory, a stone is described as illustrating "God's gift." Slotta, Bartels, Pollmann, and Lochert 1990, 566, quoting from 1596 inventory of Ferdinand II's *Kunstkammer* in Schloss Ambras. *Handsteine* also reflected the political importance of the mines to their noble collectors; see Haug 2012, 51.
38. Mathesius 1562, fol. LXXXVIII recto, quoted by Haug 2012, 53.
39. Song printed in 1530, quoted in Slotta, Rainer, Bartels, Pollmann, and Lochert 1990, 562–63.
40. Paracelsus 1941, 112.

41. Paracelsus 1941, 108–9.
42. Dym 2006, 180–81.
43. Biringuccio 1943, 91.
44. H.G., "Goldsmith's Storehouse," c. 1604, fol. 59r.
45. Anon. 1507, Part V, n.p.
46. Kertzenmacher 1613: "in the book of nature," mercury is shown to be cold in its "complexion" and effects. The metallic vapors were far more harmful than the metals and minerals themselves. For mining diseases due to *Witterung* and bad airs, see Sahmland 1988, 260 and following.
47. Paracelsus 1941, 116.
48. Daniel der Bergverstendig 1533. In 1538, Kertzenmacher, reprinted this text, beginning on fol. xxxix verso as "Von den gifftigen bösen dämpffen und räuchen der Metal." For effects of musk, see fol. xL verso. *Rechter Gebrauch* also warns about letting the smoke penetrate the body.
49. Banckes 1941, 26. For the many virtues of the variously named *alantwurtzel*, scabwort, *enula,* or *elene campana*, which was hot in the third degree, and moist in the first, see von Cuba 1515, fol. LXIII verso.
50. *Artzneybuch* 1556, 48r.
51. Banckes 1941, 12.
52. Copp 1521, fols. 21v–22r; for another use of garlic, see Smith 1968, 8–9.
53. Kertzenmacher 1613, fol. xLi verso. Among other protective substances, he lists garlic, juniper berry, and *biesam* (myrrh or amber).
54. *Artzneybuch* 1556, fol. 48r.
55. *Artzneybuch* 1556, fol. 50v.
56. *Artzneybuch* 1556, fol. 50r.
57. *Artzneybuch* 1556, fol. 53v.
58. *Artzneybuch* 1556, fol. 53v.
59. de Avila 1531, chap. XLVIII. Note that we still call a phlegm-associated illness a "cold" and still treat it with hot, dry honey. Noted also in Albala 2002, 82.
60. For the medieval prehistory of ideas about balance, see Kaye 2014.
61. *Artzneybuch* 1556, fol. 54v.
62. Albala 2002 chronicles the long debate about order of a meal, in which many thought that corruptible food should come first, such as milk or fresh cheese, and others believed butter should start a meal, because it "doth swim above in the brinks of the stomach; as the fatness doth swim above in a boiling pot, the excess of such superfice will ascend to the orifice of the stomach, and doth make eructions" (Andrew Boorde, *A compendyous Regyment or a Dyetary of Helth*, 1542, 46, quoted in Albala 2002, 109).
63. Albala 2002, 87: bread a perfectly tempered food.
64. Andrew Lacey, conversation in a bronze-casting course, led by Lacey and Francesca Bewer, July 25, 2004.
65. Knop 1957.
66. Webster and Pritchett 1924, 400–402.
67. I thank chemists Wayne Steinmetz and Nancy Hamlett for this information.
68. "A child who gets enough iron and calcium will absorb less lead. Foods with iron include eggs, lean red meat, and beans. Dairy products are high in calcium." "Lead Poisoning and Your Children," EPA Publication EPA-747-F-16-001 (December 2017), https://www.epa.gov/sites/production/files/2018-02/documents/epa_lead_brochure-posterlayout_508.pdf.
69. See Gentilcore 2004 on historiography of scholarly and popular medicine, who posits a model of overlapping spheres. In the fifteenth and sixteenth centuries, numerous medical compilations of school medicine and popular practices were published in the vernacular. See Payne 1901; and Ebel 1939.

CHAPTER 3

1. Campbell, Foister, and Roy 1997, 7. On illustrations of workshops, see Heine 1996, 20–27.
2. Taking inspiration from Sewell 1999, 58.
3. Robert Blair St. George, summarizing Jules David Prown's essay, "Mind in Matter: An Introduction to Material Culture Theory and Method," in St. George 1988, 17. See also Bernasconi 2016.
4. Ms. Fr. 640, fol. 142v, "Molding grasshoppers and things too thin." See Fu, Zhang, and Smith 2020.
5. Parts of this chapter are drawn from Smith 2014.
6. Techniques of life-casting in the sixteenth century did not always necessitate the loss of the mold, or even of the pattern (the plant or animal). See Smith and Beentjes 2010. See also Smith and Making and Knowing Project 2020; and Lacey and Lewis 2020.
7. Ms. Fr. 640, fol. 109r–v. For more on reptiles, see Demeter 2020.
8. Ms. Fr. 640, fol. 109r–v.
9. Ms. Fr. 640, fol. 105v.
10. Ms. Fr. 640, fol. 105v.
11. Ms. Fr. 640, fol. 120r–v.
12. Ms. Fr. 640, fol. 68r.
13. Ms. Fr. 640, fol. 84v.
14. Ms. Fr. 640, fols. 85v–86v.
15. Ms. Fr. 640, fols. 131r and 85r.
16. Ms. Fr. 640, fol. 85r (in the margin).
17. Ms. Fr. 640, fol. 88v (in the margin).
18. Biringuccio 1943, 143–44.
19. On Palissy, see Amico 1996; Shell 2004; Kemp 1999, 72–78; and Palissy 1988, with introduction by Keith Cameron.
20. Smith 2004, 100–110.
21. Palissy 1957.
22. Palissy 1957, 233–40.
23. Biringuccio 1943, 114.
24. Biringuccio 1943, 114. See also Crowther-Heyck 2003 on the fundamental connection in early modern Europe that existed between the study of created nature and piety. As Jacob Horst wrote in 1579: "although God is invisible, […] He can be recognized and understood through the natural creations and through the earth, which is so exquisitely made." Jakob Horst, *Von den Wunderbarlichen Geheimnissen der Natur* (Leipzig: Hansz Steinman, 1579), sig. Bvi, cited in Crowther-Heyck 2003, 263. See also Crowther-Heyck 2010.
25. Ullman 2020.
26. For examples of the spiritual depiction of lizards, see Smith 2004, 121–23.
27. See Smith 2004, 117–23. See Karin Leonhard's excellent survey of spontaneous generation in the seventeenth century, in Leonhard 2009–10, 103 and following. The seventeenth-century collection of technical recipes by J. K. (perhaps Johann Kunckel), *Der Curieusen Kunst- und Werck-Schul* (Nuremberg: Johann Ziegers, 1696), includes sections on life-casting followed by sections on spontaneous generation (J. K. 1696, 657–73).
28. Crowther-Heyck 2003, 266.
29. For living creatures in stone, see Agricola 1961, 196–97; Misson 1739, 66–67: "the Workmen that were employed to dig Stone at *Tivoli*, having cleft a great Mass, found in the Middle of it an empty Space, in which there was a living *Crayfish*, that weighed four Pounds, which they boiled and did eat. […] (Al-

exander ab Alexandro. Bapt. Fulgsus mentions a living Worm that was found in the Middle of a Flint)," quoted in Beringer 1963, 196n8. Further mentions of living crustaceans: Edward Lhwyd's *Epistola* to John Ray mentions cockles found in stone, and in the "town of Mold in Flintshire; met with several muscles [mussels] at about three foot depth in the gravel, which had living fish in them" (145), and "toads found some times in the midst of stones at land" (151).

30. The ambivalent attitude toward frogs and toads as both harmful and beneficial is illuminated by Wilson 2000, 150, 417–20.
31. Behrouzi 2002, 70–71. I thank Zohar Jolles for this reference.
32. Geckos, which can amputate their own tails at will in order to distract predators, are studied for understanding the neurons that make up central pattern generators in the spinal cord: see Fountain 2009. The all-female species of whiptail lizards reproduce without males, yet maintain a high level of genetic variation: see Bhanoo 2010, D3.
33. Albertus Magnus 1973, 104. This anonymous book of secrets apparently dates to the fourteenth century; its attribution to Albertus appeared only in fifteenth-century copies. The source for the lizard tail recipes may be the *Book of the Cow*, a ninth-century Arabic compilation of magical experiments, translated into Latin in the twelfth century from an Arabic text that is no longer extant. See van der Lugt 2009.
34. *Rechter Gebrauch der Alchimei* 1531, fol. XIII recto. See Ferguson 1959, part II, 42, for information about the variants of this recipe collection.
35. This and the next recipe: *Rechter Gebrauch der Alchimei* 1531, fol. XV verso.
36. Historians have seen such processes as encoded accounts of metallic processes. Wyckoff, in Albertus Magnus 1967, 247n2, explains the lizard and mercury process as an amalgamation technique for extracting small amounts of gold or silver from pyritic ores: "The crushed mineral was stirred with mercury, in which any free gold or silver would dissolve. After filtering off, the mercury was distilled away, leaving the precious metals." Perhaps the lizard-mercury recipe refers to a process involving marcasites, a lustrous, shiny mineral that contains sulfites, and that, according to Albertus Magnus, was viewed by alchemists as the "principal food with which Quicksilver is fed," from which gold and silver could be produced. Marcasites, like other sulfur-containing minerals—and like lizards—were regarded as generative.
37. Theophilus 1979, 119–20. For additional commentary, see Brepohl 1987.
38. Halleux (1996, 887–88) argues that it is based on Arabic recipes, perhaps part of the Jabirian corpus. Wallert (1990, 161) sees it as an encoded account of alchemical transmutation. Bucklow (1999, 143–47) views this recipe as evidence for the centrality of the vermilion-making process in providing a model for other processes of metallic transformation.
39. Vogtherr and Heilmann 2011: *Kunstbüchlein* 1538, fol. 19v.
40. Biringuccio 1943, 83.
41. Achim 2011.
42. Van Schendel 1972, 77 and following. See also Thompson 1933–34.
43. Kertzenmacher 1613, B recto–Bii verso.
44. *Rechter Gebrauch der Alchimei* 1531, X: "*Rot wasser das zü golt gehört.*"
45. Albala 2002, 80.
46. Albala 2002, 73.
47. Albala 2002, 74.
48. Albertus Magnus 1967, 19.
49. See Bynum 2007, especially chaps. 7 and 8.
50. Michael of Rhodes 2009, vol. 2, 521, fol. 184b.

51. Albertus 1967, 70; *Rechter Gebrauch der Alchimei* 1531, III verso.
52. Kertzenmacher 1613, xviii recto. Kertzenmacher (xxix verso) states that philosophers have hidden the fact that human blood is good for making silver and gold, and that menstrual blood and *"sanguis rubei collerici* [the red blood of a choleric person]" are the best of all.
53. *Rechter Gebrauch d'Alchimei* 1531, fol. III verso.
54. Cellini 1967, 123. In recounting the casting of another work, Cellini again emphasized his ability to resuscitate the dead: "owing to my thorough knowledge of the art I was here again able to bring a dead thing (un morto) to life" (125). I discussed this in Smith 2004, 106–7.
55. Cellini 1956, 343–48.
56. Perhaps for Cellini the process even mimicked the preparation of an elixir of life. See Cole 1999, 222–25. It seems apt that Cellini called the *Perseus* his "book," as noted by Cole (2002, 9). Neilson (2019, 187–91) discusses Verrocchio's interest in animation.
57. Cole 2002, 154–55.
58. Bewer 2001, 182.
59. Albertus 1967, 81.
60. Gage 1998, 39. Perhaps the cross even referred to the making of vermilion in the crucible; the root word of "crucible" in English and Italian is "cross," and it was the crucible in which sulfur and mercury underwent their own passion and transformation to produce the blood-red pigment.
61. Cennini 1960, 95; and Broecke 2015, 193.
62. Kruse 2000 and Lehmann 2008. Bohde (2007, 48) notes that Cennino was unusual in using the word *incarnazione* in fifteenth-century treatises and alone in using the verb *incarnare* to describe the use of the flesh tone.
63. Bohde 2007, 49 (blood and milk), and 51–54 (painting and incarnation).
64. For example, H. G. c. 1604, fol. 55r; Vogtherr and Heilmann 2011: *Kunstbüchlein* 1538, fol. 18v.
65. See Bucklow 1999, 145–47.
66. In the recipe for mosaic gold, Cennino calls for "sal ammoniac, tin, sulphur, quicksilver, in equal parts; except less of the quicksilver." See Cennini 1960, 101–2. The *Liber illuministarum* from Tegernsee monastery, compiled in the fifteenth and early sixteenth centuries and including much older recipes, contains a recipe for golden ink, which specifies tin, sulfur, quicksilver, sal ammoniac, and linseed oil. Only the tin and sulfur are needed to produce gold-colored stannic sulfide. Bartl, Krekel, Lautenschlager, and Oltrogge 2005, 69.
67. Bucklow 1999, 145–47.
68. See Kremnitzer and Smith 2020.
69. Bimbenet-Privat and Kugel 2007; and Grasskamp 2013.
70. Cole 2002, 155.
71. Shapin 2001.
72. Geertz 1983, chap. 4, "Common Sense as a Cultural System," 73–93. Geertz's vision of common sense is of a "thin" sort of knowledge that cannot be analyzed according to ordinary philosophical premises: "If one wants to […] suggest that common sense is a cultural system, that there is an ingenerate order to it capable of being empirically uncovered and conceptually formulated, one cannot do so by cataloguing its content, which is wildly heterogeneous, not only across societies but within them—ant-heap wisdom" (92). Many sociologists of science have studied the overlapping nature of scientific and practical knowledge, e.g., Collins 2001; Barley and Bechky 1994; and Delamont and Atkinson 2001.
73. Biringuccio 1943, 319.

1. For the transition of manuscript recipes into print, and the role of printers, see Reynolds 2019.
2. Numerous technical writings still lie unstudied in archives and libraries. See Zindel 2010; Clarke 2001; Brüning 2004; Hirsch 1950; Ferguson 1959; Darmstaedter 1926; Haberman 2002; Crossgrove 2000; Pereira 1999; Leng 2002; Werrett 2009, chap. 1; Hall 1979; and Crowther-Heyck 2003. See also Eyferth (2010), who deals with craft knowledge in China, but usefully inventories various written forms of craft knowledge. See also Löhr 2011.
3. Eamon 1994, 130, 252; and his essay in Leong and Rankin 2011b. See Celaschi and Gregori 2014.
4. Deblock 2015.
5. See Eamon 1994; Long 2001; and Long 1997. See also Glaisyer and Pennell 2003; Kavey 2007; Vérin 1993; Glatigny and Vérin 2008; Leong and Rankin 2011b; Clarke, De Munck and Dupré 2012; Córdoba 2013; Cardinal, Pérez, Spicq, and Thébaud-Sorger 2016; Wheeler and Temple 2009; Tebaux 1997; Chrisman 1982, 182 and following; Roberts, Schaffer, and Dear 2007; and Valleriani 2017. An essential treatment is Hilaire-Pérez, Nègre, Spicq, and Vermeir 2017a.
6. Ainsworth 2001, 117.
7. Borchert 2002; Borchert 2011; and Diorio 2013, 170–71.
8. Rösler 1998. See also Schotte 2019. Registers of techniques and recipes also survive from eighteenth- and nineteenth-century workshops; see, e.g., Rice 2020; and Grosjean c. 1876–83.
9. Flötner 1549. See Popplow 2004; Popplow 2002; Gerbino and Johnston 2009; and for drawings that formed 3D models, Santucci 2014.
10. De Preester 2011, 133.
11. Hilaire-Pérez and Verna 2006; Prak, Lis, Lucassen, and Soly 2006; Epstein and Prak 2008; and Davids 2008.
12. Tallis 2003, 40.
13. Alcega's *Libro de geometria, practica y traca* (first published in Madrid in 1580 and revised in 1589) was approved by two court tailors, Hernán Gutiérrez, tailor to the late Princess of Portugal, and Juan López de Burgette, tailor to the Duke of Alba. Gutiérrez had to ask Pedro de Goyenechea to sign on his behalf because he did not know how to write. De Alcega 1979, 10.
14. For techniques of modern Liberian tailors, see Lave 1977. See also Kneebone 2020.
15. Goody 1977, 142, regards the widespread circulation of recipes as related to social mobility.
16. For household texts, see Barnard and McKenzie 2002, chap. 24, "Books for daily life: household, husbandry, behavior." For conception, see Bell 1999.
17. Bergdolt 2006.
18. Neilson 2019.
19. Aruz, Benzel, and Evans 2008, 158.
20. Oppenheim, Brill, Barag, and von Saldern 1970, 24–28.
21. Clarke 2013a.
22. Oppenheim, Brill, Barag, and von Saldern 1970, 86.
23. Clunas 1998.
24. Cookbooks mentioned in Plato's *Gorgias* 518b, according to Grocock and Grainger 2006, 43.
25. Formisano 2013. See also Formisano 2001.
26. Formisano 2013, 205.
27. Formisano 2013, 208–9.
28. An interesting example can be found in Albrecht Dürer's large 1518–22 print of the Triumphal Wagon of Maximilian I, in which Dürer adds two extra horses to the original plan, which are personifications

of "experientia" and "solertia," for, as Dürer, writes, "Dann wo die erfarnuß und furrechtigkeyt nit ist mag die Reckheyt und Großmütigkeyt leycht schaden bringen" (Without experience and precision, recklessness and hubris easily cause damage).

29. Daiber 1990, 102–4.
30. Van der Lugt 2009.
31. See Hill 1991; and Abattouy and Al-Hassani 2015.
32. See Hawthorne and Smith 1974; and Clarke 2013b.
33. Oltrogge 1998.
34. Clarke 2011.
35. Ploss 1962; and the Strasbourg collection of recipes in Old Middle German of "Heinrich of Lubeck" and "Andrew of Colmar," in Neven 2016.
36. Merrifield 1967.
37. Oltrogge 1998, 90.
38. Bartl, Krekel, Lautenschlager, and Oltrogge 2005.
39. Harkness 2007.
40. See the Medici Archive Project, http://www.medici.org/.
41. Zimmermann 2014. See also Miller and Zimmermann 2005.
42. Rankin 2013; Leong 2018; Leong and Rankin 2013.
43. On Mayerne, see Boulboullé 2019. On Peiresc, Miller 2019. On other "lovers" (or *amateurs*) of art, see Leonhard 2019.
44. On the Royal Society, see Houghton Jr. 1941; Ochs 1985; Hunter 1989; Iliffe 1995; Fox 2010, chap. 1; Hunter 2013; Fransen, Reinhart, and Kusukawa 2019.
45. Cowan 2006.
46. Schneider 2007 argues that Zedler's *Lexicon* strives to bring to the reader the collective wisdom of the ages, in order to provide the lay reader the tools to make judgments (through experience)—a novel project to make writing into a tool of experience.
47. Koepp 2009 argues that Pluche intended his beautifully illustrated volumes—for which he visited the workshops of artisans—to form a compendium of general knowledge, but he also meant it as a critique of the social order that denigrated manual work and commerce.
48. Cole and Watts 1952, 15, quoting de Lalande 1764. The *Descriptions des arts et métiers* has an extremely complicated publishing history, which is recounted by Cole and Watts 1952.
49. Bernardoni 2011.
50. Chrisman 1982, 187.
51. Shelby 1997, 45.
52. Walton 2017. Büttner (2017) charts the interaction of practical experience and its abstraction in texts. See also Bennett 2003. Instruments seemed to represent for these practitioners a way of reasoning through the senses about the particularities of the physical world. See Karin J. Ekholm's analysis of Tartaglia's *Nova scientia* in Ekholm 2010, 187–88.
53. Werrett 2009, 173; a very telling example, 177–79.
54. Long 2011, 78, quoting Averlino 1965.
55. Hermens 2013; Kieffer 2014; and Neilson 2016.
56. O'Malley 2005, 120–21. Artists also found that writing down their working processes could be as appealing (and perhaps less capital intensive) to patrons as the creation of a virtuoso work: King 2007, 278–79.
57. Lukehart 2010; also Lukehart 2010–ongoing. See also Bryce 1995; and Lis and Soly 2012, on formation of academies.
58. O'Malley (2005) shows that patrons and artists discussed subject matter (163–96) and continued to

refine a painting with reference to a preliminary contract drawing (220) or a sculpture with reference to a model (221–50). See also Guthmüller, Hamm, and Tönnesmann 2006.

59. On trading zones, Long 2015, and, on the agora, Valleriani (2012, 233), who describes it as a center of exchange or marketplace where technology accumulated, was communicated, merged, organized, and generated.
60. King 2007, 279.
61. See the essays on dance in Hilaire-Pérez, Nègre, Spicq, and Vermeir 2017a.
62. Amelang 1998, 28–39. An example is found in a notebook containing accounts, recipes, and an alchemical poem in the National Art Library, V&A Museum, Spec Coll 86 SS 46, c. 1476–79. Thanks to Leon Conrad for alerting me to this notebook.
63. McHam 2013, 62–65. Petrarch was scathing about the mechanical arts and their practitioners: Lis and Soly 2012, 320.
64. Barkan 2001.
65. Quoted in King 2007, 280.
66. Gerbino and Johnston 2009, 63–64.
67. Glatigny and Vérin (2008a, 80) make the point about reflection on language and expression. Dancing masters were especially creative in putting their practice into writing: Nordera 2008.
68. Cellini 1967, 91–92; emphasis added.
69. Stahl 2009, 87–91.
70. Scully 1986, 90–91.
71. Knoop, Jones, and Hamer 1938, 104–6.
72. See Cooper 2011, chap. 2.
73. Neudörfer 1875. See Murphy 2020.
74. Nordera 2008, 274.
75. De Alcega 1979, 10.
76. See Vérin 1993.
77. Davidson and Hodson 2007, 206.
78. Quoted in Lis and Soly 2012, 362–63.
79. Gluch 2007, 1.
80. Gluch 2007, 5 and following.
81. Morrall 2006, 229. For another contemporaneous example, see Zenetti 2016.
82. Ivory turners used the prints in *Perspectiva* as models. Tarnai and Weber 2017, 552.
83. See Smith 2004, 79–80.
84. Jamnitzer 1585. See Hauschke 2014.
85. Jamnitzer 1585, vol. 1, fol. 59r: "Erklerung des Eichmasleins."
86. For an overview of mining texts, see Long 1991; for the structures within which mining books were produced and the contribution of practitioners, see Smith 2017; Asmussen 2020; and Morel 2017.
87. Niavis 1953.
88. Hannaway 1992.
89. Lefèvre 2010, 4.
90. Agricola 1530, 130. Morel (2020) convincingly argues that Agricola's account of some mining techniques was not accurate.
91. Quoted in Slotta, Bartels, Pollman, and Lochert 1990, 160, my translation.
92. Slotta and Bartels 1990, 152–54.
93. See Bingener, Bartels, and Slotta 2006. For an introduction to the *Bergbuch*, see Slotta, Bartels, Pollmann, and Lochert 1990, 146–52.

94. Plattes 1639, 59–60.
95. Glatigny and Vérin 2008, 75.
96. Kavey 2007. On women's technical writing, see Tebaux 1997. See also Deblock 2017, who argues that the *Bâtiment des recettes* fostered experimentation and exploration on the part of its readers.
97. Cormack and Mazzio 2005, 85.
98. Cormack and Mazzio 2005, 84.

CHAPTER 5

1. Anon. after 1620 (probably John Hoskins), transcribed in Murrell 1983, 3–4.
2. Rivius 1547, fol. XI recto, writes that he will leave out most of the processes involved in forming a mold because "all such things may be learned with much less effort by in-person [*gegenwertige*] teaching rather than written instruction."
3. Paré 1969, 139.
4. Paré 1951, 24.
5. Parts of this chapter are drawn from Smith 2011b.
6. For example, the 1616 German translation of the pseudonymous Alessio Piemontese's *Secreti*, or Book of Secrets, was titled *Kunstbuch des Wohlerfarnen Herrn Alexii Pedemontani, von mancherleyen nutzlichen und bewerten Secreten oder Künsten* (Book of the Art [*Kunstbuch*] […] about Many Useful and Valuable Secrets or Arts), trans. Hanß Jacob Wecker (Basel: Ludwig König).
7. The knowledge and craft ordinances of English gold- and silversmiths were known as the "mystery" of the craft. H. G. c. 1604, fol. 6v.
8. Detienne and Vernant 1978. Hunter (2013) takes artisanal intelligence as the defining characteristic of the experimental activity pursued by Robert Hooke and Christopher Wren. Wolfe (2004, Introduction) shows that notions of craft cunning were used by Renaissance writers to exemplify political dealings.
9. Biringuccio 1943, 114.
10. Al-Ghazālī 2010, 82–85.
11. Hamburger 1998, 187–90. For secrets and esotericism, see also Vermeir 2012. For an overview, see Long 2001.
12. Pagel 1982, 51. See Smith 2004, 87–88.
13. Gentilcore 1995, 309.
14. Eamon 1994; Eamon 2010.
15. Frankl 1945, 46. Roriczer was probably present at the 1459 meeting, although he did not sign the stipulations (47).
16. Garçon 2005. See Sonenscher 1987, on the invented nature of the "mysteries" of the French *compagnonnages*, informal associations of French journeymen of the seventeenth century.
17. This began, however, much before: humanist Cipriano Piccolpasso desired to extract knowledge from craftsmen in the name of the public good. See Piccolpasso 1980, vol. 2, 6–7.
18. See Reith 2008, 137; and Davids 2008, vol. 1, 235, on the ineffectiveness of Venice's bans on the transfer of knowledge from the Murano glassworks. Reith 2008, 372–75, on porcelain. See also Davids 2005.
19. Quoted in Reith 2005, 364. See also Stewart 2005; and Davids and De Munck 2014.
20. Ferguson 1959, 5.
21. Polanyi 2009, 4.
22. Collins 2010, 99.
23. Ingold 2000, 350.
24. Maines 2009.

25. Paracelsus 1928, 211.
26. Risatti 2007, 55–57.
27. Risatti 2007, 170.
28. Whiten 2005; and Whiten 2014.
29. Ingold 2000, 11.
30. Ingold 2000, 4–5.
31. Ingold 2000, 353–57.
32. Ingold 2000, 358–59.
33. Ingold 2000, 361. Ingold (175) defines the difference in skills between human and nonhuman animals: beavers always build the same kind of houses; humans build diverse houses, thus design and intentionality distinguish humans from beavers. Humans perceive the world "and the relations of which it consists […] inscribed in a separate plane of mental representations, forming a tapestry of meaning that *covers over* the world of environmental objects" (177).
34. This was echoed by Paracelsus (1929, 53): the artisan, by means of his art, makes manifest the invisible. I discuss this further in Smith 2004.
35. Dürer 1969, 295.
36. Tarule 2004, 92–99.
37. Malafouris 2013, 173.
38. Lehmann 2009; and Lehmann 2013. See also Raven 2013; Dormer 1994, 219; and Clerbois and Droth 2011.
39. Ingold 2000, 345. See also Walls 2016.
40. Biringuccio 1943, 401.
41. *Speis* also denotes the molten metal used in casting, for example, in bell founding. Dürer's outline for the painter's manual was made in 1513. He later changed the title to *Unterweysung der Messung*. Gombrich 1997, 13.
42. Paracelsus 1928, 225.
43. Nicholas of Cusa 1989, 24.
44. Ms. Fr. 640, fol. 134v.
45. Ms. Fr. 640, fol. 44v; Neven 2016.
46. Anon. 1531, fol. IX verso.
47. Theophilus 1979, 181.
48. Biringuccio 1943, 95 (vitriol), 98 (rock alum).
49. Ms. Fr. 640, fol. 145r.
50. Anon. 1531, fol. IIII.
51. For casting with lead, Ms. Fr. 640, fol. 131v.
52. Biringuccio 1943, 60 (tin), 67 (iron).
53. Theophilus 1979, 173.
54. H.G. c. 1604, fols. 22v–24r.
55. Downey 2010, 35. For a vivid account of an apprenticeship in embodied knowledge, see Wacquant 2004.
56. Tallis 2003, 28–29. See also Radman 2013.
57. Dreyfus 1992, 236–37.
58. Union with the material through engaging it with the senses was the means by which artisans gained knowledge of the powers or virtues inherent in matter. See Smith 2004, 89–93.
59. Biringuccio 1943, 198.
60. I found Sennett 2008 useful in thinking about the diverse functions of recipes.

61. Biringuccio 1943, 143–44.

62. Dupré 2018.

63. Tallis 2003, 31.

64. Ingold 2000, 24, 106.

65. Ingold 2000, 243–44.

66. Palissy 1957, 191.

67. Nicholas of Cusa 1989, 21. Polanyi (1998) regards the submission to a master as key to the transmission of skill because it can only be taught person to person.

68. Dürer 1969, 293, 296.

69. De Munck 2007, 50–53.

70. Ingold 2000, 416, referring to the work of Jean Lave.

71. Lave and Wenger 1991, 11 (Foreword by Roy Pea and John Seely Brown), 13 (Foreword by William F. Hanks).

72. Paracelsus, 1529, n.p.

73. Collins 2001, 72–73. On tacit knowledge in medicine and medical training, see Delamont and Atkinson 2001, and the sociology of science literature they cite.

74. Salomon 1993, xii (series foreword by Roy Pea and John Seely Brown). See also Tribble 2011.

75. Salomon 1993, xiv. See also Hutchins 1995; Lave 1988; Rogoff and Lave 1984; Forgas 1981.

76. Dürer 1969, 297.

77. Biringuccio 1943, 145.

78. Cellini 1967, 84.

79. Baxandall 1966, 140–42.

80. English translation by Panofsky 1955, 279–80.

81. Laudan (1984, 94) sees the community as the unit of technology transfer and the "technology generator."

82. Presenti 2004. These contracts in Stechow 1989: another about work of Dieric Bouts (13); others (78).

83. Hayward 1976, 77.

84. Doering 1894, 74, 112.

85. Dormer 1994, 18. See also Dormer 1997, 229: "the thinking is in the making."

86. Coy 1989, Introduction, xi–xii. See also Goody 1989, 255; and Marchand 2010.

87. Dormer 1994, 22–23.

88. Dormer 1994, 56–57.

89. Cennini 1960, 138.

90. Cennini 1960, 48.

91. See Smith 2004, 98. See also Fransen and Reinhart 2019, 3.

92. Bambach 1999, 130.

93. De Munck 2007, 53–5. See also De Munck, Kaplan, and Soly 2007.

94. Polanyi 1998, 62.

95. Polanyi 1998, 64.

96. Reichard 1974, 86, quoted in Keller 2001, 34.

97. Ingold 2000, 357.

98. O'Connor 2007. See also Downey 2010. In Marchand 2010b, 24 and following.

99. Heyes 2018, 122. Philosopher Hubert Dreyfus attempted to present such an account without using the mind-based vocabulary of "representations." Dreyfus 2002a; and Dreyfus 2002b.

100. Berkowitz 2018.

101. H.G. c. 1604, fols. 5v–6v.
102. Lucie-Smith 1981, 85.
103. See Schön 1983; Keller 2001; and Clark 1997.
104. Biringuccio 1943, 249.
105. Biringuccio 1943, 211.
106. Biringuccio 1943, 331.
107. Biringuccio 1943, 373.
108. Norgate 1997, 61.
109. Nicholas of Cusa 1989, book II, 43n11. On the capacity of judgment, see also Dupré and Göttler 2019; on "ingenuity," see Marr, Garrod, Marcaida, and Oosterhoff 2019.
110. Henri de Mondeville, "Chirurgie," 1306, quoted in Vérin 1993, 44 (my translation).
111. Paracelsus 1928, 211.
112. Biringuccio 1943, 114.
113. Ingold 2000, 316. Sociologists have made clear that tacit knowledge and skill are also features of scientific knowledge: Barley and Orr 1997; Pfeifer 2006.
114. Quoted in De Munck 2007, 257.
115. Quoted in De Munck 2007, 257.
116. De Munck 2007, 254-55. Suthor (2010, 184–85 and 272–73) also charts this development,.
117. On the eighteenth-century concept of "technique," see Hendriksen 2021.
118. Sudnow 1978, xiii.
119. Sudnow 1978, 12–13.
120. Sudnow 1978, 146.
121. Sudnow 1978, 151.

CHAPTER 6

1. Leong and Pennell 2017, 133-4.
2. DiMeo and Pennell 2013, 16.
3. Wall 2016.
4. Spiller 2008; and Spiller 2009.
5. Krohn 2015.
6. See also Rankin 2007; Rankin 2013; and Leong 2018. Scholarship on recipes has burgeoned in recent years and is too extensive to cite comprehensively here. Parts of this chapter are drawn from Smith 2011b.
7. Alonso-Almeida 2013, 82–86. See also Telle 2003; Elaine Leong, on "thinness" of recipes, in Leong 2017, 72; and Hagendijk 2019.
8. Neven 2016, Recipe 27, 97.
9. Theophilus 1986, 189–90.
10. See Goulding 2006; and Eamon 1994, 34, 85–87.
11. Nassau 1984, 10.
12. Albertus Magnus 1967, 70.
13. H.G. c. 1604, fol. 62v.
14. H.G. c. 1604, fol. 76v.
15. H.G. c. 1604, fol. 68r.
16. H.G. c. 1604, fol. 84v.
17. H.G. c. 1604, fol. 97v.

18. Mayerne 1620, fols. 65r–71v.
19. Dormer 1994, 57.
20. Norgate 1997, 70.
21. Paré 1951, 19–20.
22. Rankin 2013, especially chaps. 1–2; Rankin 2014; and Harkness 2007.
23. Shelby 1997, 48.
24. Lorenz Lechler, *Unterweisung*, written in 1518 for his son, discussed by Shelby and Mark 1997, 89, emphasis added.
25. J. K. 1696.
26. J. K. 1696, 290.

CHAPTER 7

1. Chartier 1994.
2. Pepys 2004, Entry for 11 June 1663.
3. Jardine and Grafton 1990; notebook of secrets on 64. For transcriptions of a sampling of Harvey's and John Dee's books, see *The Archaeology of Reading in Early Modern Europe*, https://archaeologyofreading.org/.
4. Mooney 1993.
5. Eamon 2011, 43.
6. Leong and Rankin 2011a, 1; and Eamon 2011, 34–35.
7. Mathonière 2017. See also Lemerle 2012.
8. Carruthers 2008, 336.
9. Carruthers 2008, 9.
10. Carruthers 2008, 205.
11. Carruthers 2008, 270–71, who gives a wonderful example of Boccaccio's composing a commentary on his own, self-authored text.
12. Petrucci 2011, 277–81.
13. Richardson 1999, 114.
14. Richardson 1999, 122–31.
15. Richardson 1999, 156.
16. Paracelsus 1928, 225.
17. Bacon 1999, 114.
18. Hübner 1762, fol. A3v. See Blair 2010; and Kenny 1991.
19. Chartier 2007; Sherman 2008; and Cormack and Mazzio 2005.
20. Richards 2012.
21. Aristotle (attrib.) 1970, n.p.
22. Gerbino and Johnston 2009, 46.
23. *Tectonicon* went through eighteen editions from 1556 to 1656, with the last edition in 1692. Gerbino and Johnston 2009, 49–50.
24. Sherman 2008, 4.
25. McHam 2013, 259.
26. Diderot (attrib.) 1751, "Art," vol. 1, 713–17. For English translation: Hoyt and Cassirer 2013.
27. Bate 1635. This copy is in the National Gallery of Art Library, Washington, DC, Rare Q155 .B32 1635. This book was first published in 1634. A 1634 copy held by the Library of Congress, Q155 .B32 1634, contains on the end paper two handwritten recipes for toothache and bruising.

28. Cormack and Mazzio 2005, 24, mentions co-production.
29. Chrisman 1982, 75. In his survey of library inventories in Florence, 1413–1608, Bec (1984) found, similarly, that professional and vernacular books predominated by the end of the sixteenth century.
30. Chrisman 1982, 207.
31. Richardson 1999, 109.
32. See Pelus-Kaplan 2017.
33. Chrisman 1982, 121–22.
34. Richardson 1999, 140–41.
35. *"Wer Jemandt hie der gern welt lern[n]en dütsch schriben und läsen / uß dem aller kürtzisten grundt den Jeman erdencken kan do durch / ein Jeder der vor nit ein buochstaben kan · der mag kürtzlich und bald / begriffen ein grundt do durch er mag von jm selbs lernen sin schuld / uff schribe[n] und läsen und wer es nit gelernen kan so ungeschickt / were den will ich um[m] nüt und vergeben gelert haben und gantz nüt / von jm zuo lon nem[m]en er sig wer er well burger oder handtwercks ge-/sellen frouwen und junckfrouwen wer sin bedarff der kum[m] har jn · der / wirt drüwlich gelert um[m] ein zimlichen lon · Aber die junge[n] knabe[n] / und meitlin noch den fronvasten · wie gewonheit ist · 1516."* Basel Kunstmuseum, Amerbach-Kabinett 1662, Inv. 310 and 311.
36. Hackenberg 1986.
37. Chartier 2011, 158–59.
38. Richardson 1999, 111.
39. Hackenberg 1986, 76.
40. Haage and Wegner 2007, 173. See also Eis 1960.
41. Albrecht 1514.
42. Brunschwygk 1500.
43. Brunschwygk 1500, held by the Bayerische Staatsbibliothek, 2 Inc.c.a. 3687, https://daten.digitale -sammlungen.de/~db/0003/bsb00031146/images/. On red initial letters and historiation, Carruthers 2008, 10, 337, and passim.
44. Chrisman 1982, 69.
45. Labarre 1971, 225–6.
46. Labarre 1971, 257.
47. Corbellini and Hoogvliet 2013.
48. Hackenberg 1986, 78.
49. Hackenberg 1986, 78–79.
50. Chartier 1989, 165.
51. Hackenberg 1986, 72–91.
52. Hackenberg 1986, 84–86.
53. Carl 1987, 388.
54. Papenbrock 2013, 137.
55. Beuzelin 2013, 79.
56. Bresc-Bautier 1990, 149–50.
57. Seifert 2013.
58. For an overview, see Zittel, Thimann, and Damm 2013.
59. Jorink, Lehmann, and Ramakers 2019; and Remond 2020.
60. Remond 2019, for pricking, 306–311.
61. Merrill 2017; and Marr 2013.
62. Amelang 1998, 61 and following.

63. Quoted in Mocarelli 2009, 106.
64. Lis and Soly 2012, 352–53.
65. Ramazzini, *De morbis artificum diatribe* (Modena, 1700), quoted in Mocarelli 2009, 107.
66. King 2007, 278–79, 273.
67. King 2007, 273.
68. King 2007, 279.
69. Castiglione 2002, 57.
70. Wolfe 2004, 40. See also Biow 2015, who sees discussions and use of "art" as pervasive in the Renaissance, especially chap. 1, 21–92.
71. Della Casa 2013, 60–61.
72. Della Casa 2013, 62.
73. Gerbino and Johnston 2009, 46.
74. Gerbino and Johnston 2009, 46.
75. Gerbino and Johnston 2009, 46–49.
76. Sandman 2001, 83. See also Schotte 2019.
77. Cossart 2017.
78. Rabelais 1991, 60–61.
79. Rabelais 1991, 152, 155, 156, 158.
80. Vives 1913, book 1, chap. 3, 23.
81. Vives 1913, 208.
82. Vives 1913, 229.
83. Vives 1913, 228.
84. Vives 1913, 230.
85. Vives 1913, 233. On the role of histories in medicine and empiricism, see Pomata and Siraisi 2005.
86. Vives 1913, 283.
87. Vives 1913, 209–10.
88. See Smith 2004.
89. Ruscelli, "Proem," from his *Secreti nuovi*, reprinted by Eamon and Paheau 1984, 338.
90. Eamon and Paheau 1984, 333.
91. For example, Keller 2016.
92. Glatigny and Vérin 2008b, 65–68.
93. Vérin 1993, 136.
94. Vérin 1993, 160–61. "Storehouse" was a common characterization of the method by which written accounts of process furnished the memory for action. See also Sherman 2008, 48.
95. Vérin 1993; and Glatigny and Vérin 2008b. On *methodus*, see Gilbert 1960; and Ong 1958.
96. Vérin 2008, 44. For branching diagrams, see Ong 1958.
97. Vérin 1993, 180.
98. Vérin 1993, 13 and 19.
99. Vérin 1993, 28. See also Weeks 2008. Bacon's overriding aim was to solve the problem of invention (Weeks 2008, 142–43, and passim).
100. Glatigny and Vérin 2008b, 86–87.
101. Vérin 1993.
102. Arnoux and Monnet 2004; Brioist 2013. Although discussing China, Jacob Eyferth (2010) makes this point lucidly.
103. Vérin 1993, 112.

104. Glatigny and Vérin 2008b, 52–53. From 1581 to 1621, the Dutch Republic granted about 17 percent of all printing privileges to technical texts. Buning 2017, para. 16.
105. Whalley and Barley 1997, 25.

CHAPTER 8

1. "The 'Chellini Madonna,' by Donatello, about 1450," Victoria and Albert Museum, http://www.vam.ac.uk/content/articles/t/the-chellini-madonna/.
2. Dürer and Fry 1995, 48.
3. See Russo 2020; Posada 2020; and Grasskamp 2021.
4. Göttler 2008, 43–45.
5. Leedham-Green 1986, 139.
6. Landolt and Ackermann 1991. On Amerbach's interest in Roman antiquities, Landolt 1991, 142. On goldsmith's lodging, Ackermann 1991, 53. For inventories, see Landolt 1991, 151.
7. A set of spoon blanks, recorded in his inventory, is still extant at the Historisches Museum Basel, Inv. 1870.1043.
8. Parshall, Sell, and Brodie 2001.
9. The literature is now far too voluminous to cite; some signal examples for Europe are Ago 2013; Göttler 2010; Rublack 2013; De Munck 2014; Bourgeois et al. 2018; Gerritsen and Riello 2015. A rich literature in early American material culture also exists; see the work of Jules Prown, Robert Blair St. George, and Sarah Anne Carter.
10. This chapter is drawn in part from Smith 2008b and Smith 2007.
11. Smith 1979; Blockmans and Prevenier 1999; Eichberger 2002.
12. For an overview and bibliography of the numerous works on *Kunstkammern*, see Koeppe 2019; and von Schlosser 2021. See also Grafton 2007.
13. Literature on the Habsburg and Rudolfine *Kunstkammern* is now abundant. For an overview and bibliography, see Fuciková et al. 1997; Fuciková 1988.
14. See, for example, Doering 1894.
15. Scheicher 1979; Scheicher 1977; Gries 1994.
16. Cardinal 2016.
17. Watanabe-O'Kelly 2002.
18. See Meadow 2002.
19. Quiccheberg 2013, 98–99.
20. Roth 2000, facing Latin and German text, 193–95.
21. Roth 2000, 125. In this, as at several other points, Meadow and Robertson provide a somewhat different translation of the Latin (Quiccheberg 2013, 82). Because Roth's edition includes facing Latin text of the original, the majority of my references are to that edition. Quiccheberg includes a lengthy list of nobles, patricians, merchants, scholars, citizens, and artisans who collected (Roth 2000, 164–211).
22. Roth 2000, 187.
23. Quiccheberg 2013, 74.
24. See also Quiccheberg 2013, 1–41 (Introduction by Meadows and Robertson), for more on Quiccheberg's biography and aims in this treatise.
25. Hayward 1976, 32.
26. Roth 2000, 44–47.

27. Roth 2000, 54–61.

28. Roth 2000, 60–69.

29. Roth 2000, 81 (printing press), 99 (lathe room).

30. Roth 2000, 101–3.

31. Roth 2000, 78 and following (workshops), 83 (alchemical furnace).

32. Roth 2000, 93.

33. Roth 2000, 111.

34. Hayward 1976, 148–54; and Bott 1985. In the 1607–11 inventory of the *Kunstkammer* of Rudolf II, the fountain is described as being contained in eighteen boxes marked with a particular seal. The contents of Box 8 included a small book in which the entire meaning of the fountain was neatly written out on parchment. The inventory documents are reprinted in Pechstein 1985. A description of the fountain by the seventeenth-century traveler is contained in Boesch 1888, 87–90 (Vienna, Teil II). See Miscellaneahandschrift Nr. 28722, which contains sketches ("die auf Reisen in den Jahren 1640–42 gemacht wurden," 87). The following description is from 87–88.

35. Jamnitzer 1585. For evidence of other manuals, Doering 1894.

36. Boesch 1888, 88.

37. Jamnitzer was known for his love of music. A Nuremberg composer dedicated a series of songs to him. Hauschke 2003, 136n34.

38. A casket-sized cabinet by Jamnitzer, now in Madrid, also functioned as a mirror for princes—the cabinet's ornamental program depicted the virtues necessary to a ruler. See Effmert 1989.

39. Pechstein 1974.

40. "Sum terra, Mater Omnium / Onusta caro pondere / Nascentium ex me fructuum." Quoted in Pechstein 1974, 95–96.

41. See von Schönherr 1888; and Lein (2006, 112–14), who enumerates the objects recorded in the Bavarian Wittelsbach and Habsburg collections. See Achilles-Syndram 1994, for the inventory of the Nuremberg Praun collection.

42. Diemer 1986, 159–60 (document 8). See Lein 2006, 112.

43. Fickler 2004, 130, contains inventory entries for these rarities. See also Diemer, Diemer, and Seelig 2018.

44. As I discuss in Smith 2004, chap. 4, alchemical practice was viewed as reproducing the processes of nature and alchemical theory as explaining transformation and metamorphosis. See also Smith 2008a.

45. Landolt and Ackermann 1991, 30–31, 102, 122. For inventories, see Landolt 1991, 151.

46. Jansen 1993.

47. Roth 2000, 139.

48. Neilson 2016. Hooykaas (1958, 3–8, 125) treats Petrus Ramus's visits to Nuremberg workshops. See also Rossi 1970.

49. Debuiche 2020.

50. Leibniz 2008, 49 (entry 18).

51. Roth 2000, 129.

52. Roth 2000, 117.

53. For the continued importance of artisanal skill throughout the Industrial Revolution, see among other accounts, Harris 1976; Samuel 1977; and Laudan 1984, Introduction, 10, where she pithily states, "Science owes more to the steam engine than the steam engine to science."

54. See the work of Vera Keller, especially Keller 2015.

CHAPTER 9

1. Paracelsus 1941, 91.
2. Demeter 2020.
3. See the YouTube video "Ringslangen Nieuw Wulven," https://www.youtube.com/watch?v=W6sxEf997 iM, at 0:57. I thank Joosje van Beenekom, senior metals conservator, Rijksmuseum, Amsterdam, for informing me of this film.
4. For a review, Harris 2010.
5. St. George 1988, 11.
6. Lüdtke 1995, discusses uses of *Alltagsgeschichte*; a well-known disputed case of "empathy" is Goldhagen 1996, treated by Moyn 2006.
7. Dormer 1994, 22–23.
8. Coy 1989b, 112. See also Marchand 2010b.
9. See, e.g., Keller and Keller 1993, 127.
10. Auslander, Bentley, Halevi, Sibum, and Witmore 2009, 1356–57.
11. See a special issue of *Rethinking History* 11, no. 3 (2007), especially Agnew 2007. See also Agnew, Lamb, and Tomann 2020. The literature has grown enormously, especially among historians of science and technical art historians, and is too lengthy to cite here. For recent overviews and bibliography, see Fors, Principe, and Sibum 2016; Dupré, Harris, Lulof, Kursell, and Stols-Witlox 2020; Staubermann 2011; Wrapson et al. 2012; Hagendijk 2018; Hagendijk 2020; Hendriksen 2020; Making and Knowing Project et al. 2020; Smith 2016; and Smith 2012. This chapter draws in part from Smith 2016 and Smith 2012. Many of these scholars have considered the various aims of reconstruction, and Leslie Carlyle and Maartje Stols-Witlox have made important points about methods in describing Historically Accurate Reconstruction Techniques (HART) Project. See Carlyle 2020.
12. I am grateful to Renate Keller for instruction in historical techniques in painting, and to Doug Lincoln at West Dean College for silversmithing instruction.
13. Robson 2002.
14. Martinón-Torres, Freestone, Hunt, and Rehren 2008; Martinón-Torres, Rehren, Thomas, and Mongiatti 2009.
15. Craddock and Timberlake 2004/5.
16. Young 2005, 1–2.
17. Bourgarit and Thomas 2011.
18. Stallybrass, Chartier, Mowery, and Wolfe 2004. One example of Shakespeare's use is "My Tables, my Tables; meet it is I set it downe, / That one may smile, and smile and be a Villaine" (*Hamlet*, TLN 792–93 [1.5.107–8]). See also Bostock 2020.
19. Bol 2019. Also see Bol 2014.
20. Sibum 1995.
21. Sibum 1995, 101.
22. Sibum 2000.
23. Sibum 1995, 97.
24. Sibum 1995, 101.
25. Sibum 1995, 103.
26. Sibum 1995, 102.
27. Heering 2008.
28. Principe 2016.
29. Newman and Principe 2002. On these experiments see *The Chymistry of Isaac Newton*, http://webapp1

.dlib.indiana.edu/newton/; especially the Multimedia Lab, http://webapp1.dlib.indiana.edu/newton /reference/chemLab.do.

30. Rampling 2014.
31. Dunkerton 1996–98, 292, 288. More recently, see Kneepkens 2020.
32. Lacey 2018.
33. The Making and Knowing Project has written about this phenomenon with regard to "perspective," "ephemeral art," "verre eglomisée" (painting on glass), and other categories of art historical scholarship.
34. Smith 1975, 605.
35. Fachhochschule (Köln) 2001.
36. Bucklow 1999; Bucklow 2000; and Bucklow 2001a.
37. For example, H. G. c. 1604, fol. 55r.
38. Bucklow 1999, 145.
39. Bucklow 1999, 145–47.
40. Striebel 2003, 429.
41. Stone 1981; Stone 2006; Stone 2001, 62; and Stone's contributions to Allen and Motture 2008.
42. Davidson and Hodson 2007, 206–7.
43. Rublack 2016, 12.
44. Panzanelli 2008, 2.
45. Dunkerton 1991, 174–75.
46. Wallert 2000, 268, referring to the widespread stripping of paint from polychromed statuary in the nineteenth century.
47. Panzanelli 2008.
48. Laird 1999.
49. An insight noted in Review of "Conservation Meets Its Maker," summer school, West Dean College, which investigated historical techniques of making, reported by Rogerson 2004, 17.
50. I first encountered these sketches in Amico 1996.
51. See Smith and Beentjes 2010; Smith 2012; Beentjes and Smith, 2013.
52. Palissy 1957, 192.
53. Smith (1975, 603) points out the importance of ceramics in metalworking.
54. Ms. Fr. 640, fol. 141v. This stimulated research by Davidowitz, Beentjes, van Bennekom, and Creange 2012; and Kok 2020.
55. In 2005, I collaborated with Amy Meyers and Harold Cook on a conference, "Ways of Making and Knowing: The Material Culture of Empirical Knowledge," published as Smith, Meyers, and Cook 2017.
56. For all collaborators of the project, see https://edition640.makingandknowing.org/#/content/about /credits. For sponsors, see https://edition640.makingandknowing.org/#/content/about/sponsors.
57. For more on the process, see Smith 2020a.
58. Making and Knowing Project, et al. 2020.
59. Camps and Lyautey 2020.
60. Dyon and Wacha 2020.
61. In the five years, we examined moldmaking and metalworking; color making; vernacular natural history, practical perspective, optics, and medicine; ephemeral art; and inscription, impression, and printmaking.
62. The students were required to undergo general lab safety and fire training, fire extinguisher practice, and special safety training in the Making and Knowing laboratory.
63. See Smith 2020b.

64. For essays on all these subjects, see Making and Knowing Project et al. 2020.
65. Compare vol. 3 of Michael of Rhodes 2009. Further study is merited on the use of ratios and its effects on early modern conceptions of number.
66. Fol. 145r. See Boyd, Palframan, and Smith 2020. In 1802, ovens were still measured in a similar way: "The first care is to see that the oven be sufficiently heated, yet not to such a degree as to burn the crust. If a green vegetable turns black when put in, the oven will scorch the bread; in which case it must stand open till the heat has somewhat abated." Willich 1802, vol. 1, n.p.
67. Ms. Fr. 640, fols. 124v and 72v. For further visual signs, see Boyd, Palframan, and Smith 2020. Craftspeople still rely on such visual signs today.
68. The Making and Knowing Project is preparing a research and teaching companion, with lesson plans, syllabi, skill-building exercises, and other pedagogical material, aimed at those who wish to incorporate hands-on work into their classroom teaching.
69. For a recent overview, see Fransen and Reinhart 2019, 3.

CHAPTER 10

1. Efforts to provide this vocabulary are in Lehmann 2015; see also Smith 2018; and Lehmann 2013.
2. Marchand 2010a.
3. Sigaut 1994, 438–39.
4. Fores 1994, 183n14. Sociologists of science are changing this; see Collins and Evans 2006.
5. Vega-Encabo 2016. See also Bray, 2008.
6. Lechtman 1999, 223.
7. Kuijpers 2018.
8. Cellini 1967, chap. 22, 84.
9. Biringuccio 1943, 391.
10. Kremnitzer and Smith 2020.
11. Ms. Fr. 640, fol. 98r.
12. Takeda 2011, 669, quoting Weddington 1567.
13. Gage 1998, 39.
14. See "Making Imitation Coral" about reconstructing fol. 3r in the Making and Knowing laboratory, https://vimeo.com/129811219.
15. Chessa 2020.
16. For such imitation in Ms. Fr. 640, see Lores-Chavez 2020; Kremnitzer, Shah, and Smith 2020; and Anantharaman and Smith 2020.
17. Malafouris 2013, 99.
18. Malafouris 2013, 118 and 126.
19. Malafouris 2013, 104–5.
20. All the foregoing properties of wax from Guerzoni 2012, 52.
21. Scribner 1984.
22. Liu 2020.
23. The prayer reads: "*Pater Noster, qui es in caelis, sanctificetur nomen tuum; adveniat regnum tuum; fiat voluntas tua, sicut in caelo, et in terra. Panem nostrum supersubstantialem da nobis hodie; et dimitte nobis debita nostra, sicut et nos dimittimus debitoribus nostris; et ne nos inducas in tentationem; Sed libera nos a malo. Amen.*"
24. Landsman and Rowen 2020. Other students who experimented with wax are Kang 2020; Noirot 2020; and Cataldo and van Visco 2020.

25. This is the process of "extension" in experimentation to which Chang (2011) alludes. For evidence in Ms. Fr. 640, see, among other essays, Gans 2020; Kok 2020; Fu, Zhang, and Smith 2020; Chiostrini and Palframan 2020; and Cataldo and Visco 2020.

26. On *magistra*, see Cataldo and van Visco. 2020.

27. Deanna F. Curtis, curator of woody plants, New York Botanical Garden, suggested via email that an elm hybrid cultivar known as "Pioneer" (binomial *Ulmus x hollandica*) might be more suitable for our purposes: as the Pioneer hybrid is formed by the crossing of two European species, *Ulmus glabra* (or Wych Elm) and *Ulmus minor* (the Smooth-leaved Elm), it probably more closely approximated the elms from which the author-practitioner would have sourced his roots. See Marris and Pope 2020.

28. Email communication, May 10, 2015, from Joel T. Fry, curator, Bartram's Garden, John Bartram Association, Philadelphia. As Fry made clear, the commercial uses of plants are still considered in hybridization, but they are generally not part of the taxonomic system.

29. See Cataldo and van Visco 2020.

30. Research undertaken by the Making and Knowing Project found that "impalpable" was a technical term in French, Italian, and English in the sixteenth century. Alessio Piemontese, *Secreti*, also describes the perfect sand for casting as "very soft, as if impalpable." Piemontese 1555, 206.

31. See Krekel and Lautenschlager 2005. For recipes that attempt to bring about opposite states, see Anon. 1535.

32. Landsman and Rowen 2020; and Kang 2020.

33. Wang and Smith 2020.

34. Biringuccio 1943, passim, and 46. Bernardoni (2011) sees Biringuccio's use of unctuosity as reflecting the influence of ancient writers such as Aristotle and the development of Aristotle in Albertus Magnus, but I think Biringuccio's formulations of fat and lean are too widespread among farmers, founders, and within the "health worldview" to need the influence of authorities.

35. Biringuccio 1943, 61.

36. Biringuccio 1943, 324.

37. Biringuccio 1943, 391.

38. Cellini 1967, 113.

39. For more on this, and all pertinent citations, see Smith 2015b.

40. Werrett 2009, 27 and 33.

41. Lee and Smith 2020; and Wang and Smith 2020.

42. Cennino's *Il Libro*, "Libro secondo de diversi colori e sise da mettere a oro," and Ms. Fr. 640 all use bodily products such as saliva, phlegm, urine, and ear wax, among others. See Wallert 1995.

43. Ms. Fr. 640, fol. 168v.

44. Ms. Fr. 640, fol. 121v.

45. See Soley 2020; Shi and Chang 2020; Kremnitzer, Shah, and Smith 2020; and Foyer 2020.

46. Pye 1968, 47.

47. Guthmüller 1981, 3; and Damm, Thimann, and Zittel 2013, 50.

EPILOGUE

1. This account condenses the texts of many historians, whose works are cited in Smith 2019a, from which some parts of this chapter are drawn. See especially Smith 2019b; and Smith 2015a.

2. The literature on science and empire is now vast. See Smith 2009; and more recent literature cited in Smith 2019a.

3. Gettens, Feller, and Chase 1994. See also Liu and Kuriyama 2020.

4. Albertus 1967, 212.

5. Albertus 1967, 112, 153.

6. Halleux 1996, 890–91.

7. Anawati 1996; and Halleux 1996.

8. Waller 1990, 155.

9. Albertus 1967, 201.

10. Thurston 1915. My thanks to Robert Goulding for his lecture "Snakes in a Flame," containing this reference. See Van der Lugt 2009.

11. My thanks to Dorothy Ko for this information.

12. See Grafit 2020; and Smith, DeVinney, Grafit, and Liu 2019.

13. Ms. Fr. 640, fol. 55r–v, "Horrible venom which kills if one steps on a board or a stirrup."

14. See Gengzhe 2006, 191–93. With deep thanks to Xiaomeng Liu for his contribution to this research.

15. Feng and Shryock 1935.

16. Anthropologist Cecil H. Brown coined the term "wug" from the words worm and bug. Brown (1979) notes that "the wug life-form of Chinese encompasses small reptiles other than snakes."

17. From de Groot 1892, 847.

18. Fox 2016.

19. Su 1985.

20. Tao 1819, 4b–5b.

21. Yinglong 1985, 16. For the ancient significance of silkworms, see Kuhn 1988, 301.

22. de Groot 1892, 851.

23. Diamond 1988.

24. Historian of alchemy Robert Halleux, who states that European alchemy "owes more or less everything" to Arabic alchemical literature, notes that the manuscript Palermo 40 Qq A 10 (early fourteenth century), which uses "animal substances, especially eggs and worms for transmutation, seems to have been translated from an Arabic original, probably belonging to the Jabirian corpus." Halleux 1996, 888.

References

Abattouy, Mohammed, and Salim Al-Hassani. 2015. *The Corpus of Al-Isfizārī in the Sciences of Weights and Mechanical Devices: New Arabic Texts in Theoretical and Practical Mechanics from the Early XIIth Century.* London: Al-Furqān Islamic Heritage Foundation.

Achilles-Syndram, Katrin, ed. 1994. *Die Kunstsammlung des Paulus Praun. Die Inventare von 1616 und 1719.* Nuremberg: Im Selbstverlag des Stadtrats zu Nürnberg.

Achim, Miruna. 2011. "From Rustics to Savants: Indigenous *Materia Medica* in Eighteenth-century Mexico." *Studies in History and Philosophy of Biological and Biomedical Sciences* 42, no. 3:275–84.

Ackermann, Felix. 1991. "Plaketten im Amerbach-Kabinett." In *Sammeln in der Renaissance: Das Amerbach-Kabinett. Beiträge zu Basilius Amerbach*, 51–72. Basel: Historisches Museum.

Adamson, Glenn. 2010. *The Craft Reader.* Oxford: Berg.

Adamson, Glenn. 2018. "The Case for Material Intelligence." *Aeon Magazine*, November.

Agnew, Vanessa. 2007. "History's Affective Turn: Historical Reenactment and Its Work in the Present." *Rethinking History* 11, no. 3:299–312.

Agnew, Vanessa, Jonathan Lamb, and Juliane Tomann, eds. 2020. *The Routledge Handbook of Reenactment Studies: Key Terms in the Field.* Abingdon, UK: Routledge.

Ago, Renata. 2013. *Gusto for Things: A History of Objects in Seventeenth-century Rome.* Trans. Bradford Bouley and Corey Tazzara, with Paula Findlen. Chicago: University of Chicago Press.

Agricola, Georgius. 1530. *Bermannus, sive de re metallica.* Basil: Froben.

Agricola, Georgius. 1961. *De animantibus subterraneis liber* (1549, 1556). In *Ausgewählte Werke.* Trans. Georg Fraustadt and Rolf Hertel. Berlin: VEB Deutscher Verlag der Wissenschaften.

Ainsworth, Maryan W. 2001. "Commentary: An Integrated Approach." In *Early Netherlandish Painting at the Crossroads. A Critical Look at Current Methodologies*, ed. Maryan W. Ainsworth, 106–21. New Haven: Yale University Press.

Albala, Ken. 2002. *Eating Right in the Renaissance.* Berkeley: University of California Press.

Albertus Magnus. 1967. *Book of Minerals.* Trans. Dorothy Wyckoff. Oxford: Clarendon Press.

Albertus Magnus (attrib.). 1973. *The Book of Secrets of Albertus Magnus of the Virtues of Herbs, Stones and Certain Beasts; Also a Book of the Marvels of the World* (1530). Trans. and ed. Michael R. Best and Frank H. Brightman. Oxford: Clarendon Press.

Albrecht, Meister. 1514. *Diß büechlin sagt wie man pferdt artzneyen und ein yegliches roß erkennen und probieren soll.* Strasbourg: Mathias Hüpffuff.

al-Ghazālī, Abu Hamid Muhammad ibn Muhammad. 2010. *The Rescuer from Error*, ed. Muhammed Ali Khalidi. In *Medieval Islamic Philosophical Writings*. Cambridge: Cambridge University Press.

Allen, Denise, and Peta Motture, eds. 2008. *Andrea Riccio: Renaissance Master of Bronze*. New York: Frick Collection.

Alonso-Almeida, Francisco. 2013. "Genre Conventions in English Recipes, 1600–1800." In *Reading and Writing Recipe Books*, ed. Michelle DiMeo and Sara Pennell, 68–90. Manchester, UK: Manchester University Press.

Amelang, James S. 1998. *Flight of Icarus: Artisan Autobiography in Early Modern Europe*. Stanford: Stanford University Press.

Ames-Lewis, Francis. 2000. *The Intellectual Life of the Early Renaissance Artist*. New Haven: Yale University Press.

Amico, Leonard M. 1996. *Bernard Palissy: In Search of Earthly Paradise*. Paris: Flammarion.

Anantharaman, Divya, and Pamela H. Smith. 2020. "Animals Dried in an Oven." In *Secrets of Craft and Nature*, ed. Making and Knowing Project et al. https://edition640.makingandknowing.org/#/essays/ann_502_ad_20.

Anawati, Georges C. 1996. "Arabic Alchemy." In *Encyclopedia of the History of Arabic Science, Vol. 3*, ed. Roshdi Rashed, 853–85. London: Routledge.

Anon. 1507. *Küchenmeisterey*. Strasbourg: Hüpffuff.

Anon. 1531. *Rechter Gebrauch der Alchimei, mitt vil bißher verborgenen, unnd lustigen Künsten, nit allein den fürwitzigen Alchimisten, sonder allen kunstbaren Werckleutten, in und ausserhalb feurs; Auch sunst aller menglichen inn vil wege zugebrauchen.*

Anon. 1535. *Künstbüchlin gerechten gründlichen gebrauches aller kunstbaren Werckleüt*. Augsburg: Heinrich Steyner.

Anon. after 1620. "The Art of Limning either by the Life, Landscape, or Histories." British Library, Harl. 6376.

Aristotle. 1933. *Metaphysics*. Trans. Hugh Tredennick. Cambridge, MA: Harvard University Press.

Aristotle. 1944. *Politics*. Trans. H. Rackham. Cambridge, MA: Harvard University Press.

Aristotle (attrib.). 1970. *Secreta secretorum*. London, 1528. Amsterdam, NY: Da Capo Press.

Aristotle. 1999. *Nicomachean Ethics*. Trans. Terence Irwin. 2nd ed. Indianapolis: Hackett Publishing.

Arnoux, Mattieu, and Piere Monnet, eds. 2004. *Le technicien dans la cité en Europe occidentale, 1250–1650*. Rome: EFR.

Artzneybuch. 1556. [Compiled by physician Tarquinius Schnellenberg and printer Johann Dautman.] Königsberg: Johann Daubman.

Aruz, Joan, Kim Benzel, and Jean M. Evans, eds. 2008. *Beyond Babylon: Art, Trade, and Diplomacy in the Second Millenium B.C.* New Haven: Yale University Press.

Asmussen, Tina, ed. 2020. "The Cultural and Material Worlds of Mining in Early Modern Europe." Special issue. *Renaissance Studies* 34, no. 1:1–148.

Auslander, Leora, Amy Bentley, Leor Halevi, H. Otto Sibum, and Christopher Witmore. 2009. "AHR Conversation: Historians and the Study of Material Culture." *American Historical Review* 114:1355–1404.

Averlino, Antonio di Piero. 1965. *Treatise on Architecture; Being the Treatise by Antonio di Piero Averlino, known as Filarete*. Trans. John R. Spencer. New Haven: Yale University Press.

Bacon, Francis. 1999. "Of Studies." In *The Essays or Counsels Civil and Moral* (1625), ed. Brian Vickers. Oxford: Oxford University Press.

Bambach, Carmen C. 1999. *Drawing and Painting in the Italian Renaissance Workshop: Theory and Practice, 1300–1600*. Cambridge: Cambridge University Press.

Banckes, Richard. 1941. *An Herbal* (1525). Ed. Sanford V. Larkey and Thomas Pyles. Delmar, NY: Scholars' Facsimiles & Reprints.

Barclay, Robert. 1992. *The Art of the Trumpet-Maker: The Materials, Tools, and Techniques of the Seventeenth and Eighteenth Centuries in Nuremberg*. Oxford: Clarendon Press.

Barkan, Leonard. 2001. *Unearthing the Past: Archaeology and Aesthetics in the Making of Renaissance Culture*. Princeton, NJ: Princeton University Press.

Barley, Stephen R., and Beth A. Bechky. 1994. "In the Backrooms of Science: The Work of Technicians in Science Labs." *Work and Occupations* 21, no. 1:85–126.

Barley, Stephen R., and Julian E. Orr, eds. 1997. *Between Craft and Science: Technical Work in U.S. Settings*. Ithaca, NY: Cornell University Press.

Barnard, John, and D. F. McKenzie, eds. 2002. *The Cambridge History of the Book in Britain, Vol. 4: 1557–1695*. Cambridge: Cambridge University Press.

Barney, Rachel. 2009. "Twenty Questions about Protagorean Wisdom." Paper presented in the conference "Wisdom in Ancient Thought," Columbia University, March 2009.

Bartl, Anna, Christoph Krekel, Manfred Lautenschlager, and Doris Oltrogge. 2005. *Der "Liber illuministarum" aus Kloster Tegernsee*. Stuttgart: Franz Steiner Verlag.

Bate, John. 1635. *The Mysteries of Nature and Art*. London: Ralph Mabb.

Baxandall, Michael. 1966. "Hubert Gerhard and the Altar of Christoph Fugger: The Sculpture and Its Making." *Münchner Jahrbuch der bildenden Kunst* 3, no. 17:127–44.

Bec, Christian. 1984. *Les livres des Florentins (1413–1608)*. Florence: L. S. Olschki.

Beentjes, Tonny, and Pamela H. Smith. 2013. "Sixteenth-century Life-casting Techniques: Experimental Reconstructions Based on a Preserved Manuscript." In *The Renaissance Workshop*, ed. David Saunders, Marika Spring, and Andrew Meek, 144–51. London: Archetype.

Behrouzi, Nitza. 2002. *The Hand of Fortune: Khamsas from the Gross Family Collection and the Eretz Israel Museum Collection*. Tel Aviv: Eretz Israel Museum.

Belhoste, Bruno. 2012. "A Parisian Craftsman among the Savants: The Joiner André-Jacob Roubo (1739–1791) and His Works." *Annals of Science* 69, no. 3:395–411.

Bell, Rudolf M. 1999. *How to Do It: Guides to Good Living for Renaissance Italians*. Chicago: University of Chicago Press.

Bennett, James A., and Scott Mandelbrote. 1998. *The Garden, the Ark, the Tower, the Temple: Biblical Metaphors of Knowledge in Early Modern Europe*. Oxford: Museum of the History of Science.

Bennett, James. 2003. "Presidential Address: Knowing and Doing in the Sixteenth Century: What Were Instruments For?" *British Journal for the History of Science* 36:129–50.

Bergdolt, Klaus. 2006. "Der Künstler als Literat—Das Beispiel Ghiberti." In *Künstler und Literat: Schrift- und Buchkultur in der europäischen Renaissance*, ed. Bodo Guthmüller, Berndt Hamm, and Andreas Tönnesmann, 13–30. Wiesbaden: Harrassowitz Verlag.

Beringer, Johann Bartholomew Adam. 1963. *The Lying Stones of Dr. Johann Bartholomew Adam Beringer being his Lithographiae Wirceburgensis*. Trans. and annot. Melvin E. Jahn and Daniel J. Woolf. Berkeley: University of California Press.

Berkowitz, Aaron L. 2018. "The Cognitive Neuroscience of Improvisation." In *The Oxford Handbook of Critical Improvisation Studies*, ed. George E. Lewis and Benjamin Piekut, 1:63–64.

Bernardoni, Andrea. 2011. *La conoscenza del fare: Ingegneria, arte, scienza nel "De la pirotechnia" di Vannoccio Biringuccio*. Rome: L'Erma di Bretschneider.

Bernasconi, Gianenrico. 2016. "L'objet comme document." In *L' Europe technicienne: XVe–XVIIIe siècle*, ed. Catherine Cardinal, Liliane Pérez, Delphine Amélie Spicq, and Marie Thébaud-Sorger. Special issue. *Artefact: Technique, histoire et sciences humaines* 4:31–47.

Bertucci, Paola. 2017. *Artisanal Enlightenment: Science and the Mechanical Arts in Old Regime France*. Chicago: University of Chicago Press.

Beuzelin, Cécile. 2013. "Jacopo Pontormo: A Scholarly Craftsman." In *The Artist as Reader: On Education and Non-education of Early Modern Artists*, ed. Heiko Damm, Michael Thimann, and Claus Zittel, 71–105. Leiden: Brill.

Bewer, Francesca G. 2001. "The Sculpture of Adriaen de Vries: A Technical Study." In *Small Bronzes in the Renaissance*, ed. Debra Pincus, 159–93. Studies in the History of Art 62, Symposium Papers 39. Washington, DC: Center for Advanced Study in the Visual Arts.

Bhanoo, Sindya N. 2010. "Puzzle Solved: How a Fatherless Lizard Species Maintains Its Genetic Diversity." *New York Times*, February 23.

Bimbenet-Privat, Michèle, and Alexis Kugel. 2007. "La Daphné d'argent et de corail par Wenzel Jamnitzer au Musée National de la Renaissance." *La revue du Louvre et des musées de France* 4:62–74.

Bingener, Andreas, Christoph Bartels, and Rainer Slotta, eds. 2006. *Das Schwazer Bergbuch*. 3 vols. Bochum: Deutsches Bergbau-Museum.

Biow, Douglas. 2015. *On the Importance of Being an Individual in Renaissance Italy: Men, Their Professions, and Their Beards*. Philadelphia: University of Pennsylvania Press.

Biringuccio, Vannoccio. 1943. *The Pirotechnia* (1540). Trans. Cyril Stanley Smith and Martha Teach Gnudi. New York: Basic Books.

Blair, Ann. 1999. "Authorship in the Popular 'Problemata Aristotelis.'" *Early Science and Medicine* 4, no. 3:189–227.

Blair, Ann. 2010. *Too Much to Know: Managing Scholarly Information before the Modern Age*. New Haven: Yale University Press.

Blockmans, Wim, and Walter Prevenier. 1999. *The Promised Lands: The Low Countries under Burgundian Rule, 1369–1530*. Trans. Elizabeth Fackelman. Philadelphia: University of Pennsylvania Press.

Boesch, Hans, ed. 1888. "Urkunden und Auszüge aus dem Archiv und der Bibliothek des Germanischen Museums in Nürnberg." *Jahrbuch des kunsthistorischen Sammlungen der allerhöchsten Kaiserhauses* 7.

Bohde, Daniela. 2007. "'Le tinte delle carni': Zur Begrifflichkeit für Haut und Fleisch in Italienischen Kunsttraktaten des 15. bis 17. Jahrhunderts." In *Weder Haut noch Fleisch: Das Inkarnat in der Kunstgeschichte*, ed. Daniela Bohde and Mechthild Fend, 41–63. Berlin: Gebr. Mann.

Bol, Marjolijn. 2014. "Coloring Topaz, Crystal and Moonstone: Gems and the Imitation of Art and Nature, 300–1500." In *Fakes!? Hoaxes, Counterfeits and Deception in Early Modern Science*, ed. Marco Beretta and Maria Conforti, 108–29. Sagamore Beach: Science History Publications.

Bol, Marjolijn. 2019. "The Emerald and the Eye: On Sight and Light in the Artisan's Workshop and the Scholar's Study." In *Perspective as Practice: Renaissance Cultures of Optics*, ed. Sven Dupré. Turnhout: Brepols.

Bondioli, Mauro. 2009. "Early Shipbuilding Records and the Book of Michael of Rhodes." In *The Book of Michael of Rhodes*, ed. Pamela O. Long, David McGee, and Alan M. Stahl, 3:243–80. Cambridge, MA: MIT Press.

Borchert, Till-Holger, ed. 2002. *De eeuw van Van Eyck 1430–1530: De Vlaamse Primitieven en het Zuiden*. Ghent: Ludion.

Borchert, Till-Holger, ed. 2011. *Van Eyck to Dürer: The Influence of Early Netherlandish Painting on European Art, 1430–1530*. London: Thames & Hudson.

Bostock, Sahar. 2020. "Tablets." In *Secrets of Craft and Nature*, ed. Making and Knowing Project et al. https://edition640.makingandknowing.org/#/essays/ann_068_fa_18.

Bott, Gerhard, ed. 1985. *Wenzel Jamnitzer und die Nürnberger Goldschmiedekunst 1500–1700*. Exh. cat., Germanisches Nationalmuseum, Nuremberg. Munich: Klinkhardt & Bierman.

Bouillon, Didier, André Guillerme, Martine Mille, and Gersende Piernas, eds. 2017. *Gestes techniques, techniques du geste: Approches pluridisciplinaires*, ed. Villeneuve d'Ascq: Presses Universitaires du Septentrion.

Boulboullé, Jenny. 2019. "Drawn Up by a Learned Physician from the Mouths of Artisans: The *Mayerne* Manuscript Revisited." In *Lessons in Art: Art, Education, and Modes of Instruction since 1500*, ed. Eric Jorink, Ann-Sophie Lehmann, Bart Ramakers, 204–49. Leiden: Brill.

Bourgarit, David, and Nicolas Thomas. 2011. "From Laboratory to Field Experiments: Shared Experience in Brass Cementation." *Historical Metallurgy* 45, no. 1:8–16.

Bourgeois, Luc, Danièle Alexandre-Bidon, Laurent Feller, Perrine Mane, Catherine Verna, and Mickaël Wilmart, eds. 2018. *La culture matérielle: Un objet en question. Anthropologie, archéologie et histoire. Actes du colloque international de Caen (9 et 10 octobre 2015)*. Caen: Presses Universitaires de Caen.

Boyd, Emily, Jef Palframan, and Pamela H. Smith. 2020. "Making Gold Run for Casting." In *Secrets of Craft and Nature*, ed. Making and Knowing Project et al. https://edition640.makingandknowing.org/#/essays/ann_505_ad_20.

Brandl, Rainer. 1986. "Zwischen Kunst und Handwerk: Kunst und Künstler im mittelalterlichen Nürnberg." In *Nürnberg 1300–1550: Kunst der Gotik und Renaissance*. Munich: Prestel-Verlag, 51–60.

Bray, Francesca. 2008. "Science, Technique, Technology: Passages between Matter and Knowledge in Imperial Chinese Agriculture." *British Journal for the History of Science* 4, no. 3: 319–44.

Brepohl, Erhard. 1987. *Theophilus Presbyter und die Mittelalterliche Goldschmiedekunst*. Vienna: Bohlau.

Bresc-Bautier, Geneviève. 1990. "Jean Séjourné, sculpteur-fontainier d'Henri IV, mort au Louvre en 1614." *Bulletin de la Société Nationale des Antiquaires de France* 1992:140–56.

Brioist, Pascal. 2013. *Léonard homme de guerre*. Paris: Alma.

Broecke, Lara. 2015. *Cennino Cennini's "Il libro dell'arte": A New English translation and Commentary with Italian Transcription*. London: Archetype.

Brown, Cecil H. 1979. "Folk Zoological Life-Forms: Their Universality and Growth." *American Anthropologist* n.s. 81, no. 4:791–817.

Brüning, Volker Fritz. 2004. *Bibliographie der alchemistischen Literatur, Band 1: Die alchemistischen Druckwerke von der Erfindung der Buchdruckerkunst bis zum Jahr 1690*. Munich: K. G. Saur.

Brunschwygk, Hieronymus. 1500. *Liber de arte distillandi; De simplicibus; Das buch rechten kunst zu distilieren die eintzigen dinge*. Strasbourg: Johannes Grüeninger.

Bryce, Judith. 1995. "The Oral World of the Early Accademia Fiorentina." *Renaissance Studies* 9:77–103.

Bucklow, Spike. 1999. "Paradigms and Pigment Recipes: Vermilion, Synthetic Yellows and the Nature of Egg." *Zeitschrift für Kunsttechnologie und Konservierung* 13: 140–49.

Bucklow, Spike. 2000. "Paradigms and Pigment Recipes: Natural Ultramarine." *Zeitschrift für Kunsttechnologie und Konservierung* 14:5–14.

Bucklow, Spike. 2001a. "Paradigms and Pigment Recipes: Silver and Mercury Blues." *Zeitschrift für Kunsttechnologie und Konservierung* 15:25–33.

Bucklow, Spike. 2001b. "The Use of Metals in a Fourteenth-century East Anglian Painters' Workshop: The Thornham Parva Retable." In *New Offerings, Ancient Treasures: Studies in Medieval Art for George Henderson*, ed. Paul Binski and William Noel. Phoenix Mill, Thrupp, Stroud, Gloucestershire, UK: Sutton Publishing.

Buning, Marius. 2017. "Promoting Technical Knowledge: Printing Privileges and Technical Literature in the Early Dutch Republic." In *Le livre technique avant le XXe siècle: À l'échelle du monde*, ed. Liliane Hilaire-Pérez, Valérie Nègre, Delphine Spicq, and Koen Vermeir, 453–62. Paris: CNRS éditions. http://books.openedition.org/editionscnrs/27787.

Büttner, Jochen. 2017. "Shooting with Ink." In *Structures of Practical Knowledge*, ed. Matteo Valleriani, 115–66. Cham: Springer International.

Bynum, Caroline Walker. 2007. *Wonderful Blood: Theology and Practice in Late Medieval Northern Germany and Beyond*. Philadelphia: University of Pennsylvania Press.

Campbell, Lorne, Susan Foister, and Ashok Roy, eds. 1997. "The Methods and Materials of Northern European Painting, 1400–1550." *National Gallery Technical Bulletin* 18:6–55.

Camps, Celine, and Margot Lyautey. 2020. "Ma<r>king and Knowing: Encoding BnF Ms. Fr. 640." In *Secrets of Craft and Nature*, ed. Making and Knowing Project et al. https://edition640.makingandknowing.org/#/essays/ann_335_ie_19.

Cardinal, Catherine. 2016. "Le banc d'orfèvre du prince électeur Auguste de Saxe (1526–1586) au Musée National de la Renaissance." In *L'Europe technicienne: XVe–XVIIIe siècle*, ed. Catherine Cardinal, Liliane Pérez, Delphine Amélie Spicq, and Marie Thébaud-Sorger. Special issue. *Artefact: Technique, histoire et sciences humaines* 4:367–70.

Cardinal, Catherine, Liliane Pérez, Delphine Amélie Spicq, and Marie Thébaud-Sorger, eds. 2016. *L'Europe technicienne: XVe—XVIIIe siècle*. Special issue. *Artefact: Technique, histoire et sciences humaines* 4.

Carl, Doris. 1987. "Das Inventar der Werkstatt von Filippino Lippi aus dem Jahre 1504." *Mitteilungen des Kunsthistorischen Institutes in Florenz* 31:373–91.

Carlson, Raymond, and Jordan Katz. 2020. "Casting in Frames." In *Secrets of Craft and Nature*, ed. Making and Knowing Project et al. https://edition640.makingandknowing.org/#/essays/ann_010_fa_14.

Carlyle, Leslie. 2020. "Reconstructions of Oil Painting Materials and Techniques: The HART Model for Approaching Historical Accuracy." In *Reconstruction, Replication and Re-enactment in the Humanities and Social Sciences*, ed. Sven Dupré, Anna Harris, Patricia Lulof, and Julia Kursell. Amsterdam: Amsterdam University Press.

Carruthers, Mary. 2008. *The Book of Memory: A Study of Memory in Medieval Culture*. 2nd ed. Cambridge: Cambridge University Press.

Castiglione, Baldesar. 2002. *The Book of the Courtier*. Trans. Charles Singleton; ed. Daniel Javitch. New York: W. W. Norton.

Cataldo, Emogene, and Julianna van Visco. 2020a. "*Eau Magistra*: Investigating Binders for Sand Molds." In *Secrets of Craft and Nature*, ed. Making and Knowing Project et al. https://edition640.makingandknowing.org/#/essays/ann_011_fa_14.

Cataldo, Emogene, and Julianna van Visco. 2020b. "Wax and Tallow: Material Explorations." In *Secrets of Craft and Nature*, ed. Making and Knowing Project et al. https://edition640.makingandknowing.org/#/essays/ann_001_fa_14.

Celaschi, Massimiliano, and Angonella Gregori. 2014. *Da Girolamo Ruscelli a Alessio Piemontese: "I Secreti" in Italia e in Europa dal Cinque al Settecento*. Rome: Vecchiarelli Editore.

Cellini, Benvenuto. 1956. *The Autobiography of Benvenuto Cellini*. Trans. George Bull. London: Penguin.

Cellini, Benvenuto. 1967. *The Two Treatises on Goldsmithing and Sculpture*. Trans. C. R. Ashbee. New York: Dover.

Cennini, Cennino d'Andrea. 1960. *The Craftsman's Handbook: Il libro dell'Arte*. Trans. Daniel V. Thompson Jr. New York: Dover.

Chang, Hasok. 2011. "How Historical Experiments Can Improve Scientific Knowledge and Science Education: The Cases of Boiling Water and Electrochemistry." *Science and Education* 20, nos. 3–4:317–41.

Chaplin, Joyce E. 2008. "Knowing the Ocean: Benjamin Franklin and the Circulation of Atlantic Knowledge." In *Science and Empire in the Atlantic World*, ed. James Delbourgo and Nicholas Dew, 73–96. New York: Routledge.

Chartier, Roger. 1989. "The Practical Impact of Writing." In *The Book History Reader*. Cambridge, MA: Belknap Press of Harvard University Press.

Chartier, Roger. 1994. *The Order of Books: Readers, Authors, and Libraries in Europe between the Fourteenth and Eighteenth Centuries*. Trans. Lydia G. Cochrane. Stanford: Stanford University Press.

Chartier, Roger. 2007. "The *Order of Books* Revisited." *Modern Intellectual History* 4, no. 3:509–19.

Chartier, Roger. 2011. "The Practical Impact of Writing." In *The History of Reading: A Reader*, ed. Shafquat Towheed, Rosalind Crone, and Katie Halsey. London: Routledge.

Chartier, Roger, and Peter Stallybrass. 2013. "What Is a Book?" In *The Cambridge Companion to Textual Scholarship*, ed. Neil Fraistat and Julia Flanders, 188–204. Cambridge: Cambridge University Press.

Chessa, Maria Alessandra. 2020. "Counterfeit Coral." In *Secrets of Craft and Nature*, ed. Making and Knowing Project et al. https://edition640.makingandknowing.org/#/essays/ann_015_sp_15.

Chiostrini, Giulia, and Jef Palframan. 2020. "Molding a Rose." In *Secrets of Craft and Nature*, ed. Making and Knowing Project et al. https://edition640.makingandknowing.org/#/essays/ann_022_sp_15.

Chrisman, Miriam Usher. 1982. *Lay Culture, Learned Culture: Books and Social Change in Strasbourg, 1488–1599*. New Haven: Yale University Press.

Cifoletti, Giovanna. 2006a. "From Valla to Viète: The Rhetorical Reform of Logic and Its Use in Early Modern Algebra." *Early Science and Medicine* 11, no. 4:390–423.

Cifoletti, Giovanna. 2006b. "Mathematics and Rhetoric: Introduction." *Early Science and Medicine* 11, no. 4:369–89.

Clark, Andy. 1997. *Being There: Putting Brain, Body, and World Together Again*. Cambridge, MA: MIT Press.

Clarke, Mark. 2001. *The Art of All Colours: Mediaeval Recipe Books for Painters and Illuminators*. London: Archetype.

Clarke, Mark. 2011. *Medieval Painters' Materials and Techniques: The Montpellier* Liber diversarum arcium. London: Archetype.

Clarke, Mark. 2013a. "The Earliest Technical Recipes: Assyrian Recipes, Greek Chemical Treatises and the *Mappae Clavicula* Text Family." In *Craft Treatises and Handbooks: The Dissemination of Technical Knowledge in the Middle Ages*, ed. Ricardo Córdoba, 9–31. Turnhout: Brepols.

Clarke, Mark. 2013b. "Late Medieval Artists' Recipes Books (14th–15th Centuries). In *Craft Treatises and Handbooks: The Dissemination of Technical Knowledge in the Middle Ages*, ed. Ricardo Córdoba, 33–53. Turnhout: Brepols.

Clarke, Mark, Bert De Munck, and Sven Dupré, eds. 2012. *Transmission of Artists' Knowledge*. Brussels: Koninklijke Vlaamse Academie van Belgie voor Wetenschappen en Kunsten.

Clerbois, Sébastian, and Martina Droth, eds. 2011. *Revival and Invention: Sculpture through Its Material Histories*. Oxford: Peter Lang.

Clunas, Craig. 1998. "Luxury Knowledge: The *Xiushilu* ('Records of Lacquering') of 1625." *Techniques & Culture* 29:27–40.

Cole, Arthur H., and George B. Watts. 1952. *The Handicrafts of France*. Boston: Harvard Graduate School of Business Administration.

Cole, Michael W. 1999. "Cellini's Blood." *Art Bulletin* 81, no. 2:215–35.

Cole, Michael W. 2002. *Cellini and the Principles of Sculpture*. Cambridge: Cambridge University Press.

Collins, Harry M. 2001. "Tacit Knowledge, Trust and the Q of Sapphire." *Social Studies of Science* 31, no. 1:71–85.

Collins, Harry M. 2010. *Tacit and Explicit Knowledge*. Chicago: University of Chicago Press.

Collins, Harry M., and Robert Evans. 2006. "The Third Wave of Science Studies: Studies of Expertise and Experience." In *The Philosophy of Expertise*, ed. Evan Selinger and Robert P. Crease, 39–110. New York: Columbia University Press.

Conner, Clifford D. 2005. *A People's History of Science: Miners, Midwives, and "Low Mechanicks."* New York: Nation Books.

Cooper, Lisa H. 2011. *Artisans and Narrative Craft in Late Medieval England*. Cambridge: Cambridge University Press.

Copp, Joan. 1521. *Ein nutzlich Regiment*. Erfurt: Matthes Maler.

Corbellini, Sabrina, and Margriet Hoogvliet. 2013. "Artisans and Religious Reading in Late Medieval Italy and Northern France (ca. 1400–ca. 1520)." *Journal of Medieval and Early Modern Studies* 43, no. 3:521–44.

Córdoba, Ricardo, ed. 2013. *Craft Treatises and Handbooks: The Dissemination of Technical Knowledge in the Middle Ages*. Turnhout: Brepols.

Cormack, Bradin, and Carla Mazzio. 2005. *Book Use, Book Theory, 1500–1700*. Chicago: University of Chicago Library.

Cossart, Brice. 2017. "Traités d'artillerie et écoles d'artilleurs: Intéractions entre pratiques d'enseignement et livres techniques à l'époque de Philippe II d'Espagne." In *Le livre technique avant le xxe siècle: À l'échelle du monde*, ed. Liliane Hilaire-Pérez, Valérie Nègre, Delphine Spicq, and Koen Vermeir, 341–53. Paris: CNRS Éditions. http://books.openedition.org/editionscnrs/27763.

Cowan, Brian. 2006. "Art and Connoisseurship in the Auction Market of Later Seventeenth-Century London." In *Mapping Markets for Paintings in Europe 1450–1750*, ed. Neil De Marchi and Hans J. van Miegroet, 263–84. Turnhout: Brepols.

Coward, Fiona. 2016. "Scaling Up: Material Culture as Scaffold for the Social Brain." *Quaternary International* 405:78–90.

Coy, Michael W., ed. 1989a. *Apprenticeship: From Theory to Method and Back Again*. Albany: State University of New York Press.

Coy, Michael W. 1989b. "To Method." In *Apprenticeship: From Theory to Method and Back Again*, ed. Michael W. Coy, 107–13. Albany: State University of New York Press.

Craddock, Paul, and Simon Timberlake. 2004/5. "Smelting Experiments at Butser." *HMS News* 58 (Winter): 1–3.

Crawford, Matthew B. 2010. *Shop Class as Soulcraft: An Inquiry into the Value of Work*. New York: Penguin.

Crossgrove, William. 2000. "The Vernacularization of Science, Medicine, and Technology in Late Medieval Europe: Broadening Our Perspectives." *Early Science and Medicine* 5, no. 1:47–63.

Crowther-Heyck, Kathleen. 2003. "Wonderful Secrets of Nature: Natural Knowledge and Religious Piety in Reformation Germany," *Isis* 94, no. 2:253–73.

Crowther-Heyck, Kathleen. 2010. *Adam and Eve in the Protestant Reformation*. Cambridge: Cambridge University Press.

Cuomo, Serafina. 2007. *Technology and Culture in Greek and Roman Antiquity*. New York: Cambridge University Press.

Daiber, Hans. 1990. "Qostā ibn Lūqā (9. Jh.) über die Einteilung der Wissenschaften." *Zeitschrift der arabisch-islamischen Wissenschaften* 6:93–125.

Damm, Heiko, Michael Thimann, and Claus Zittel. 2013. "Introduction: Close and Extensive Reading among Artists in the Early Modern Period." In *The Artist as Reader: On Education and Non-education of Early Modern Artists*, ed. Heiko Damm, Michael Thimann, and Claus Zittel, 1–68. Leiden: Brill.

Daniel der Bergverstendig [pseud.]. 1533. *Bergwerck vnd Probirbüchlin: Für die Bergk vnd Feuerwercker/Goltschmid/Alchimisten vnd Künstner*. Frankfurt am Main: Christian Egenolff.

Darmstaedter, Ernst. 1926. *Berg-, Probir- und Kunstbüchlein*. Munich: Verlag der Münchner Drucke.

Davidowitz, T., T. Beentjes, J. van Bennekom, and S. Creange. 2012. "Identifying 16th-Century Paints on Silver Using a Contemporary Manuscript." In *The Artist's Process: Technology and Interpretation: Proceedings of the 4th Symposium of the Art Technological Source Research Working Group*, ed. S. Eyb-Green, 72–78. London: Archetype.

Davids, Karel. 2005. "Craft Secrecy in Europe in the Early Modern Period: A Comparative View." *Early Science and Medicine* 10:341–48.

Davids, Karel. 2008. *The Rise and Decline of Dutch Technological Leadership: Technology, Economy, and Culture in the Netherlands, 1350–1800*. 2 vols. Leiden: Brill.

Davids, Karel, and Bert De Munck, eds. 2014. *Innovation and Creativity in Late Medieval and Early Modern European Cities*. Farnham, Surrey, UK: Ashgate.

Davidson, Adam. 2012. "It Ain't Just Pickles." *New York Times Magazine*, February, 14–17.

Davidson, Hilary, and Anna Hodson. 2007. "Joining Forces: The Intersection of Two Replica Garments." In *Textiles and Text*, ed. Maria Hayward and Elizabeth Kramer, 204–10. London: Archetype.

De Alcega, Juan. 1979. *Tailor's Pattern Book, 1589*. Trans. Jean Pain and Cecilia Bainton. Carlton, UK: Ruth Bean.

De Avila, Ludovicus. 1531. *Ein Nutzlich Regiment*. Trans. Michael Krautwadel. Augsburg: Heynrich Steyner.

Deblock, Geneviève. 2015. *Le Bâtiment des recettes: Présentation et annotation de l'édition Jean Ruelle, 1560*. Rennes: Presses Universitaires de Rennes.

Deblock, Geneviève. 2017. "Le *Bâtiment des recettes*, un ouvrage technique français de la Bibliothèque bleue." In *Le livre technique avant le xxᵉ siècle: À l'échelle du monde*, ed. Liliane Hilaire-Pérez, Valérie Nègre, Delphine Spicq, and Koen Vermeir, 195–201. Paris: CNRS Éditions. http://books.openedition.org/editionscnrs/27724.

Debuiche, Colin. 2020. "Le Ms. Fr. 640 et la collection Béthune." In *Secrets of Craft and Nature*, ed. Making and Knowing Project et al. https://edition640.makingandknowing.org/#/essays/ann_316_ie_19.

De Groot, Jakob Jan Maria. 1892. *The Religious System of China: Its Ancient Forms, Evolution, History and Present Aspect, Manners, Custom and Social Institutions Connected Therewith*. Vol. 5. Leiden: Brill.

De Lalande, M. 1764. *Art du tanneur*. Paris: H. L. Guérin & L. F. Delatour.

Delamont, Sara, and Paul Atkinson. 2001. "Doctoring Uncertainty: Mastering Craft Knowledge." *Social Studies of Science* 31, no. 1:87–107.

Della Casa, Giovanni. 2013. *Galateo, or The Rules of Polite Behavior (1558)*. Trans. M. F. Rusnak. Chicago: University of Chicago Press.

Demeter, Béla. 2020. "Mr Toad's Wild Ride: From Bestiary to Shop Manual." In *Secrets of Craft and Nature*, ed. Making and Knowing Project et al. https://edition640.makingandknowing.org/#/essays/ann_500_ad_20.

De Munck, Bert. 2007. *Technologies of Learning: Apprenticeship in Antwerp Guilds from the Fifteenth Century to the End of the Ancien Régime*. Turnhout: Brepols.

De Munck, Bert. 2014. "Artisans, Products, and Gifts: Rethinking the History of Material Culture in Early Modern Europe." *Past and Present* 224:39–74.

De Munck, Bert. 2019. "Artisans as Knowledge Workers: Craft and Creativity in a Long Term Perspective." *Geoforum* 99 (February): 227–37.

De Munck, Bert, Steven L. Kaplan, and Hugo Soly. 2007. *Learning on the Shop Floor: Historical Perspectives on Apprenticeship*. New York: Berghahn Books.

De Preester, Helena. 2011. "Technology and the Body: The (Im)possibilities of Re-embodiment." *Foundations of Science* 16:119–37.

Detienne, Marcel, and Jean-Pierre Vernant. 1978. *Cunning Intelligence in Greek Culture and Society*. Trans. Janet Lloyd. Atlantic Highlands, NJ: Humanities Press.

Diamond, Norma. 1988. "The Miao and Poison: Interactions on China's Southwest Frontier." *Ethnology* 27, no. 1:1–25.

Diderot, Denis [attrib.]. 1751. "Art." *Encyclopédie ou Dictionnaire raisonné des sciences, des arts et des métiers*, 1:713–17. Paris.

Diderot, Denis. 2003. "Art." *The Encyclopedia of Diderot & d'Alembert Collaborative Translation Project*. Trans. Nelly S. Hoyt and Thomas Cassirer. Ann Arbor: MPublishing, University of Michigan Library. https://quod.lib.umich.edu/d/did/did2222.0000.139/.

Diemer, Dorothea. 1986. "Bronzeplastik um 1600 in München: Neue Quellen und Forschungen." In *Jahrbuch des Zentralinstituts für Kunstgeschichte* 2:107–77.

Diemer, Dorothea, Peter Diemer, and Lorenz Seelig, eds. 2018. *Die Münchner Kunstkammer*. 3 vols. Munich: C. H. Beck.

DiMeo, Michelle, and Sara Pennell. 2013. *Reading and Writing Recipe Books, 1550–1800*. Manchester, UK: Manchester University Press.

Diorio, Jennifer. 2013. "Neri di Bicci and the Diffusion of Cartoons between Fifteenth-century Florentine Workshops." In *The Renaissance Workshop*, ed. David Saunders, Marika Spring, and Andrew Meek, 170–71. London: Archetype.

Doering, Oscar. 1894. "Des Augsburger Patriziers Philipp Hainhofer Beziehungen zum Herzog Philipp II von Pommern-Stettin: Correspondenzen aus den Jahren 1610–1619." *Quellenschriften für Kunstgeschichte und Kunsttechnik des Mittelalters und der Neuzeit*, NF 6. Vienna: Carl Graeser.

Dormer, Peter. 1994. *The Art of the Maker: Skill and Its Meaning in Art, Craft and Design*. London: Thames & Hudson.

Dormer, Peter. 1997. *The Culture of Craft: Status and Future*. Manchester, UK: Manchester University Press.

Downey, Greg. 2010. "'Practice without Theory': A Neuroanthropological Perspective on Embodied Learning." In *Making Knowledge: Explorations of the Indissoluble Relation between Mind, Body and Environment*, ed. Trevor H. J. Marchand, 21–38. Chichester, West Sussex, UK: Wiley-Blackwell.

Dreyfus, Hubert L. 1992. *What Computers Still Can't Do: A Critique of Artificial Reason*. Cambridge, MA: MIT Press.

Dreyfus, Hubert L. 2002a. "Intelligence without Representation: The Relevance of Phenomenology to Scientific Explanation." *Phenomenology and the Cognitive Sciences* 1, no. 4:367–83.

Dreyfus, Hubert L. 2002b. "Refocusing the Question: Can There Be Skillful Coping without Propositional Representations or Brain Representations?" *Phenomenology and the Cognitive Sciences* 1, no. 4:413–25.

Druart, Therese-Anne. 2020. "al-Farabi." In *The Stanford Encyclopedia of Philosophy*, ed. Edward N. Zalta. https://plato.stanford.edu/archives/fall2020/entries/al-farabi/.

Dunkerton, Jill. 1991. *Giotto to Dürer: Early Renaissance Painting in the National Gallery*. New Haven: Yale University Press.

Dunkerton, Jill. 1996–98. "Observations on the Handling Properties of Binding Media Identified in European Painting from the Fifteenth to the Seventeenth Centuries." *Bulletin de l'Institut Royal du Patrimoine Artistique* 27:287–92.

Dupré, Sven. 2018. "The Role of Judgment in the Making of Glass Colors in the Seventeenth Century." *Ferrum: Nachrichten aus der Eisenbibliothek, Stiftung der Georg Fischer AG* 90.

Dupré, Sven, and Christine Göttler, eds. 2019. *Knowledge and Discernment in the Early Modern Arts*. London: Routledge.

Dupré, Sven, Anna Harris, Patricia Lulof, Julia Kursell, and Maartje Stols-Witlox, eds. 2020. *Reconstruction, Replication and Re-enactment in the Humanities and Social Sciences*. Amsterdam: Amsterdam University Press.

Dürer, Albrecht. 1969. *Schriftlicher Nachlass*. Vol. 3. Ed. Hans Rupprich. Berlin: Deutscher Verlag für Kunstwissenschaft.

Dürer, Albrecht, and Roger Fry. 1995. *Dürer's Record of Journeys to Venice and the Low Countries*. New York: Dover.

Dym, Warren. 2006. "Mineral Fumes and Mining Spirits: Popular Beliefs in the Sarepta of Johann Mathesius (1504–1565)." *Renaissance and Reformation Review* 8, no. 2:161–85.

Dyon, Soersha, and Heather Wacha. 2020. "Turning Turtle: The Process of Translating BnF Ms. Fr. 640." In *Secrets of Craft and Nature*, ed. Making and Knowing Project et al. https://edition640.makingandknowing.org/#/essays/ann_318_ie_19.

Eamon, William. 1994. *Science and the Secrets of Nature: Books of Secrets in Medieval and Early Modern Culture*. Princeton, NJ: Princeton University Press.

Eamon, William. 2010. *The Professor of Secrets: Mystery, Medicine, and Alchemy in Renaissance Italy*. Washington, DC: National Geographic Society.

Eamon, William. 2011. "How to Read a Book of Secrets." In *Secrets and Knowledge in Medicine and Science, 1500–1800*, ed. Elaine Leong and Alisha Rankin, 23–46. Farnham, Surrey, UK: Ashgate.

Eamon, William, and Françoise Paheau. 1984. "The Accademia Segreta of Girolamo Ruscelli: A Sixteenth-Century Italian Scientific Society." *Isis* 75, no. 2:327–42.

Ebel, Heinrich. 1939. "Der 'Herbarius communis' des Hermannus de Sancto Portu (ca. 1284) und das 'Arzneibüchlein' (1484) des Claus von Metry." Diss., Friedrich-Wilhelms-Universität, Berlin.

Edgerton, David. 1999. "From Innovation to Use: Ten Eclectic Theses on the Historiography of Technology." *History and Technology* 16, no. 2:111–36.

Effmert, Viola 1989. "'... ein schön kunstlich silbre vergult truhelein ...': Wenzel Jamnitzers Prunkkassette in Madrid." *Anzeiger des Germanischen Nationalmuseums*: 131–58.

Egmond, Florike. 1999. "Natuurlijke historie en savoir prolétaire." In *Kometen, monsters en muilezels: Het veranderende natuurbeeld en de natuurwetenschap in de zeventiende eeuw*, ed. Florike Egmond and Eric Jorink. Arcadia: R. H. Vermij.

Eichberger, Dagmar. 2002. *Leben mit Kunst: Wirken durch Kunst. Sammelwesen und Hofkunst unter Margarete von Österreich, Regentin der Niederlande*. Turnhout: Brepols.

Eis, Gerhard. 1960. *Meister Albrants Rossarzneibuch: Verzeichnis der Handschriften, Text der ältesten Fassung, Literaturverzeichnis*. Constance: Terra-Verlag.

Ekholm, Karin J. 2010. "Tartaglia's *Ragioni*: A *Maestro d'Abaco's* Mixed Approach to the Bombardier's Problem." *British Journal for the History of Science* 43:181–207.

EPA. 2017. "Lead Poisoning and Your Children." EPA Publication EPA-747-F-16–001 (December). Accessed August 3, 2020. https://www.epa.gov/sites/production/files/2018–02/documents/epa_lead_brochure-posterlayout_508.pdf.

Epstein, Stephan R., and Maarten Prak. 2008. *Guilds, Innovation, and the European Economy, 1400–1800*. Cambridge: Cambridge University Press.

Eyferth, Jacob. 2010. "Craft Knowledge at the Interface of Written and Oral Cultures." *East Asian Science, Technology and Society* 4:185–205.

Fachhochschule (Köln). 2001. Fachbereich Restaurierung und Konservierung von Kunst- und Kulturgut. "Die Farben des Mittelalters." In *Restaurieren heißt verstehen: Zerstörungsfreie Untersuchung und Restaurierung in der Studienrichtung Restaurierung und Konservierung von Schriftgut, Graphik und Buchmalerei*. Cologne: Fachhochschule Köln.

Falchetta, Piero. 2009. "The Portolan of Michael of Rhodes." In *The Book of Michael of Rhodes*, ed. Pamela O. Long, David McGee, and Alan M. Stahl, 3:193–210. Cambridge, MA: MIT Press.

Fehrenbach, Frank. 1997. *Licht und Wasser: Zur Dynamik naturphilosophischer Leitbilder im Werk Leonardo da Vincis*. Tübingen: E. Wasmuth.

Felfe, Robert. 2015. *Naturform und bildnerische Prozesse: Elemente einer Wissensgeschichte in der Kunst des 16. und 17. Jahrhunderts*. Berlin: Walter De Gruyter.

Feng, H. Y., and J. K. Shryock. 1935. "The Black Magic in China Known as Ku." *Journal of the American Oriental Society* 55, no. 1:1–30.

Ferguson, John K. 1959. *Bibliographical Notes on Histories of Inventions and Books of Secrets* (1898), 2 vols. London: Holland Press.

Fickler, Johann Baptist. 2004. *Das Inventar der Münchner herzoglichen Kunstkammer von 1598. Bayerische Akademie der Wissenschaften, Philosophisch-Historische Klasse Abhandlungen*, n.s. 125, ed. Peter Diemer. Munich: C. H. Beck.

Floyd-Wilson, Mary. 2013. *Occult Knowledge, Science, and Gender on the Shakespearean Stage*. Cambridge: Cambridge University Press.

Flötner, Peter. 1549. *Kunstbuch*. Zurich: Rudolff Wyssenbach.

Forgas, Joseph P., ed. 1981. *Social Cognition: Perspectives on Everyday Understanding*. London: Academic Press.

Formisano, Marco. 2001. *Tecnica e scrittura: Le letterature tecnico-scientifiche nello spazio letterario tardolatino*. Rome: Carocci.

Formisano, Marco. 2013. "Late Latin Encyclopedism: Toward a New Paradigm of Practical Knowledge." In *Encyclopaedism from Antiquity to the Renaissance*, ed. Jason König and Greg Woolf, 197–215. New York: Cambridge University Press.

Fors, Hjalmar, Lawrence Principe, and Otto Sibum. 2016. "From the Library to the Laboratory and Back Again: Experiment as a Tool for Historians of Science." *Ambix* 63:85–97.

Fountain, Henry. 2009. "The Tail of a Gecko Has Life of Its Own." *New York Times*, September 10.

Fox, Ariel. 2016. "Precious Bodies: Money Transformation Stories from Medieval to Late Imperial China." *Harvard Journal of Asiatic Studies* 76, nos. 1–2:43–85.

Fox, Celina. 2010. *The Arts of Industry in the Age of the Enlightenment*. New Haven: Yale University Press.

Foyer, Emilie. 2020. "Color of Gold without Gold on Silver." In *Secrets of Craft and Nature*, ed. Making and Knowing Project et al. https://edition640.makingandknowing.org/#/essays/ann_032_fa_15.

Franci, Raffaela. 2009. "Mathematics in the Manuscript of Michael of Rhodes." In *The Book of Michael of Rhodes*, ed. Pamela O. Long, David McGee, and Alan M. Stahl, 3:115–46. Cambridge, MA: MIT Press.

Frankl, Paul. 1945. "The Secret of Medieval Masons." *Art Bulletin* 27, no. 1:46–60.

Fransen, Sietske, and Katherine M. Reinhart, eds. 2019. "The Practice of Copying in Making Knowledge in Early Modern Europe." Special issue. *Word & Image* 35, no. 3:211–22.

Fransen, Sietske, Katherine M. Reinhart, and Sachiko Kusukawa. 2019. "Copying Images in the Archives of the Early Royal Society." Special issue. *Word & Image*, 35, no. 3:256–76.

Fu, Shiye, Zhiqi Zhang, and Pamela H. Smith. 2020. "Molding Grasshoppers and Things Too Thin." In *Secrets of Craft and Nature*, ed. Making and Knowing Project et al. https://edition640.makingandknowing.org/#/essays/ann_013_sp_15.

Fuciková, Eliska, ed. 1988. *Prag um 1600: Beiträge zur Kunst und Kultur am Hofe Rudolfs II* Freren, Germany: Luca.

Fuciková, Eliska, et al., eds. 1997. *Rudolf II and Prague: The Court and the City*. London: Thames & Hudson.

Gage, John. 1998. "Colour Words in the High Middle Ages." In *Looking through Paintings: The Study of Painting Techniques and Materials in Support of Art Historical Research*, ed. Erma Hermens, 35–48. Leids Kunsthistorisch Jaarboek 11. Baarn, Netherlands: Uitgeverij de Prom.

Gans, Sofia. 2020. "Excellent Sand from Alabaster." In *Secrets of Craft and Nature*, ed. Making and Knowing Project et al. https://edition640.makingandknowing.org/#/essays/ann_016_sp_15.

Garçon, Anne-Françoise. 2005. "Les dessous des métiers: Secrets, rites et soustraitance dans la France du XVIIIe siècle." *Early Science and Medicine* 10, no. 3:378–91.

Geertz, Clifford. 1983. *Local Knowledge: Further Essays in Interpretive Anthropology*. New York: Basic Books.

Gengzhe, Yu. 2006. *Xu gu zhi di: yi xiang wenhua qishi fuhao de qianzhuan liuyi* [Land of Sorcery: The Shift of a Symbol of Cultural Discrimination]. *Zhongguo shehui kexue* [Social Sciences in China] 2.

Gentilcore, David. 1995. "'Charlatans, Mountebanks and Other Similar People': The Regulation and Role of Itinerant Practitioners in Early Modern Europe." *Social History* 20, no. 3:297–314.

Gentilcore, David. 2004. "Was There a Popular Medicine in Early Modern Europe?" *Folklore* 115, no. 2:151–66.

Geoghegan, Arthur. 1945. *The Attitude toward Labor in Christianity and Ancient Culture*. Washington, DC: Catholic University of America Press.

Gerbino, Anthony, and Stephen Johnston 2009. *Compass and Rule: Architecture as Mathematical Practice in England, 1500–1750*. New Haven: Yale University Press.

Gerritsen, Anne, and Giorgio Riello, eds. 2015. *Writing Material Culture History*. London: Blooms-bury Academic.

Gettens, Rutherford J., Robert L. Feller, and W. T. Chase. 1994. "Vermilion and Cinnabar." In *Artists' Pigments: A Handbook of Their History and Characteristics*, ed. Ashok Roy. Vol. 2. Oxford: Oxford University Press.

Geurts, Kathryn Linn. 2005. "Consciousness as 'Feeling in the Body': A West African Theory of Embodiment, Emotion and the Making of Mind." In *Empire of the Senses: The Sensual Culture Reader*, ed. David Howes, 164–78. New York: Berg.

G. H. 1744. *Neu-eröfnetes Probier-Buch Darinnen nicht nur Alle Geheimnisse der Probierkunst*. Lübeck: Johann Benj. Rüdiger.

Gilbert, Neal W. 1960. *Renaissance Concepts of Method*. New York: Columbia University Press.

Glaisyer, Natasha, and Sara Pennell, eds. 2003. *Didactic Literature in England 1500–1800: Expertise Constructed*. Aldershot, UK: Ashgate.

Glatigny, Pascal Dubourg. 2017. "Réduction en art et *Erkenntnissteuerung*: Deux tendances historio-graphiques actuelles sur l'écriture des savoirs à l'époque modern." In *Penser la technique autre-ment: En homage à l'oeuvre d'Hélène Vérin*, 31–41. Paris: Classiques Garnier.

Glatigny, Pascal Dubourg, and Hélène Vérin. 2008a. "La réduction en art, un phénomène culturel." In *Réduire en art: La technologie de la Renaissance aux Lumières*, ed. Pascal Dubourg Glatigny and Hélène Vérin, 59–94. Paris: Éditions de la Maison des Sciences de l'Homme.

Glatigny, Pascal Dubourg, and Hélène Vérin, eds. 2008b. *Réduire en art: La technologie de la Renais-sance aux Lumières*. Paris: Éditions de la Maison des Sciences de l'Homme.

Gluch, Sibylle. 2007. "The Craft's Use of Geometry in 16th c. Germany: A Means of Social Advance-ment? Albrecht Dürer and After." *Anistoriton Journal* 10, no. 3:1–16.

Goldhagen, Daniel. 1996. *Hitler's Willing Executioners: Ordinary Germans and the Holocaust*. New York: Alfred A. Knopf.

Gombrich, Ernst H. 1997. *Speis der Malerknaben: Zu den technischen Grundlagen von Dürers Kunst*. Vienna: WUV-Universitätsverlag.

Goody, Esther N. 1989. "Learning, Apprenticeship and the Division of Labor." In *Apprenticeship: From Theory to Method and Back Again*, ed. Michael W. Coy, 233–56. Albany: State University of New York Press.

Goody, Jack. 1977. *The Domestication of the Savage Mind*. Cambridge: Cambridge University Press.

Goody, Jack. 1987. *The Interface between the Written and the Oral*. Cambridge: Cambridge University Press.

Goody, Jack. 2000. *Power of Written Tradition*. Washington, DC: Smithsonian Institution Press.

Göttler, Christine. 2008. "Fire, Smoke and Vapour: Jan Brueghel's 'Poetic Hells': 'Ghespoock' in Early Modern European Art." In *Spirits Unseen: The Representation of Subtle Bodies in Early Modern European Culture*, ed. Christine Göttler and Wolfgang Neuber. Leiden: Brill.

Göttler, Christine. 2010. *Last Things: Art and the Religious Imagination in the Age of Reform*. Turn-hout: Brepols.

Goulding, Robert. 2006. "Deceiving the Senses in the Thirteenth Century: Trickery and Illusion in the *Secretum Philosophorum*." In *Magic and the Classical Tradition*, ed. Charles S. F. Burnett and William F. Ryan, 135–62. London: Warburg Institute.

Grafit, Sasha. 2020. "Silkworms and the Work of Algiers." In *Secrets of Craft and Nature*, ed. Making and Knowing Project et al. https://edition640.makingandknowing.org/#/essays/ann_059_sp_17.

Grafton, Anthony. 2007. "Renaissance Histories of Art and Nature." In *The Artificial and the Natural:*

An Evolving Polarity, ed. Bernadette Bensaude-Vincent and William R. Newman, 185–210. Cambridge, MA: MIT Press.

Grasskamp, Anna. 2013. "Metamorphose in Rot: Die Inszenierung von Korallen-fragmenten in Kunstkammern des 16. und 17. Jahrhunderts." In *Metamorphosen*, ed. Jessica Ullrich and Antonia Ulrich. Special issue. *Tierstudien* 4:13–24.

Grasskamp, Anna. 2021. "Unpacking Foreign Ingenuity: The German Conquest of Artful Objects with 'Indian' Provenance around 1600." In *Ingenuity in the Making: Matter and Technique in Early Modern Europe*, ed. Richard J. Oosterhoff, José Ramón Marcaida, and Alexander Marr. Pittsburgh: University of Pittsburgh Press.

Gries, Christian. 1994. "Erzherzog Ferdinand II: Von Tirol und die Sammlungen auf Schloß Ambras." *Frühneuzeit-Info* 5:7–37.

Grocock, Christopher, and Sally Grainger. 2006. *Apicius*. Blackawton, Totnes, Devon, UK: Prospect Books.

Grosjean, Charles T. ca. 1876–83. Diaries, Working Notebooks, and Day Book. Brown University Library.

Guerzoni, Guido Antonio. 2012. "Use and Abuse of Beeswax in the Early Modern Age: Two Apologues and a Taste." In *Waxing Eloquent: Italian Portraits in Wax*, ed. A. Daninos, 43–59. Milan: Officina Libraria.

Guthmüller, Bodo. 1981. *Ovidio meta-morphoseos vulgare: Formen und Funktionen der volkssprachlichen Wiedergabe klassischer Dichtung in der italienischen Renaissance*. Veröffentlichungen zur Humanismusforschung 3. Boppard am Rhein.

Guthmüller, Bodo, Berndt Hamm, and Andreas Tönnesmann, eds. 2006. *Künstler und Literat: Schrift- und Buchkultur in der europäischen Renaissance*. Wiesbaden: Harrassowitz Verlag.

Haage, Bernhard Dietrich, and Wolfgang Wegner. 2007. *Deutsche Fachliteratur der Artes in Mittelalter und Früh Neuzeit*. Berlin: Erich Schmidt Verlag.

Haberman, Mechthild. 2002. *Deutsche Fachtexte der frühen Neuzeit: Naturkundlich-medizinische Wissensvermittlung im Spannungsfeld von Latein und Volkssprache*. Berlin: de Gruyter.

Hackenberg, Michael. 1986. "Books in Artisan Homes of Sixteenth-Century Germany." *Journal of Library History* 21:72–91.

Hackett, Jeremiah, ed. 1997. *Roger Bacon and the Sciences*. Leiden: Brill.

Hacking, Ian. 1983. *Representing and Intervening: Introductory Topics in the Philosophy of Natural Science*. Cambridge: Cambridge University Press.

Hagendijk, Thijs. 2018. "Learning a Craft from Books: Historical Re-enactment of Functional Reading in Gold- and Silversmithing." *Nuncius* 33:198–235.

Hagendijk, Thijs. 2019. "Unpacking Recipes and Communicating Experience: The *Ervarenissen* of Simon Eikelenberg (1663–1738) and the Art of Painting." *Early Science and Medicine* 24:248–82.

Hagendijk, Thijs. 2020. "Reworking Recipes: Reading and Writing Practical Texts in the Early Modern Arts." PhD diss., University of Amsterdam.

Hall, Bert S. 1979. "'Der Meister sol auch kennen schreiben und lesen': Writings about Technology ca. 1400–ca. 1600 A.D. and their Cultural Implications." In *Early Technologies*, ed. Denise Schmandt-Besserat, 47–58. Malibu. CA: Undena Publications.

Halleux, Robert. 1996. "The Reception of Arabic Alchemy in the West." In *Encyclopedia of the History of Arabic Science*, vol. 3, ed. Roshdi Rashed. London: Routledge.

Halleux, Robert. 2009. *Le savoir de la main: Savants et artisans dans l'Europe pré-industrielle*. Paris: Colin.

Hamburger, Jeffrey. 1998. *The Visual and the Visionary*. New York: Zone Books.

Hannaway, Owen. 1992. "Georgius Agricola as Humanist." *Journal of the History of Ideas* 53:553–60.

Harkness, Deborah E. 2007. *The Jewel House: Elizabethan London and the Scientific Revolution*. New Haven: Yale University Press.

Harris, J. R. 1976. "Skills, Coal and British Industry in the Eighteenth Century." *History* 61:167–82.

Harris, William V. 2010. "History, Empathy and Emotions." In *Antike und Abendland: Beiträge zum Verständnis der Griechen und Römer und ihres Nachlebens*, ed. Werner von Koppenfels, Christoph Riedweg, and Helmut Krasser, 1–23. Berlin: De Gruyter.

Haug, Henrike. 2012. "'Wünderbarliche Gewechse': Bergbau und Goldschmeidekunst in 16. Jahrhundert." *Kritische Berichte* 3:49–63.

Hauschke, Sven. 2003. "Wenzel Jamnitzer im Porträt: Der Künstler als Wissenschaftler." *Anzeiger des Germanischen Nationalmuseums*: 127–36. Nuremberg: Germanisches Nationalmuseum Nürnberg.

Hauschke, Sven. 2014. "Wenzel Jamnitzers Beschreibung seines 'kunstlichen und wolgetzierten Schreibtischs' und die Erfindung des Kunstschranks." *Jahrbuch der Coburger Landesstiftung* 58:1–164.

Hawthorne, John G., and Cyril Stanley Smith, eds. 1974. *Mappae Clavicula: A Little Key to the World of Medieval Techniques*. Transactions of the American Philosophical Society, n.s. 64, no. 4.

Hayward, J. F. 1976. *Virtuoso Goldsmiths and the Triumph of Mannerism, 1540–1620*. London: Sotheby Parke Bernet Publications.

Heering, Peter. 2008. "The Enlightened Microscope: Re-enactment and Analysis of Projections with Eighteenth-century Solar Microscopes." *British Journal for the History of Science* 41:345–67.

Heine, Günther. 1996. "Fact or Fancy? The Reliability of Old Pictorial Trade Representations." *Tools & Trades* 9:20–27.

Hendriksen, Marieke M. A., ed. 2020. *Rethinking Performative Methods in the History of Science*. Special issue. *Berichte zur Wissenschaftsgeschichte* 43, no. 3.

Hendriksen, Marieke M. A. 2021. "From *Ingenuity* to *Genius* and *Technique*: Shifting Concepts in Eighteenth-century Theories of Art and Craft." In *Ingenuity in the Making: Matter and Technique in Early Modern Europe*, ed. Richard J. Oosterhoff, José Ramón Marcaida, and Alexander Marr. Pittsburgh: University of Pittsburgh Press.

Hermens, Erma. 2013. "The *Botteghe degli Artisti*: Artistic Enterprise at the della Rovere and Medici Courts in the Late Sixteenth Century." In *The Renaissance Workshop*, ed. David Saunders, Marika Spring, and Andrew Meek, 105–13. London: Archetype.

Heyes, Cecilia. 2018. *Cognitive Gadgets: The Cultural Evolution of Thinking*. Cambridge, MA: Harvard University Press.

H. G. c. 1604. "The Goldsmith's Storehouse." Folger Shakespeare Library, V.a. 179.

Hilaire-Pérez, Liliane. 2013. *La pièce et le geste: Artisans, marchands et savoir technique à Londres au xviiie siècle*. Paris: Éditions Albin Michel.

Hilaire-Pérez, Liliane, Valérie Nègre, Delphine Spicq, and Koen Vermeir, eds. 2017a. *Le livre technique avant le xxe siècle: À l'échelle du monde*. Paris: CNRS Éditions. http://books.openedition.org/editionscnrs/27664.

Hilaire-Pérez, Liliane, Valérie Nègre, Delphine Spicq, and Koen Vermeir. 2017b. "Regards croisés sur le livre et les techniques avant le xxe siècle." In *Le livre technique avant le xxe siècle: À l'échelle du monde*, ed. Liliane Hilaire-Pérez, Valérie Nègre, Delphine Spicq, and Koen Vermeir, 5–39. Paris: CNRS Éditions. http://books.openedition.org/editionscnrs/27676.

Hilaire-Pérez, Liliane, Fabien Simon, and Marie Thébaud-Sorger, eds. 2016. *L'Europe des sciences et des techniques: Un dialogue des savoirs, XVᵉ–XVIIIᵉ siècle*. Rennes: Presses Universitaires de Rennes. http://books.openedition.org/pur/45861.

Hilaire-Pérez, Liliane, and Catherine Verna. 2006. "Dissemination of Technical Knowledge in the Middle Ages and the Early Modern Era." *Technology and Culture* 47, no. 3:536–65.

Hill, Donald. 1991. "Arabic Mechanical Engineering: Survey of the Historical Sources." *Arabic Sciences and Philosophy* 1:67–186.

Hirsch, Rudolf. 1950. "The Invention of Printing and the Diffusion of Alchemical and Chemical Knowledge." *Chymia* 3:115–41.

Hooykaas, Reijer. 1958. *Humanisme, science et réforme: Pierre de la Ramee*. Leiden: Brill.

Houghton, Walter E., Jr. 1941. "The History of Trades: Its Relation to Seventeenth-Century Thought: As Seen in Bacon, Petty, Evelyn, and Boyle." *Journal of the History of Ideas* 2, no. 1:33–60.

Howes, David. 2005. "Skinscapes: Embodiment, Culture, and Environment." In *The Book of Touch*, ed. Constance Classen, 27–39. Oxford: Berg.

Hoyt, Nelly S., and Thomas Cassirer. 2013. "Art." In *The Encyclopedia of Diderot & d'Alembert Collaborative Translation Project*. Ann Arbor: MPublishing, University of Michigan Library. Accessed February 16, 2012. http://hdl.handle.net/2027/spo.did2222.0000.139.

Huber, Peter. 1995. "'Die schöneste Stuffe': Handsteine aus fünf Jahrhunderten." *ExtraLapis* 8:58–67.

Hübner, Johann. 1762. *Curieuses und reales Natur-Kunst-Berg-Gewerck- und Handlungs-Lexicon*. 2nd ed. Leipzig: J. F. Gleditschens Handlung.

Hughes, Thomas P. 1983. *Networks of Power: Electrification in Western Society, 1880–1930*. Baltimore: Johns Hopkins University Press.

Hunter, Matthew C. 2013. *Wicked Intelligence: Visual Art and the Science of Experiment in Restoration London*. Chicago: University of Chicago Press.

Hunter, Michael. 1989. *Establishing the New Science: The Experience of the Early Royal Society*. Woodbridge, Suffolk, UK: Boydell Press.

Hutchins, Edwin. 1995. *Cognition in the Wild*. Cambridge, MA: MIT Press.

Iliffe, Rob. 1995. "Material Doubts: Hooke, Artisan Culture and the Exchange of Information in 1670s London." *British Journal for the History of Science* 28:285–318.

Iliopoulos, Antonis, and Duilio Garofoli. 2016. "The Material Dimensions of Cognition: Reexamining the Nature and Emergence of the Human Mind." *Quaternary International* 405:1–7.

Ingham, Patricia Clare. 2015. *The Medieval New: Ambivalence in an Age of Innovation*. Philadelphia: University of Pennsylvania Press.

Ingold, Tim, ed. 1994. *Companion Encyclopedia of Anthropology: Humanity, Culture and Social Life*. London: Routledge.

Ingold, Tim. 2000. *The Perception of the Environment: Essays in Livelihood, Dwelling and Skill*. New York: Routledge.

Jamnitzer, Wenzel. 1585. "Ein gar Künstlicher und wohlgezierter Schreibtisch sampt allehant Künstlichen Silbern und vergulten newerfunden Instrumenten … Zum Gebrauch der Geometrischen und Astronomischen, auch andern schönen und nützlichen Künsten. Alles durch Wentzel Jamnitzer, Bürger und Goltschmidt in Nürmberg auffs new verfertigt." 2 vols. MSL 1893/1600 and MSL 1893/1601, National Art Library, Victoria and Albert Museum.

Jansen, Dirk Jacob. 1993. "Samuel Quiccheberg 'Inscriptiones': De encyclopedische verzameling als hulpmiddel voor de wetenschap." In *Verzamelen: Van rariteitenkabinet tot kunstmuseum*, ed. Ellinoor Bergvelt, Debora J. Meijers, and Mieke Rijnders, 56–76. Houten, Netherlands: Gaade.

Jardine, Lisa, and Anthony Grafton. 1990. "'Studied For Action': How Gabriel Harvey Read His Livy." *Past and Present* 129, no. 1:30–78.

J. K. 1696. *Curieusen Kunst- und Werck-Schul: Erster Theil*. Nuremberg: Johann Ziegers.

Jones, Matthew L. 2016. *Reckoning with Matter: Calculating Machines, Innovation, and Thinking about Thinking from Pascal to Babbage*. Chicago: University of Chicago Press.

Jorink, Eric, Ann-Sophie Lehmann, and Bart Ramakers, eds. 2019. *Lessons in Art: Art, Education, and Modes of Instruction since 1500*. Leiden: Brill.

Jütte, Robert. 1988. "Aging and the Body Image in the Sixteenth Century: Hermann Weinsberg's (1518–97) Perception of the Aging Body." *European History Quarterly* 18, no. 3:259–90.

Kaliardos, Katharina Pilaski. 2013. *The Munich Kunstkammer: Art, Nature, and the Representation of Knowledge in Courtly Contexts*. Tübingen: Mohr Siebeck.

Kamil, Neil D. 2005. *Fortress of the Soul: Violence, Metaphysics, and Material Life in the Huguenots' New World, 1517–1751*. Baltimore: Johns Hopkins University Press.

Kang, Charles. 2020. "Black Sulfured Wax." In *Secrets of Craft and Nature*, ed. Making and Knowing Project et al. https://edition640.makingandknowing.org/#/essays/ann_051_fa_16.

Kastan, David Scott. 2002. "Plays into Print: Shakespeare to His Earliest Readers." In *Books and Readers in Early Modern England*, ed. Jennifer Andersen and Elizabeth Sauer, 23–41. Material Studies. Philadelphia: University of Pennsylvania Press.

Kavey, Alison. 2007. *Books of Secrets: Natural Philosophy in England, 1550–1600*. Urbana: University of Illinois Press.

Kaye, Joel. 2014. *A History of Balance, 1250–1375: The Emergence of a New Model of Equilibrium and Its Impact on Thought*. Cambridge: Cambridge University Press.

Keller, Charles M. 2001. "Thought and Production: Insights of the Practitioner." In *Anthropological Perspectives on Technology*, ed. Michael Brian Schiffer, 33–45. Albuquerque: University of New Mexico Press.

Keller, Charles, and Janet Dixon Keller. 1993. "Thinking and Acting with Iron." In *Understanding Practice: Perspectives on Activity and Context*, ed. Seth Chaiklin and Jean Lave, 125–43. Cambridge: Cambridge University Press.

Keller, Vera. 2015. *Knowledge and the Public Interest, 1574–1725*. Cambridge: Cambridge University Press.

Keller, Vera. 2016. "Art Lovers and Scientific Virtuosi? The *Philomathia* of Erhard Weigel (1625–1699) in Context." In *Nuncius* 31:523–48.

Kemp, Martin. 1998–99. "Palissy's Philosophical Pots: Ceramics, Grottoes and the Matrices of the Earth." In *Le origini della modernità*, ed. W. Tega, 2:72–78. Florence: L. S. Olschki.

Kenny, Neil. 1991. *The Palace of Secrets: Béroalde de Verville and Renaissance Conceptions of Knowledge*. Oxford: Clarendon Press.

[Kertzenmacher] Kärtzenmacher, Petrus. 1538. *Alchimia: Wie mann alle farben/wasser/olea/salia und alumina damit mann alle corpora/spiritus und calces preparirt/sublimirt und fixirt machen sol. Und wie mann diese ding nutze-auff das Sol und Luna werden mög. Auch von solviren uund schaidung aller metal/Polirung aller handt edel gestain/fürtrefflichen wassern zum etzen/schaiden und solviren/undt zletst wie die gifftige dämpf züverhüten ein kurtzer begrif*. Strasbourg: Jacob Cammerlander.

Kertzenmacher, Petrus. 1613. *Alchimia: Das ist alle Farben, Wasser, Olea, Salia, und Alvmina, damit man alle Corpora Spiritus und Calces praeparirt, sublimirt und fixirt zubereyten. Und wie man diese Ding nutze, auff dass Sol und Lvna werden möge. Auch von Solviren und Scheidung aller*

Metall, Polierung allerhand Edelgestein, fürtrefflichen Wassern zum etzen, scheiden und solviren: und zuletzt wie die gifftige Dämpff zu verhüten ein kurtzer Bericht. Frankfurt am Main: Steinmeyer.

Kheirandish, Elaheh. 2007. "Quṣṭā ibn Lūqā al-Baʿlabakk ī." In *The Biographical Encyclopedia of Astronomers*, ed. Thomas Hockey et al., 948–49. New York: Springer.

Kieffer, Fanny. 2014. "The Laboratories of Art and Alchemy at the Uffizi Gallery in Renaissance Florence: Some Material Aspects." In *Laboratories of Art: Alchemy and Art Technology from Antiquity to the 18th Century*, ed. Sven Dupré, 105–27. Cham, Switzerland: Springer.

King, Catherine. 2007. "Making Histories, Publishing Theories." In *Making Renaissance Art*, ed. Kim. W. Woods, 251–80. New Haven: Yale University Press.

Kneebone, Roger. 2020. *Expert: Understanding the Path to Mastery.* London: Viking.

Kneepkens, Indra. 2020. "Masterful Mixtures: Practical Aspects of Fifteenth- and Early Sixteenth-century Oil Paint Formulation." PhD diss., University of Amsterdam.

Knoop, Douglass, G. P. Jones, and Douglas Hamer, eds. 1938. *The Two Earliest Masonic Manuscripts.* Manchester, UK: Manchester University Press.

Knop, Wilhelm. 1957. *Arbeitsschutz im Metallbetrieb: Handbuch für den Umgang mit Eisen und NE-Metallen bei ihrer Erzeugung, Verwendung und Bearbeitung.* Berlin: Metall-Verlag.

Koepp, Cynthia J. 2009. "Advocating for Artisans: The Abbé Pluche's *Spectacle de la nature* (1732–51)." In *The Idea of Work in Europe from Antiquity to Modern Times*, ed. Josef Ehmer and Catharina Lis, 245–73. Burlington, VT: Ashgate.

Koeppe, Wolfram, ed. 2019. *Making Marvels: Science and Splendor at the Courts of Europe.* New Haven: Yale University Press.

Kok, Cindy. 2020. "Colors for Green Leaves and Painting on Metal." In *Secrets of Craft and Nature*, ed. Making and Knowing Project et al. https://edition640.makingandknowing.org/#/essays/ann_030_fa_15.

Korn, Peter. 2015. *Why We Make Things, and Why It Matters.* Boston: David R. Godine.

Krekel, Christoph, and Manfred Lautenschlager. 2005. "Bearbeitung von Glas, Edelstein, Bein und Horn." In *Der "Liber illuministarum" aus Kloster Tegernsee*, ed. Anna Bartl, Christoph Krekel, Manfred Lautenschlager, and Doris Oltrogge, 673–78. Stuttgart: Franz Steiner Verlag.

Kremnitzer, Kathryn, Siddhartha Shah, and Pamela H. Smith. 2020. "Gemstones and Imitation." In *Secrets of Craft and Nature*, ed. Making and Knowing Project et al. https://edition640.makingandknowing.org/#/essays/ann_029_fa_15.

Kremnitzer, Kathryn, and Pamela H. Smith. 2020. "Imitation Rubies and Failure." In *Secrets of Craft and Nature*, ed. Making and Knowing Project et al. https://edition640.makingandknowing.org/#/essays/ann_082_fa_15.

Krohn, Deborah. 2015. *Food and Knowledge in Renaissance Italy: Bartolomeo Scappi's Paper Kitchens.* Burlington, VT: Ashgate.

Kruse, Christiane. 2000. "Fleisch werden—Fleisch malen: Malerei als 'incarnazione': Mediale Verfahren des bildwerdens in Libro dell'Arte von Cennino Cennini." *Zeitschrift für Kunstgeschichte* 63:305–25.

Kuhn, Dieter. 1988. *Chemistry and Chemical Technology, Vol. 5, Part IX: Textile Technology: Spinning and Reeling.* In *Science and Civilization in China*, ed. Joseph Needham. Cambridge: Cambridge University Press.

Kuijpers, Maikel H. G. 2018. *An Archaeology of Skill: Metalworking Skill and Material Specialization in Early Bronze Age Central Europe.* New York: Routledge.

Kuijpers, Maikel H. G. 2019. *The Future Is Handmade*. Video. https://www.youtube.com/watch?time
_continue=424&v=jOMujtoicTk.

Labarre, Albert. 1971. *Le Livre dans la vie Amiénoise du seizième siècle: L'enseignement des inven-
taires après décès 1503–1576*. Paris: Beatrice-Nauwelaerts.

Lacey, Andrew. 2018. "The Sculptor at Work: Recreating the Rothschild Bronzes." In *Michelangelo
Sculptor in Bronze: The Rothschild Bronzes*, ed. Victoria Avery, 201–25. London: Philip Wilson
Publishers.

Lacey, Andrew, and Siân Lewis. 2020. "In Pursuit of Magic." In *Secrets of Craft and Nature*, ed.
Making and Knowing Project et al. https://edition640.makingandknowing.org/#/essays/ann_501
_ad_20.

Laird, Mark. 1999. *The Flowering of the Landscape Garden: English Pleasure Grounds 1720–1800*.
Philadelphia: University of Pennsylvania Press.

Landolt, Elisabeth. 1991."Das Amerbach-Kabinett und seine Inventare." In *Sammeln in der
Renaissance: Das Amerbach-Kabinett: Beiträge zu Basilius Amerbach*, 73–206. Basel: Historisches
Museum.

Landolt, Elisabeth, and Felix Ackermann. 1991. *Sammeln in der Renaissance: Das Amerbach-
Kabinett: Die Objekte im Historischen Museum Basel*. Basel: Historisches Museum.

Landsman, Rozemarijn, and Jonah Rowen. 2020. "Uses of Sulfur in Casting." In *Secrets of Craft and
Nature*, ed. Making and Knowing Project et al. https://edition640.makingandknowing.org/#
/essays/ann_007_fa_14.

Laudan, Rachel. 1984a. "Cognitive Change in Technology and Science." In *The Nature of Technological Knowledge. Are Models of Scientific Change Relevant?*, ed. Rachel Laudan, 83–104. Dordrecht:
D. Reidel Publishing.

Laudan, Rachel, ed. 1984b. *The Nature of Technological Knowledge: Are Models of Scientific Change
Relevant?* Boston: D. Reidel Publishing.

Lave, Jean. 1977. "Cognitive Consequences of Traditional Apprenticeship Training in West Africa."
Anthropology and Education Quarterly 8, no. 3:170–80.

Lave, Jean. 1988. *Cognition in Practice: Mind, Mathematics and Culture in Everyday Life*. Cambridge: Cambridge University Press.

Lave, Jean, and Etienne Wenger. 1991. *Situated Learning: Legitimate Peripheral Participation*. Cambridge: Cambridge University Press.

Lavin, Irving. 1977–78. "The Sculptor's Last Will and Testament." *Allen Memorial Art Museum Bulletin* 35:4–39.

Lecain, Timothy J. 2017. *The Matter of History: How Things Create the Past*. Cambridge: Cambridge
University Press.

Lechtman, Heather. 1999. "Afterword." In *The Social Dynamics of Technology: Practice, Politics, and
World Views*, ed. Marcia-Anne Dobres and Christopher R. Hoffman, 223–32. Washington, DC:
Smithsonian Institution Press.

Lee, Michelle, and Pamela H. Smith. 2020. "Lean Sands and Fat Binders." In *Secrets of Craft and
Nature*, ed. Making and Knowing Project et al. https://edition640.makingandknowing.org/#
/essays/ann_009_fa_14.

Leedham-Green, E. S. 1986. *Books in Cambridge Inventories: Book-lists from Vice-Chancellor's Court
Probate Inventories in the Tudor and Stuart Periods*. New York: Cambridge University Press.

Lefèvre, Wolfgang. 2010. "Picturing the World of Mining in the Renaissance: The *Schwazer Bergbuch*
(1556)." Max-Planck-Institut für Wissenschaftsgeschichte preprint 407.

Le Goff, Jacques. 1980. *Time, Work, and Culture in the Middle Ages*. Trans. Arthur Goldhammer. Chicago: University of Chicago Press.

Lehmann, Ann-Sophie. 2008. "Fleshing Out the Body: The 'Colours of the Naked' in Workshop Practice and Art Theory, 1400–1600." In *Body and Embodiment in Netherlandish Art*, ed. Ann-Sophie Lehmann and Herman Roodenburg, 86–109. Zwolle, Netherlands: Waanders.

Lehmann, Ann-Sophie. 2009. "Wedging, Throwing, Dipping and Dragging—How Motions, Tools and Materials Make Art." In *Folded Stones*, ed. Barbara Baert and Trees de Mits, 41–60. Ghent: Institute for Practice-based Research in the Arts.

Lehmann, Ann-Sophie. 2013. "How Materials Make Meaning." In *Meaning in Materials, 1400–1800*, ed. Ann-Sophie Lehmann, Frits Scholten, and Perry Chapman, 7–26. Netherlands Yearbook for History of Art 62. Leiden: Brill.

Lehmann, Ann-Sophie. 2015. "The Matter of the Medium: Some Tools for an Art-Theoretical Interpretation of Materials." In *The Matter of Art*, ed. Christy Anderson, Anne Dunlop, and Pamela H. Smith, 21–41. Manchester, UK: Manchester University Press.

Leibniz, Gottfried Wilhelm. 2008. *Protogaea*. Ed. and trans. Claudine Cohen and Andre Wakefield. Chicago: University of Chicago Press.

Lein, Edgar. 2006. "'Wie man allerhand Insecta, als Spinnen, Fliegen, Käfer, Eydexen, Frösche und auch ander zart Laubwerck scharff abgiessen solle, als wann sie natürlich also gewachsen wären'—Die Natur als Modell in Johann Kunckels Beschreibungen des Naturabgusses von Tieren und Pflanzen." In *Das Modell in der bildenden Kunst des Mittelalters und der Neuzeit—Festschrift für Herbert Beck*, ed. Peter C. Bol and Heike Richter, 103–19. Petersberg, Germany: M. Imhof.

Lemerle, Frédérique. 2012. "Vitruve, Vignole, Palladio au xviiᵉ siècle: Traductions, abrégés et augmentations." In *Architecture et théorie: L'héritage de la Renaissance: Actes de colloque*, ed. Jean-Philippe Garric et al. Paris: Publications de l'Institut National d'Histoire de l'Art. http://books.openedition.org/inha/3328.

Leng, Rainer. 2002. *Ars belli: Deutsche taktische und kriegstechnische Bilderhandschriften und Traktate im 15. und 16. Jahrhundert*. 2 vols. Wiesbaden: Ludwig Reichert Verlag.

Leong, Elaine. 2017. "Brewing Ale and Boiling Water in 1651." In *The Structures of Practical Knowledge*, ed. Matteo Valeriani, 55–75. Berlin: Springer.

Leong, Elaine. 2018. *Recipes and Everyday Knowledge: Medicine, Science, and the Household in Early Modern England*. Chicago: University of Chicago Press.

Leong, Elaine, Angela N. H. Creager, and Mathias Grote, eds. 2020. *Learning by the Book: Manuals and Handbooks in the History of Science*. Special issue. *BJHS Themes* 5.

Leong, Elaine, and Sara Pennell. 2017. "Recipe Collections and the Currency of Medical Knowledge in the Early Modern 'Medical Marketplace.'" In *The Medical Marketplace and Its Colonies c. 1450–c. 1850*, ed. Mark Jenner and Patrick Wallis, 133–52. London: Palgrave Macmillan.

Leong, Elaine, and Alisha Rankin. 2011a. "Introduction." In *Secrets and Knowledge in Medicine and Science, 1500–1800*, ed. Elaine Leong and Alisha Rankin, 1–20. Farnham, UK: Ashgate.

Leong, Elaine, and Alisha Rankin, eds. 2011b. *Secrets and Knowledge in Medicine and Science, 1500–1800*. Farnham, UK: Ashgate.

Leong, Elaine, and Alisha Rankin. 2013. "Testing Drugs and Trying Cures: Experiment and Medicine in Medieval and Early Modern Europe." *Bulletin of the History of Medicine* 92, no. 2:157–82.

Leonhard, Karin. 2009–10. "Pictura's Fertile Field: Otto Marsus van Schrieck and the Genre of Sottobosco Painting." *Simiolus* 34:95–118.

Leonhard, Karin. 2013. *Bildfelder. Stilleben und Naturstücke des 17. Jahrhunderts*. Berlin.

Leonhard, Karin. 2019. "'The Various Natures of Middling Colours We May Learne of Painters': Sir Kenelm Digby Looks at Rubens and Van Dyck." In *Knowledge and Discernment in the Early Modern Arts*, ed. Sven Dupré and Christine Göttler, 163–86. London: Routledge.

Lis, Catharina. 2009. "Perceptions of Work in Classical Antiquity: A Polyphonic Heritage." In *The Idea of Work in Europe from Antiquity to Modern Times*, ed. Josef Ehmer and Catharina Lis, 33–68. Burlington, VT: Ashgate.

Lis, Catharina, and Hugo Soly. 2012. *Worthy Efforts: Attitudes to Work and Workers in Pre-Industrial Europe*. Leiden: Brill.

Liu, Xiaomeng. 2020. "An Excellent Salve for Burns." In *Secrets of Craft and Nature*, ed. Making and Knowing Project et al. https://edition640.makingandknowing.org/#/essays/ann_080_sp_17.

Liu, Yan, and Shigehisa Kuriyama. 2020. "Fluid Being: Mercury in Chinese Medicine and Alchemy." In *Fluid Matter(s): Flow and Transformation in the History of the Body*, ed. Natalie Köhle and Shigehisa Kuriyama. Asian Studies Monograph Series 14. Canberra: ANU Press. Accessed August 27, 2020.

Lloyd, Geoffrey, and Nathan Sivin. 2002. *The Way and the Word: Science and Medicine in Early China and Greece*. New Haven: Yale University Press.

Löhr, Wolf-Dietrich. 2008. "Handwerk und Denkwerk des Malers: Kontexte für Cenninis Theorie der Praxis." In *"Fantasie und Handwerk": Cennino Cennini und die Tradition der toskanischen Malerei von Giotto bis Lorenzo Monaco*, ed. Wolf-Dietrich Löhr and Stefan Weppelmann. Munich: Hirmer.

Löhr, Wolf-Dietrich. 2011. "Die Rede der Hand." In *Autorschaft: Ikonen—Stile—Institutionen*, ed. Christel Meier and Martina Wagner-Egelhaaf, 164–93. Munich: Akademie Verlag.

Long, Pamela O. 1991. "The Openness of Knowledge: An Ideal and Its Context in Sixteenth-century Writings on Mining and Metallurgy." *Technology and Culture*: 318–55.

Long, Pamela O. 1997. "Power, Patronage, and the Authorship of Ars: From Mechanical Know-how to Mechanical Knowledge in the Last Scribal Age." *Isis* 88:1–41.

Long, Pamela O. 2000. *Technology and Society in the Medieval Centuries: Byzantine, Islam, and the West, 500–1300*. Washington, DC: AHA Publications.

Long, Pamela O. 2001. *Openness, Secrecy, Authorship: Technical Arts and the Culture of Knowledge from Antiquity to the Renaissance*. Baltimore: Johns Hopkins University Press.

Long, Pamela O. 2009. "Introduction: The World of Michael of Rhodes." In *The Book of Michael of Rhodes*, ed. Pamela O. Long, David McGee, and Alan M. Stahl, 3:1–33. Cambridge, MA: MIT Press.

Long, Pamela O. 2011. *Artisan/Practitioners and the New Sciences 1400–1600*. Corvallis: Oregon State University Press.

Long, Pamela O. 2015. "Trading Zones in Early Modern Europe." *Isis* 106, no. 4:840–47.

Long, Pamela O., David McGee, and Alan M. Stahl, eds. 2009. *The Book of Michael of Rhodes: A Fifteenth-Century Maritime Manuscript*. 3 vols. Cambridge, MA: MIT Press.

Lores-Chavez, Isabella. 2020. "Imitating Raw Nature." In *Secrets of Craft and Nature*, ed. Making and Knowing Project et al. https://edition640.makingandknowing.org/#/essays/ann_045_fa_16.

Lucie-Smith, Edward. 1981. *The Story of Craft: The Craftsman's Role in Society*. Oxford: Phaidon.

Lüdtke, Alf, ed. 1995. *The History of Everyday Life: Reconstructing Historical Experiences and Ways of Life*. Trans. William Templer. Princeton, NJ: Princeton University Press.

Lukehart, Peter, ed. 2010. *The Accademia Seminars: The Early History of the Accademia di San Luca in Rome, c. 1590–1635*. New Haven: Yale University Press.

Lukehart, Peter, dir. 2010–ongoing. *The Early History of the Accademia di San Luca, c. 1590–1635*. National Gallery of Art, Washington, DC. https://www.nga.gov/research/casva/research-projects/early-history-of-the-accademia-di-san-luca.html.

Maines, Rachel P. 2009. *Hedonizing Technologies: Paths to Pleasure in Hobbies*. Baltimore: Johns Hopkins University Press.

Making and Knowing Project, Pamela H. Smith, Naomi Rosenkranz, Tianna Helena Uchacz, Tillmann Taape, Clément Godbarge, Sophie Pitman, Jenny Boulboullé, Joel Klein, Donna Bilak, Marc Smith, and Terry Catapano, eds. 2020. *Secrets of Craft and Nature in Renaissance France: A Digital Critical Edition and English Translation of BnF Ms. Fr. 640*. New York: Making and Knowing Project. https://edition640.makingandknowing.org.

Malafouris, Lambros. 2013. *How Things Shape the Mind: A Theory of Human Engagement*. Cambridge, MA: MIT Press.

Marchand, Trevor H. J. 2010a. "Embodied Cognition and Communication: Studies with British Fine Woodworkers." In *Making Knowledge: Explorations of the Indissoluble Relation between Mind, Body and Environment*, ed. Trevor H. Marchand, 95–114. Oxford: John Wiley & Sons.

Marchand, Trevor H. J., ed. 2010b. *Making Knowledge: Explorations of the Indissoluble Relation between Mind, Body and Environment*. Chichester: Wiley-Blackwell.

Marr, Alexander. 2013. "The Architect-Engineer as Reader." In *The Artist as Reader*, ed. Heiko Damm, Michael Thimann, and Claus Zittel, 421–46. Leiden: Brill.

Marr, Alexander, Raphaële Garrod, José Ramón Marcaida, and Richard J. Oosterhoff. 2019. *Logodaedalus: Word Histories of Ingenuity in the Early Modern Period*. Pittsburgh: University of Pittsburgh Press.

Marris, Caroline, and Stephanie Pope. 2020. "Sand Molds of Ox Bone, Wine, and Elm Root." In *Secrets of Craft and Nature*, ed. Making and Knowing Project et al. https://edition640 .makingandknowing.org/#/essays/ann_020_sp_15.

Martinón-Torres, Marcos, I. C. Freestone, A. Hunt, and T. Rehren. 2008. "Mass-produced Mullite Crucibles in Medieval Europe: Manufacture and Material Properties." *Journal of the American Ceramic Society* 91, no. 5:2071–74.

Martinón-Torres, Marcos, T. Rehren, Nicolas Thomas, and Aude Mongiatti. 2009. "Identifying Materials, Recipes and Choices: Some Suggestions for the Study of Archaeological Cupels." In *Archaeometallurgy in Europe 2007*, 435–45. Milan: Associazione Italiana di Metallurgia.

Mathesius, Johannes. 1562. *Sarepta, oder, Bergpostill: Sampt der Jochimssthalischen kurtzen Chroniken*. Nuremberg: J. vom Berg & U. Newbar.

Mathonière, Jean-Michel. 2017. "Vignole et les compagnons du tour de France." In *Le livre technique avant le xxᵉ siècle: À l'échelle du monde*, ed. Liliane Hilaire-Pérez, Valérie Nègre, Delphine Spicq, and Koen Vermier. Paris: CNRS Éditions. http://books.openedition.org/editionscnrs/27709.

Mauss, Marcel. 1990. *The Gift: The Form and Reason for Exchange in Archaic Society*. Trans. W. D. Halls. London: Routledge.

Mayerne, Theodore Turquet de. 1620. "Pictoria Sculptoria & quae subalternarum artium." Sloane Ms. 2052. British Library.

McGee, David. 2009. "The Shipbuilding Text of Michael of Rhodes." In *The Book of Michael of Rhodes*, ed. Pamela O. Long, David McGee, and Alan M. Stahl, 3:211–41. Cambridge, MA: MIT Press.

McHam, Sarah Blake. 2013. *Pliny and the Artistic Culture of the Italian Renaissance*. New Haven: Yale University Press.

Meadow, Mark A. 2002. "Merchants and Marvels: Hans Jacob Fugger and the Origins of the Wunderkammer." In *Merchants and Marvels: Commerce, Science, and Art in Early Modern Europe*, ed. Pamela H. Smith and Paula Findlen, 182–200. New York: Routledge.

Merrifield, Mary P. 1967. *Original Treatises on the Arts of Painting* (1849). 2 vols. New York: Dover.

Merrill, Elizabeth M. 2017. "Pocket-Size Architectural Notebooks and the Codification of Practical

Knowledge." In *The Structures of Practical Knowledge*, ed. Matteo Valleriani, 21–54. Cham, Switzerland: Springer.

Merton, Robert K. 1961. "Singletons and Multiples in Scientific Discovery: A Chapter in the Sociology of Science." *Proceedings of the American Philosophical Society* 105, no. 5:470–86.

Merton, Robert K. 1973. *The Sociology of Science: Theoretical and Empirical Investigations*. Chicago: University of Chicago Press.

Miller, Matthias, and Karin Zimmermann. 2005. *Die Codices Palatini germanici in der Universitätsbibliothek Heidelberg*. Wiesbaden: Harrassowitz.

Miller, Peter N. 2019. "Peiresc in the Parisian Jewel House." In *Knowledge and Discernment in the Early Modern Arts*, ed. Sven Dupré and Christine Göttler, 213–35. London: Routledge.

Misson, Maximilian. 1739. *A New Voyage to Italy*. 5th ed. London: J. & J. Bonwick.

Mocarelli, Luca. 2009. "The Attitude of Milanese Society to Work and Commercial Activities." In *The Idea of Work in Europe from Antiquity to Modern Times*, ed. Josef Ehmer and Catharina Lis, 101–21. Burlington, VT: Ashgate.

Mooney, Linne R. 1993. "A Middle English Text on the Seven Liberal Arts." *Speculum* 68, no. 4:1027–52.

Morel, Thomas. 2017. "Bringing Euclid into the Mines: Classical Sources and Vernacular Knowledge in the Development of Subterranean Geometry." In *Translating Early Modern Science*, ed. Sietske Fransen, Niall Hodson, and K. A. E. Enenkel, 154–81. Leiden: Brill.

Morel, Thomas. 2020. "De Re Geometrica: Writing, Drawing, and Preaching Mathematics in Early-Modern Mines." *Isis* 111, no. 1:22–45.

Morrall, Andrew. 2006. "Entrepreneurial Craftsmen in Late Sixteenth-Century Augsburg." In *Mapping Markets for Paintings in Europe 1450–1750*, ed. Neil De Marchi and Hans J. van Miegroet, 211–36. Turnhout: Brepols.

Motture, Peta. 2001. *Bells & Mortars and Related Utensils*. London: V&A Publications.

Moxon. Joseph. 1970. *Mechanick Exercises or the Doctrine of Handy-Works* (1683, 1693, 1703). Ed. Charles F. Montgomery. New York: Praeger.

Moyn, Samuel. 2006. "Empathy in History, Empathizing with Humanity." *History and Theory* 45:379–415.

Mukerji, Chandra. 2006. "Tacit Knowledge and Classical Technique in Seventeenth-Century France." *Technology and Culture* 47, no. 4:713–33.

Murphy, Hannah. 2020. "Artisanal 'Histories' in Early Modern Nuremberg." In *Knowledge and the Early Modern City: A History of Entanglements*, ed. Bert De Munck and Antonella Romano, 58–78. New York: Routledge.

Murrell, Jim. 1983. *The Way Howe to Lymne: Tudor Miniatures Observed*. London: Victoria and Albert Museum.

Myers, Natasha. 2008. "Molecular Embodiments and the Body-Work of Modeling in Protein Crystallography." *Social Studies of Science* 38, no. 2:163–99.

Nassau, Kurt. 1984. *Gemstone Enhancement*. London: Butterworths.

Nègre, Valérie. 2016. "Histoire de l'art, histoire de l'architecture et histoire des techniques (Europe, xve–xviiie siècle)." In *L' Europe technicienne: XVe–XVIIIe siècle*, ed. Catherine Cardinal, Liliane Pérez, Delphine Amélie Spicq, and Marie Thébaud-Sorger. Special issue. *Artefact: Technique, histoire et sciences humaines* 4:49–61.

Neilson, Christina. 2016. "Demonstrating Ingenuity: The Display and Concealment of Knowledge in Renaissance Artists' Workshops." In *Shared Spaces and Knowledge Transactions in the Italian Renaissance City*. Special issue. *I Tatti Studies in the Italian Renaissance* 19, no. 1:63–91.

Neilson, Christina. 2019. *Practice and Theory in the Italian Renaissance Workshop: Verrocchio and the Epistemology of Art Making*. Cambridge: Cambridge University Press.

Neudörfer, Johann. 1875. "Nachrichten von Künstlern und Werkleuten … aus dem Jahre 1547." Transcribed and annotated by G. W. K. Lochner. *Quellenschriften für Kunstgeschichte und Kunsttechnik des Mittelalters und der Renaissance* 10. Vienna: Wilhelm Braumüller.

Neven, Sylvie. 2016. *The Strasbourg Manuscript: A Medieval Tradition of Artists' Recipe Collections (1400–1570)*. London: Archetype.

Newman, William R. "The Alchemy of Isaac Newton." Accessed June 2, 2011. http://webapp1.dlib .indiana.edu/newton/reference/chemLab.do.

Newman, William R., and Lawrence M. Principe. 2002. *Alchemy Tried in the Fire: Starkey, Boyle, and the Fate of Helmontian Chemistry*. Chicago: University of Chicago Press.

Niavis, Paul. 1953. *Iudicium Iovis oder das Gericht der Götter über den Bergbau*. Trans. Paul Krenkel. Freiberger Forschungshefte D3, "Kultur und Technik." Berlin.

Nicholas of Cusa. 1989. *The Layman on Wisdom and the Mind*. Trans. M. L. Führer. Ottawa: Doverhouse Editions.

Noirot, Fabien. 2020. "Molding, Modeling, and Repairing: Lifecast Snakes Modeled in Black Wax." In *Secrets of Craft and Nature*, ed. Making and Knowing Project et al. https://edition640 .makingandknowing.org/#/essays/ann_504_ad_20.

Nordera, Marina. 2008. "La réduction de la danse en art (xve–xviiie siècle)." In *Réduire en art: La technologie de la Renaissance aux Lumières*, ed. Pascal Dubourg Glatigny and Hélène Vérin. Paris: Éditions de la Maison des Sciences de l'Homme.

Norgate, Edward. 1997. *Miniatura, or, The Art of Limning: New Critical Edition*. Ed. Jeffrey M. Muller and Jim Murrell. New Haven: Yale University Press.

Ochs, Kathleen H. 1985. "The Royal Society of London's History of Trades Programme: An Early Episode in Applied Science." *Notes and Records of the Royal Society of London* 39, no. 2:129–58.

O'Connor, Erin. 2007. "Embodied Knowledge in Glassblowing: The Experience of Meaning and the Struggle towards Proficiency." *Sociological Review* 55: 126–41.

Oltrogge, Doris. 1998. "'Cum sesto et rigula': L'organisation du savoir technologique dans le *Liber diversarum artium* de Montpellier et dans le *De diversis artibus* de Théophile." In *Discours et savoirs: Encyclopédies médiévales*, ed. Bernard Baillaud, Jérôme de Gramont, and Denis Hüe, 67–99. Cahiers Diderot 10. Rennes: Presses Universitaires de Rennes.

O'Malley, Michelle. 2005. *The Business of Art: Contract and the Commissioning Process in Renaissance Italy*. New Haven: Yale University Press.

Ong, Walter J. 1958. *Ramus, Method, and the Decay of Dialogue: From the Art of Discourse to the Art of Reason*. Cambridge, MA: Harvard University Press.

Ong, Walter J. 1986. "Writing Is a Technology That Restructures Thought." In *The Written Word: Literacy in Transition*, ed. Gerd Bauman, 23–50. Oxford: Clarendon Press.

Oppenheim, A. Leo, Robert H. Brill, Dan Barag, Axel von Saldern. 1970. *Glass and Glassmaking in Ancient Mesopotamia*. Corning, NY: Corning Museum of Glass Press.

Ovitt, George. 1983. "The Status of the Mechanical Arts in Medieval Classifications of Learning." *Viator* 14: 89–105.

Ovitt, George. 1987. *The Restoration of Perfection: Labor and Technology in Medieval Culture*. New Brunswick, NJ: Rutgers University Press.

Pacey, Arnold. 1992. *The Maze of Ingenuity*. Cambridge, MA: MIT Press.

Pagel, Walter. 1982. *Paracelsus: An Introduction to Philosophical Medicine in the Era of the Renaissance*. 2nd ed. Basel: Karger.

Palissy, Bernard. 1957. *The Admirable Discourses of Bernard Palissy*. Trans. Aurele La Rocque. Urbana: University of Illinois Press.

Palissy, Bernard. 1988. *Recepte Veritable* (1563). Geneva: Librairie Droz.

Panofsky, Erwin. 1955. *The Life and Art of Albrecht Dürer*. Princeton, NJ: Princeton University Press.

Panzanelli, Roberta, ed. 2008. *The Color of Life: Polychromy in Sculpture from Antiquity to the Present*. With Eike D. Schmidt and Kenneth Lapatin. Los Angeles: Getty Research Institute.

Papenbrock, Martin. 2013. "Gillis van Coninxloo: Der Künstler als Leser." In *The Artist as Reader: On Education and Non-education of Early Modern Artists*, ed. Heiko Damm, Michael Thimann, and Claus Zittel, 129–53. Leiden: Brill.

Pappano, Margaret A., and Nicole R. Rice, eds. 2013. "Medieval and Early Modern Artisan Culture." *Journal of Medieval and Early Modern Studies* 43, no. 3:473–85.

Paracelsus (Theophrastus von Hohenheim). 1529. *Von der frantzösischen Kranckheit Drey Bücher*. Nuremberg: Peypus, 1529.

Paracelsus (Theophrastus von Hohenheim). 1928. *Die große Wundarznei* (1536). In *Sämtliche Werke: Medizinische, naturwissenschaftliche und philosophische Schriften*, ed. Karl Sudhoff, vol. 10. Munich: Oldenbourg.

Paracelsus (Theophrastus von Hohenheim). 1929. *Astronomia Magna: oder die gantze Philosophia sagax der großen und kleinen Welt/des von Gott hocherleuchten/erfahrnen/ und bewerten teutschen Philosophi und Medici* (finished 1537/38; 1st published 1571). In *Sämtliche Werke: Medizinische, naturwissenschaftliche und philosophische Schriften*, ed. Karl Sudhoff, vol. 12. Munich: Oldenbourg.

Paracelsus (Theophrastus von Hohenheim). 1941. "On the Miners' Sickness and Other Miners' Diseases." Trans. George Rosen. In *Four Treatises of Theophrastus von Hohenheim called Paracelsus*, ed. Henry E. Sigerist. Baltimore: Johns Hopkins University Press.

Paré, Ambroise. 1951. *The Apologie and Treatise, containing the Voyages made into Divers Places*. Ed. Geoffrey Keynes. London: Falcon Educational Books.

Paré, Ambroise. 1969. *Ten Books of Surgery with the Magazine of the Instruments Necessary for It*. Trans. Robert White Linker and Nathan Womack. Athens: University of Georgia Press.

Parry, Richard. 2014. "Episteme and Techne." In *The Stanford Encyclopedia of Philosophy*, ed. Edward N. Zalta. https://plato.stanford.edu/archives/fall2014/entries/episteme-techne/.

Parshall, Peter, Stacey Sell, and Judith Brodie. 2001. *The Unfinished Print*. Washington, DC: National Gallery of Art.

Paster, Gail Kern. 2004. *Humoring the Body: Emotions and the Shakespearean Stage*. Chicago: University of Chicago Press.

Payne, Joseph Frank. 1901. *On the "Herbarius" and "Hortus Sanitatis."* London: Blades, East & Blades.

Pechstein, Klaus. 1974. "Der Merkelsche Tafelaufsatz von Wenzel Jamnitzer." *Mitteilungen des Vereins für Geschichte der Stadt Nürnberg* 61:90–121.

Pechstein, Klaus. 1985. "Der Goldschmied Wenzel Jamnitzer." In *Wenzel Jamnitzer und die Nürnberger Goldschmiedekunst 1500–1700*, 67–70. Munich: Klinkhardt & Biermann.

Pelus-Kaplan, Marie-Louise. 2017. "Le livre et la langue comme clés de contact: Deux manuels d'apprentissage des langues slaves dans les villes hanséatiques (fin xvie siècle, début du xviie siècle): Entre livres techniques et codes de civilité." In *Le livre technique avant le xxe siècle: À l'échelle du monde*, ed. Liliane Hilaire-Pérez, Valérie Nègre, Delphine Spicq, and Koen Vermeir, 183–92. Paris: CNRS Éditions. http://books.openedition.org/editionscnrs/27718.

Pepys, Samuel. 2004. *Diary of Samuel Pepys—Complete*. Ed. Henry B. Wheatley. Project Gutenberg. https://www.gutenberg.org/ebooks/4200.

Pereira, Michela. 1999. "Alchemy and the Use of Vernacular Languages in the Late Middle Ages." *Speculum* 74:336–56.

Petrucci, Armando. 2011. "Reading in the Middle Ages." In *The History of Reading: A Reader*, ed. Shafquat Towheed, Rosalind Crone, and Katie Halsey, 275–82. London: Routledge.

Pfeifer, Rolf. 2006. *How the Body Shapes the Way We Think*. Cambridge, MA: MIT Press.

Piccolpasso, Cipriano. 1980. *The Three Books of the Potter's Art: I tre libri dell'arte del vasaio* (ca. 1558). Trans. Ronald Lightbown and Alan Caiger-Smith. 2 vols. London: Scolar Press.

Pickstone, John V. 1997. "Thinking over Wine and Blood: Craft-products, Foucault, and the Reconstruction of Enlightenment Knowledges." *Social Analysis: The International Journal of Social and Cultural Practice* 41, no. 1:97–105.

Piemontese, Alessio. 1616. *Kunstbuch des Wohlerfarnen Herrn Alexii Pedemontani, von mancherleyen nutzlichen und bewerten Secreten oder Künsten*. Trans. Hanß Jacob Wecker. Basel: Ludwig Königs.

Piemontese, Alessio. 1555. *Secreti del Reverendo Donno Alessio Piemontese*. Venice: Sigismondo Bondogna.

Plato. 1953. *The Dialogues of Plato*. Trans. and ed. Benjamin Jowett. 4th ed. Oxford: Oxford University Press.

Plattes, Gabriel. 1639. *Discovery of Subterraneall Treasure … London*.

Pliny. 1952. *Natural History*. Trans. H. Rackham. Vol. 9. Cambridge, MA: Harvard University Press.

Ploss, Emil Ernst. 1962. *Ein Buch von alten Farben: Technologie der Textilfarben in Mittelalter mit einem Ausblick auf die festen Farben*. Heidelberg: Impuls Verlag Heinz Moos.

Polanyi, Michael. 1998. *Personal Knowledge: Towards a Post-Critical Philosophy* (1962). London: Routledge.

Polanyi, Michael. 2009. *The Tacit Dimension* (1966). Chicago: University of Chicago Press.

Pomata, Gianna, and Nancy Siraisi, eds. 2005. *Historia: Empiricism and Erudition in Early Modern Europe*. Cambridge, MA: MIT Press.

Popplow, Marcus. 2004. "Why Draw Pictures of Machines? The Social Contexts of Early Modern Machine Drawings." In *Picturing Machines*, ed. Wolfgang Lefévre, 17–48. Cambridge, MA: MIT Press.

Popplow, Marcus. 2002. "Models of Machines: A Missing Link between Early Modern Engineering and Mechanics." Max-Planck-Institut für Wissenschaftsgeschichte preprint 225.

Posada, Andrés Vélez. 2020. "Genius, as *Ingenium*." In *Encyclopedia of Early Modern Philosophy and the Sciences*, ed. D. Jalobeanu and C. T. Wolfe. Cham, Switzerland: Springer Nature. Accessed August 27, 2020.

Prag um 1600: Kunst und Kultur am Hofe Rudolfs II. 1988. Essen: Kulturstiftung Ruhr.

Prak, Maarten. 2011. "Mega-structures of the Middle Ages: The Construction of Religious Buildings in Europe and Asia, c. 1000–1500." *Journal of Global History* 6, no. 3:381–406.

Prak, Maarten, Catharina Lis, Jan Lucassen, and Hugo Soly, eds. 2006. *Craft Guilds in the Early Modern Low Countries: Work, Power, and Representation*. Burlington, VT: Ashgate.

Presenti, Allegra. 2004. "Communicating Design: Drawings for the Patron in Italy 1400–1600." Paper presented at V&A Seminar, February 3.

Principe, Lawrence. 2016. "Chymical Exotica in the Seventeenth Century, or, How to Make the Bologna Stone." *Ambix* 63:118–44.

Prown, Jonathan, and Richard Miller. 1996. "The Rococo, the Grotto, and the Philadelphia High

Chest." In *American Furniture*, ed. Luke Beckerdite. Philadelphia: Chipstone Foundation. Accessed October 12, 2021.

Pye, David. 1968. *The Nature and Art of Workmanship*. Cambridge: Cambridge University Press.

Quiccheberg, Samuel. 2013. *The First Treatise on Museums: Samuel Quiccheberg's Inscriptiones, 1565*. Trans. and ed. Mark A. Meadow and Bruce Robertson. Los Angeles: Getty Publications.

Rabelais, François. 1991. *The Complete Works of François Rabelais*. Trans. Donald M. Frame. Berkeley: University of California Press.

Radman, Zdravko. 2013. *The Hand, an Organ of the Mind: What the Manual Tells the Mental*. Cambridge, MA: MIT Press.

Rampling, Jennifer M. 2014. "Transmuting Sericon: Alchemy as 'Practical Exegesis' in Early Modern England." *Osiris* 29, no. 1:19–34.

Rankin, Alisha. 2007. "Becoming an Expert Practitioner: Court Experimentalism and the Medical Skills of Anna of Saxony (1532–1585)." *Isis* 98:23–53.

Rankin, Alisha. 2013. *Panaceia's Daughters: Noblewomen as Healers in Early Modern Germany*. Chicago: University of Chicago Press.

Rankin, Alisha. 2014. "How to Cure the Golden Vein: Medical Remedies as Wissenschaft in Renaissance Germany." In *Ways of Making and Knowing: The Material Culture of Empirical Knowledge*, ed. Pamela H. Smith, Amy Meyers, and Harold J. Cook, 113–37. Ann Arbor: University of Michigan Press.

Raven, Diederick. 2013. "Artisanal Knowledge." *Acta Baltica Historiae Philosophiae Scientarum* 1, no. 1:5–34.

Reichard, Gladys. 1974. *Weaving a Navajo Blanket* (1936). New York: Dover.

Reith, Reinhold. 2005. "Know-how, Technologietransfer und die *Arcana Artis* im Mitteleuropa der Frühen Neuzeit." *Early Science and Medicine* 10, no. 3:349–77.

Reith, Reinhold. 2008. "The Circulation of Skilled Labor in Late Medieval and Early Modern Central Europe." In *Guilds, Innovation, and the European Economy, 1400–1800*, ed. Stephan R. Epstein and Maarten Prak, 114–42. Cambridge: Cambridge University Press.

Remond, Jaya. 2019. "'Draw Everything That Exists in the World': *'t Light der Teken en Schilderkonst* and the Shaping of Art Education in Early Modern Northern Europe." In *Lessons in Art: Art, Education, and Modes of Instruction since 1500*, ed. Erick Jorink et al., 287–321. Leiden: Brill.

Remond, Jaya. 2020. "Artful Instruction: Pictorializing and Printing Artistic Knowledge in Early Modern Germany." *Word & Image* 36, no. 2:101–34.

Reynolds, Melissa. 2019. "'Here Is a Good Boke to Lerne': Practical Books, the Coming of the Press, and the Search for Knowledge, ca. 1400–1560." *Journal of British Studies* 58:259–88.

Ribard, Dinah. 2010. "Le travail intellectuel: Travail et philosophie, xviie–xixe siècles." *Annales: Histoire, sciences sociales* 65, no. 3:715–42.

Ribard, Dinah. 2018. "Savoirs des mots, savoir du livre, savoirs livresques: Réflexions artisanes sur le passage de l'écrit au livre." Paper presented at the conference "Du manuscrit au livre, l'écriture des savoir-faire à la Renaissance," Université Toulouse-Jean Jaurès, Toulouse, 15–16 March.

Rice, Silas. 2020. "Instructions for Jewelry Making, ca. 1802–25." Microfilm 3019, Winterthur Museum, Gardens, and Library.

Richards, Jennifer. 2012. "Useful Books: Reading Vernacular Regimens in Sixteenth-century England." *Journal of the History of Ideas* 73, no. 2:247–71.

Richardson, Brian. 1999. *Printing, Writers, and Readers in Renaissance Italy*. Cambridge: Cambridge University Press.

Risatti, Howard. 2007. *A Theory of Craft: Function and Aesthetic Expression*. Chapel Hill: University of North Carolina Press.

Rivius, Gualtherus H. (Walther Ryff). 1547. *Der furnembsten notwendigsten der gantzen Architectur angehörigen Mathematischen und Mechanischen künst eygentlicher bericht und vast klare verstendliche unterrichtung zu rechtem verstandt der lehr Vitruvij*. Nuremberg.

Roberts, Lissa L., Simon Schaffer, and Peter Dear, eds. 2007. *The Mindful Hand: Inquiry and Invention from the Late Renaissance to Early Industrialisation*. Amsterdam: University Presses.

Robson, Eleanor. 2002. "Technology in Society: Three Textual Case Studies from Late Bronze Age Mesopotamia." In *The Social Context of Technological Change. Egypt and the Near East, 1650–1550 BC*, ed. Andrew J. Shortland, 39–57. Oxford: Oxbow Books.

Rogerson, Cordelia. 2004. "Conservation Meets Its Maker." *Conservation News* 88 (January).

Rogoff, Barbara, and Jean Lave, eds. 1984. *Everyday Cognition: Its Development in Social Context*. Cambridge, MA: Harvard University Press.

Rösler, Irmtraud. 1998. "'De Seekarte Ost Vnd West To Segelen…': On Northern European Nautical 'Fachliteratur' in the Late Middle Ages." *Early Science and Medicine* 3:103–18.

Rossi, Paolo. 1970. *Philosophy, Technology, and the Arts in the Early Modern Era*. Trans. Salvator Attanasio. New York: Harper & Row.

Roth, Harriet, ed. 2000. *Der Anfang der Museumslehre in Deutschland: Das Traktat "Inscriptiones vel Tituli Theatri Amplissimi" von Samuel Quiccheberg*. Berlin: Akademie Verlag.

Rublack, Ulinka. 2013. "Matter in the Material Renaissance." *Past and Present* 219: 41–85.

Rublack, Ulinka. 2016. "Renaissance Dress, Cultures of Making, and the Period Eye." In *New Directions in Making and Knowing*, ed. Pamela H. Smith. Special issue. *West 86th: A Journal of Decorative Arts, Design History, and Material Culture* 23, no. 1:6–34.

Russo, Alessandra. 2020. "Lights on the Antipodes: Francisco de Holanda and an Art History of the Universal." *Art Bulletin* 102:4, 37–65.

Sahmland, Irmtraut. 1988. "Gesundheitsschädigung der Bergleute: Die Bedeutung der Bergpredigten des 16. bis frühen 18. Jahrhunderts als Quelle arbeitsmedizingeschichtlicher Fragestellungen." *Medizinisches Journal* 23, nos. 3–4:240–76.

St. George, Robert Blair, ed. 1988. *Material Life in America 1600–1860*. Boston: Northeastern University Press.

Salomon, Gavriel, ed. 1993. *Distributed Cognitions: Psychological and Educational Considerations*. Cambridge: Cambridge University Press.

Samuel, Raphael. 1977. "Workshop of the World: Steam Power and Hand Technology in Mid-Victorian Britain." *History Workshop Journal* 3, no. 1:6–72.

Sandman, Alison. 2001. "Mirroring the World: Sea Charts, Navigation, and Territorial Claims in Sixteenth-Century Spain." In *Merchants and Marvels: Commerce, Science, and Art in Early Modern Europe*, ed. Pamela H. Smith and Paula Findlen, 83–108. New York: Routledge.

Santucci, Giovanni. 2014. "Federico Brandani's Paper Model for the Chapel of the Dukes of Urbino at Loreto." *Burlington Magazine* 156:4–11.

Sarton, George. 1956. "East and West" (1931). In George Sarton, *The History of Science and the New Humanism*, 59–110. New York: George Braziller.

Sawyer, Keith. 2007. *Group Genius: The Creative Power of Collaboration*. New York: Basic Books.

Schatzberg, Eric. 2018. *Technology: Critical History of a Concept*. Chicago: University of Chicago Press.

Scheicher, Elisabeth. 1977. *Die Kunstkammer (Schloss Ambras)*. Innsbruck: Kunsthistorisches Museum.

Scheicher, Elisabeth. 1979. *Die Kunst- und Wunderkammern der Habsburger*. Vienna: Molden.

Schneider, Ulrich Johannes. 2007. "Die Enzyklopädie als Medizin: Aufklärung durch das größte Lexikon des 18. Jahrhunderts." *Leipziger Jahrbuch zur Buchgeschichte* 16:285–96.

Schneider, Ulrich Johannes, ed. 2009. *In Pursuit of Knowledge: 600 Years of Leipzig University*. Leipzig: Universitätsbibliothek.

Schön, Donald A. 1983. *The Reflective Practitioner: How Professionals Think in Action*. New York: Basic Books.

Schotte, Margaret E. 2019. *Sailing School: Navigating Science and Skill, 1550–1800*. Baltimore: Johns Hopkins University Press.

Schreiber, Georg. 1962. *Der Bergbau in Geschichte, Ethos und Sakralkultur, Wissenschaftliche Abhandlungen der Arbeitsgemeinschaft für Forschung des Landes Nordrhein-Westfalen*. Cologne: Westdeutscher Verlag.

Schultz, Eva. 1990. "Notes on the History of Collecting and of Museums in the Light of Selected Literature of the Sixteenth to the Eighteenth Century." *Journal of History of Collections* 2:205–18.

Schürer, Ralf. 1994. "Ein erbar handwerckh von goldschmiden." In *Silber und Gold: Augsburger Goldschmiedekunst für die Höfe Europas*, ed. Reinhold Baumstark and Helmut Seling, 1:57–65. 2 vols. Munich: Hirmer Verlag.

Scribner, Bob. 1984. "Cosmic Order and Daily Life: Sacred and Secular in Pre-Industrial German Society." In *Religion and Society in Early Modern Europe 1500–1800*, ed. Kaspar von Greyerz. London: Allen & Unwin for the German Historical Institute.

Scully, Terence, ed. and trans. 1986. *Chiquart's "On Cookery": A Fifteenth-Century Savoyard Culinary Treatise*. New York: Peter Lang.

Seifert, Christian Tico. 2013. "Pieter Lastman als Leser: Eine Künstlerbibliothek und ihre Nutzung." In *The Artist as Reader*, ed. Heiko Damm, Michael Thimann, and Claus Zittel, 155–94. Leiden: Brill.

Sennett, Richard. 2008. *The Craftsman*. New Haven: Yale University Press.

Sewell, William H., Jr. 1999. "The Concept(s) of Culture." In *Beyond the Cultural Turn: New Directions in the Study of Society and Culture*, ed. Victoria E. Bonnell and Lynn Hunt, 35–61. Los Angeles: University of California Press.

Shapin, Steven. 2001. "Proverbial Economies: How an Understanding of Some Linguistic and Social Features of Common Sense Can Throw Light on More Prestigious Bodies of Knowledge, Science for Example." *Social Studies of Science* 31, no. 5:731–69.

Shelby, Lon R. 1997. "The Geometrical Knowledge of Medieval Master Masons." In *The Engineering of Medieval Cathedrals*, ed. Lynn T. Courtenay, 27–61. Aldershot, UK: Ashgate.

Shelby, Lon R., and R. Mark. 1997. "Late Gothic Structure Design in the 'Instructions' of Lorenz Lechler." In *The Engineering of Medieval Cathedrals*, ed. Lynn T. Courtenay. Aldershot, UK: Ashgate.

Shell, Hanna Rose. 2004. "Casting Life, Recasting Experience: Bernard Palissy's Occupation between Maker and Nature." *Configurations* 12:1–40.

Sherman, William. 2008. *Used Books: Marking Readers in Renaissance England*. Philadelphia: University of Pennsylvania Press.

Shi, Yuanxie, and Amy Chang. 2020. "Rouge Clair: Glass or Paint?" In *Secrets of Craft and Nature*, ed. Making and Knowing Project et al. https://edition640.makingandknowing.org/#/essays/ann_034_sp_16.

Sibum, Heinz Otto. 1995. "Reworking the Mechanical Value of Heat: Instruments of Precision and

Gestures of Accuracy in Early Victorian England." *Studies in History and Philosophy of Science* 26:73–106.

Sibum, H. Otto. 2000. "Experimental History of Science." In *Museums of Modern Science: Nobel Symposium 112*, ed. Svante Lindqvist, Marika Hedin, and Ulf Larsson, 77–86. Canton, MA: Science History Publications.

Sigaut, François. 1994. "Technology." In *Companion Encyclopedia of Anthropology: Humanity, Culture and Social Life*, ed. Tim Ingold, 420–59. London: Routledge.

Sigerist, Henry E. 1941. "Medieval Medicine." In *Studies in the History of Science*. Philadelphia: University of Pennsylvania Press.

Sisco, Anneliese Grünhaldt, and Cyril Stanley Smith. 1949. *Bergwerk- und Probierbüchlein: A Translation from the German of the Bergbüchlein, a Sixteenth-century Book on Mining Geology, by Anneliese Grünhaldt Sisco; and of the Probierbüchlein, a Sixteenth-century Work on Assaying, by Anneliese Grünhaldt Sisco and Cyril Stanley Smith*. New York: American Institute of Mining & Metallurgical Engineers.

Skemer, Don C. 2006. *Binding Words: Textual Amulets in the Middle Ages*. University Park: Pennsylvania State University Press.

Slotta, Rainer, Christoph Bartels, Heinz Pollmann, and Martin Lochert, eds. 1990. *Meisterwerke bergbaulicher Kunst vom 13. bis 19. Jahrhundert*. Bochum: Deutsches Bergbau-Museum Bochum.

Smith, Cyril Stanley. 1943. "Introduction: Life of Biringuccio." In *The Pirotechnia*, trans. Cyril Stanley Smith and Martha Teach Gnudi, ix–xxiii. New York: Basic Books.

Smith, Cyril Stanley, ed. 1968. *Von Stahel und Eisen* (Nuremberg, 1532). In *Sources for the History of the Science of Steel, 1532–1796*, trans. Anneliese G. Sisco, 7–19. Cambridge, MA: MIT Press.

Smith, Cyril Stanley. 1975. "Metallurgy as Human Experience." *Metallurgical Transactions* 6A:603–23.

Smith, Jeffrey Chipps. 1979. "The Artistic Patronage of Philip the Good, Duke of Burgundy (1419–1467)." PhD diss., Columbia University.

Smith, Pamela H. 2004. *The Body of the Artisan: Art and Experience in the Scientific Revolution*. Chicago: University of Chicago Press.

Smith, Pamela H. 2007. "Making and Knowing in a Sixteenth-century Goldsmith's Workshop." In *The Mindful Hand: Inquiry and Invention between the Late Renaissance and Early Industrialization*, ed. Lissa L. Roberts, Simon Schaffer, and Peter Dear, 20–37. Amsterdam: KNAW Press.

Smith, Pamela H. 2008a. "Alchemy as the Imitator of Nature." In *Glass of the Alchemists*, ed. Dedo von Kerssenbrock-Krosigk, 22–33. Corning, NY: Corning Museum of Glass.

Smith, Pamela H. 2008b. "Collecting Nature and Art: Artisans and Knowledge in the *Kunstkammer*." In *Engaging with Nature: Essays on the Natural World in Medieval and Early Modern Europe*, ed. Barbara Hannawalt and Lisa Kiser, 115–36. Notre Dame, IN: University of Notre Dame Press.

Smith, Pamela H. 2009. "Science on the Move: Recent Trends in the History of Early Modern Science." *Renaissance Quarterly* 62: 345–375.

Smith, Pamela H. 2010. "Why Write a Book? From Lived Experience to the Written Word in Early Modern Europe." *Bulletin of the German Historical Institute* 47:25–50.

Smith, Pamela H. 2011a. "Science." In *The Oxford Companion to History*, ed. Ulinka Rublack, 268–97. Oxford: Oxford University Press.

Smith, Pamela H. 2011b. "What Is a Secret? Secrets and Craft Knowledge in Early Modern Europe." In *Secrets and Knowledge in Medicine and Science, 1500–1800*, ed. Elaine Leong and Alisha Rankin, 47–66. Farnham, UK: Ashgate.

Smith, Pamela H. 2012. "In the Workshop of History: Making, Writing, and Meaning." *West 86th: A Journal of Decorative Arts, Design History, and Material Culture* 19:4–31.

Smith, Pamela H. 2014. "Making as Knowing: Craft as Natural Philosophy." In *Ways of Making and Knowing: The Material Culture of Empirical Knowledge*, ed. Pamela H. Smith, Amy Meyers, and Harold J. Cook, 17–47. Ann Arbor: University of Michigan Press.

Smith, Pamela H. 2015a. "Itineraries of Materials and Knowledge in the Early Modern World." In *The Global Lives of Things*, ed. Anne Gerritsen and Giorgio Riello, 31–61. London: Routledge.

Smith, Pamela H. 2015b. "The Matter of Ideas in the Working of Metals in Early Modern Europe." In *The Matter of Art: Materials, Practices, Cultural Logics, c. 1250–1750*, ed. Christy Anderson, Anne Dunlop, and Pamela H. Smith, 42–67. Manchester, UK: Manchester University Press.

Smith, Pamela H., ed. 2016. "New Directions in Making and Knowing." *West 86th: A Journal of Decorative Arts, Design History, and Material Culture* 23, no. 1:3–101.

Smith, Pamela H. 2017. "The Codification of Vernacular Theories of Metallic Generation in Sixteenth-century European Mining and Metalworking." In *The Structures of Practical Knowledge*, ed. Matteo Valeriani, 371–92. Cham, Switzerland: Springer.

Smith, Pamela H. 2018. "Artisanal Epistemology." In *Encyclopedia of Renaissance Philosophy*, ed. Marco Sgarbi. Dordrecht: Springer.

Smith, Pamela H., ed. 2019a. *Entangled Itineraries of Materials, Practices, and Knowledges across Eurasia*. Pittsburgh: University of Pittsburgh Press.

Smith, Pamela H. 2019b. "Introduction: Nodes of Convergence, Material Complexes, and Entangled Itineraries across Eurasia." In *Entangled Itineraries*, ed. Pamela H. Smith, 5–24. Pittsburgh: University of Pittsburgh Press.

Smith, Pamela H. 2020a. "Making the Edition of Ms. Fr. 640." In *Secrets of Craft and Nature*, ed. Making and Knowing Project et al. https://edition640.makingandknowing.org/#/essays/ann_329_ie_19.

Smith, Pamela H. 2020b. "Reconstruction Insights." In *Secrets of Craft and Nature*, ed. Making and Knowing Project et al. https://edition640.makingandknowing.org/#/content/resources/reconstruction-insights.

Smith, Pamela H., and Tonny Beentjes. 2010. "Nature and Art, Making and Knowing: Reconstructing Sixteenth-Century Life Casting Techniques." *Renaissance Quarterly* 63:128–79.

Smith, Pamela H., Joslyn DeVinney, Sasha Grafit, and Xiaomeng Liu. 2019. "Smoke and Silkworms: Itineraries of Material Complexes across Eurasia." In *Entangled Itineraries of Materials, Practices, and Knowledges across Eurasia*, ed. Pamela H. Smith, 165–81. Pittsburgh: University of Pittsburgh Press.

Smith, Pamela H. 2020. "Lifecasting in Ms. Fr. 640." In *Secrets of Craft and Nature*, ed. Making and Knowing Project et al. https://edition640.makingandknowing.org/#/essays/ann_511_ad_20.

Smith, Pamela H., Amy R. W. Meyers, and Harold J. Cook, eds. 2017. *Ways of Making and Knowing: The Material Culture of Empirical Knowledge*. Chicago: University of Chicago Press.

Soley, Teresa. 2020. "Imitation Marble." In *Secrets of Craft and Nature*, ed. Making and Knowing Project et al. https://edition640.makingandknowing.org/#/essays/ann_040_sp_16.

Sonenscher, Michael. 1987. "Mythical Work: Workshop Production and the *Campagnonnages* of Eighteenth-century France." In *The Historical Meanings of Work*, ed. Patrick Joyce, 31–63. Cambridge: Cambridge University Press.

Spiller, Elizabeth. 2008. *Seventeenth-Century English Recipe Books: Cooking, Physic, and Chirurgery in the Worlds of Elizabeth Talbot Grey and Aletheia Talbot Howard*. Aldershot, UK: Ashgate.

Spiller, Elizabeth. 2009. "Recipes for Knowledge: Maker's Knowledge, Traditions, Paracelsian Recipes and the Invention of the Cookbook, 1600–1660." In *Renaissance Food from Rabelais to Shakespeare*, ed. Joan Fitzpatrick. Aldershot, UK: Ashgate.

Stahl, Alan M. 2009. "Michael of Rhodes: Mariner in Service to Venice." In *The Book of Michael of Rhodes*, ed. Pamela O. Long, David McGee, and Alan M. Stahl, 3:35–98. Cambridge, MA: MIT Press.

Stallybrass, Peter, Roger Chartier, J. Franklin Mowery, and Heather Wolfe. 2004. "Hamlet's Tables and the Technologies of Writing in Renaissance England." *Shakespeare Quarterly* 55:379–419.

Staubermann, Klaus, ed. 2011. *Reconstructions: Recreating Science and Technology of the Past*. Edinburgh: National Museums Scotland.

Stechow, Wolfgang. 1989. *Northern Renaissance Art 1400–1600: Sources and Documents*. Evanston, IL: Northwestern University Press.

Stewart, Larry. 2005. "Science, Instruments, and Guilds in Early Modern Britain." *Early Science and Medicine* 10, no. 3:392–410.

Stone, Richard E. 1981. "Antico and the Development of Bronze Casting in Italy at the End of the Quattrocento." *Metropolitan Museum Journal* 16:87–116.

Stone, Richard E. 2001. "A New Interpretation of the Casting of Donatello's *Judith and Holofernes*." In *Small Bronzes in the Renaissance*, ed. Debra Pincus, 55–67. Studies in the History of Art 62, Symposium Papers 39. Washington, DC: Center for Advanced Study in the Visual Arts.

Stone, Richard E. 2006. "Severo Calzetta da Ravenna and the Indirectly Cast Bronze." *Burlington Magazine* 148, no. 1245:810–19.

Striebel, Ernst. 2003. "Über das Färben von Holz, Horn und Bein: Das Augsburger Kunstbüchlin von 1535." *Restauro* 6:424–30.

Su, E. 1985. *Duyang za bian* [Compiling Miscellanies at Duyang]. Beijing: Zhonghua shuju.

Sudhoff, Karl. 1908. "Deutsche medizinische Inkunabeln." *Studies in the History of Medicine* 2–3. Leipzig: J. A. Barth.

Sudnow, David. 1978. *Ways of the Hand: The Organization of Improvised Conduct*. Cambridge, MA: Harvard University Press.

Summers, David. 1987. *The Judgment of Sense: Renaissance Naturalism and the Rise of Aesthetics*. Cambridge: Cambridge University Press.

Suthor, Nicola. 2010. *Bravura: Virtuosität und Mutwilligkeit in der Malerei der frühen Neuzeit*. Munich: Wilhelm Fink.

Sutton, John. 2007. "Batting, Habit and Memory: The Embodied Mind and the Nature of Skill." *Sport in Society* 10, no. 5:763–86.

Takeda, Reiko. 2011. "Literacy in a Business Context: Literacy Practices of Some Bristol Merchants in the Sixteenth Century." *History of Education* 40, no. 5: 651–71.

Tallis, Raymond. 2003. *The Hand: A Philosophical Inquiry into Human Being*. Edinburgh: Edinburgh University Press.

Tao, Cai. 1819. *Tie wei shan con tan* [Miscellaneous Records at Mount Tiewei]. Hangzhou: Zhi buzu zhai.

Tarnai, Tibor, and Dirk Weber. 2017. "Turned Geometry: Two Masterpieces by Georg Friedel." *Burlington Magazine* 159:544–52.

Tarule, Robert. 2004. *The Artisan of Ipswich: Craftsmanship and Community in Colonial New England*. Baltimore: Johns Hopkins University Press.

Tebaux, Elizabeth. 1997. "Women and Technical Writing, 1475–1700: Technology, Literacy and

Development of a Genre." In *Women, Science and Medicine, 1500–1700: Mothers and Sisters of the Royal Society*, ed. Lynette Hunter and Sarah Hutton, 29–62. Thrupp, Stroud, Gloucestershire, UK: Sutton Publishing.

Tell, Håkan P. 2011. *Plato's Counterfeit Sophists*. Cambridge, MA: Harvard University Press.

Telle, Joachim. 2003. "Das Rezept als literarische Form." *Berichte zur Wissenschaftsgeschichte* 26:251–74.

Theophilus. 1979. *On Divers Arts*. Trans. John G. Hawthorne and Cyril Stanley Smith. New York: Dover.

Theophilus. 1986. *The Various Arts. De Diversis Artibus*. Ed. and trans. C. R. Dodwell. Oxford: Clarendon Press.

Thompson, Daniel V., Jr. 1933–34. "Artificial Vermilion in the Middle Ages." *Technical Studies in the Field of the Fine Arts* 2:62–70.

Thurston, Edgar. 1915. *Omens and Superstitions of Southern India*. New York: McBride, Nast & Co.

Tribble, Evelyn B. 2011. *Cognition in the Globe: Attention and Memory in Shakespeare's Theatre*. New York: Palgrave Macmillan.

Tribble, Evelyn B., and John Sutton. 2012. "Minds in and out of Time: Memory, Embodied Skill, Anachronism, and Performance." *Textual Practice* 26, no. 4:587–607.

Truitt, Elly. 2015. *Medieval Robots: Mechanism, Magic, Nature, and Art*. Philadelphia: University of Pennsylvania Press.

Ullman, Reut. 2020. "Artificial Grottos." In *Secrets of Craft and Nature*, ed. Making and Knowing Project et al. https://edition640.makingandknowing.org/#/essays/ann_063_fa_17.

Valleriani, Matteo. 2012. "The Knowledge Agora: The Role of the Officials." In *Cultures of Knowledge: Technology in Chinese History*, ed. Dagmar Schäffer, 233–46. Leiden, Boston: Brill.

Valleriani, Matteo, ed. 2017. *The Structures of Practical Knowledge*. Cham, Switzerland: Springer.

Van der Lugt, Maaike. 2009. "'Abominable Mixtures': The *Liber Vaccae* in the Medieval West, or the Dangers and Attractions of Natural Magic." *Traditio* 64:229–77.

Van Schendel, A. F. E. 1972. "Manufacture of Vermilion in 17th-Century Amsterdam, The Pekstok Papers." *Studies in Conservation* 17, no. 2:77–78.

Vega-Encabo, Jesus. 2016. "Artisanal Knowledge: Some Preliminary Observations." Paper presented at the conference "Artisanal Knowledge," University of Lisbon, 4–7 May.

Vérin, Hélène. 1993. *Le gloire des ingénieurs: L'intelligence technique du XVIe au XVIIIe siècle*. Paris: Albin Michel.

Vérin, Hélène. 2008. "Rédiger et réduire en art: Un projet de rationalisation des pratiques." In *Réduire en art*, ed. Pascal Dubourg Glatigny and Hélène Vérin, 17–58. Paris: Éditions de la Maison des Sciences de l'Homme.

Vermeir, Koen. 2012. "Openness versus Secrecy? Historical and Historiographical Remarks." *British Journal for the History of Science* 45, no. 2:165–88.

Vives, Juan Luis. 1913. *Vives: On Education*. Trans. Foster Watson. Cambridge: Cambridge University Press.

Vogtherr, Heinrich, and Maria Heilmann. 2011. *Kunstbüchlein* (1538). Heidelberg: Universitätsbibliothek der Universität Heidelberg.

Von Cuba, Johannes. 1515. *In disem Buch ist der Herbary: Oder Kreuterbuch genant der Gart der Gesuntheit: mit merern Figuren und Register*. Thiergarten: Renatus Heck.

Von Schlosser, Julius. 2021. *Art and Curiosity Cabinets of the Late Renaissance*. Ed. Thomas DaCosta Kaufmann; trans. Jonathan Blower. Malibu, CA: Getty Publications.

Von Schönherr, David. 1888. "Wenzel Jamnitzers Arbeiten für Erzherzog Ferdinand." *Mitteilungen des Instituts für Oesterreichische Geschichtsforschung*, ed. Ritter von Sickel, H. Ritter von Zeissberg, and E. Mühlbacher, 9:289–305.

Wacquant, Loïc. 2004. *Body and Soul: Notebooks of an Apprentice Boxer.* Oxford: Oxford University Press.

Walker, Katherine. 2015. "Early Modern Almanacs and The Witch of Edmonton." *Early Modern Literary Studies* 18, nos. 1–2:1–25.

Wall, Wendy. 2016. *Recipes for Thought: Knowledge and Taste in the Early Modern English Kitchen.* Philadelphia: University of Pennsylvania Press.

Waller, Arie. 1990. "Alchemy and Medieval Art Technology." In *Alchemy Revisited*, ed. Z. R. W. M. von Martels, 154–61. Leiden: Brill.

Wallert, Arie. 1995. "'Libro Secondo de Diversi Colori e Sise da Mettere a Oro': A Fifteenth-Century Technical Treatise on Manuscript Illumination." In *Historical Painting Techniques, Materials, and Studio Practice*, ed. Arie Wallert, Erma Hermens, and Marja Peek, 380–47. Marina Del Rey, CA: Getty Conservation Institute.

Wallert, Arie. 2000. "Makers, Materials and Manufacture." In *Netherlandish Art in the Rijksmuseum 1400–1600*, ed. Henk van Os, Jan Piet Filedt Kok, Ger Luijten, and Frits Scholten, 253–69. Amsterdam: Waanders.

Wallis, Faith. 2009. "Michael of Rhodes and Time Reckoning: Calendar, Almanac, Prognostication." In *The Book of Michael of Rhodes*, ed. Pamela O. Long, David McGee, and Alan M. Stahl, 3:281–319. Cambridge, MA: MIT Press.

Walls, Matthew. 2016. "Making as a Didactic Process: Situated Cognition and the Chaîne Opératoire." *Quaternary International* 405:21–30.

Walton, Steven A. 2017. "Technologies of Pow(d)er: Military Mathematical Practitioners' Strategies and Self-Presentation." In *Mathematical Practitioners and the Transformation of Natural Knowledge in Early Modern Europe*, ed. Lesley Cormack, John A. Schuster, and Steven A. Walton. Cham, Switzerland: Springer.

Wang, Yijun, and Pamela H. Smith. 2020. "Fat, Lean, Sweet, Sour: Sand of Ox Bone and Rock Salt." In *Secrets of Craft and Nature*, ed. Making and Knowing Project et al. https://edition640.makingandknowing.org/#/essays/ann_012_fa_14.

Watanabe-O'Kelly, Helen. 2002. *Court Culture in Dresden: From Renaissance to Baroque.* New York: Palgrave.

Watson-Verran, Helen, and David Turnbull. 1995. "Science and Other Indigenous Knowledge Systems." In *Handbook of Science and Technology Studies*, ed. Sheila Jasanoff, Gerald E. Markle, James C. Peterson, and Trevor Pinch, 115–39. Thousand Oaks, CA: Sage Publications.

Webster, L. T., and I. W. Pritchett. 1924. "Microbic Virulence and Host Susceptibility on Paratyphoid-enteritidis Infection of White Mice. V: The Effect of Diet on Host Resistance." *Journal of Experimental Medicine* 40, no. 3:397–404.

Weddington, John. 1567. *A Breffe Instruction, and Manner, Howe to Kepe Marchantes Bokes, of Accomptes.* Antwerp: Petter van Keerberghen. Facsimile ed. London: Scolar Press.

Weeks, Sophie. 2008. "The Role of Mechanics in Francis Bacon's *Great Instauration*." In *Philosophies of Technology*, ed. Claus Zittel, 133–95. Leiden: Brill.

Werrett, Simon. 2009. *Fireworks: Pyrotechnic Arts and Science in European History.* Chicago: University of Chicago Press.

Whalley, Peter, and Stephen R. Barley. 1997. "Technical Work in the Division of Labor: Stalking the

Wily Anomaly." In *Between Craft and Science: Technical Work in U.S. Settings*, ed. Stephen R. Barley and Julian E. Orr. Ithaca, NY: Cornell University Press.

Wheeler, Jo, with Katy Temple. 2009. *Renaissance Secrets, Recipes and Formulas*. London: V&A.

Whiten, Andrew. 2005. "The Second Inheritance System of Chimpanzees and Humans." *Nature* 437, no. 1 (September): 52–55.

Whiten, Andrew. 2014. "Incipient Tradition in Wild Chimpanzees." *Nature* 514, no. 9 (October): 178–79.

Whitney, Elsbeth. 1990. "Paradise Restored: The Mechanical Arts from Antiquity through the Thirteenth Century." *Transactions of the American Philosophical Society* 80.

Willich, A. F. M. 1802. *Domestic Encyclopedia, or A Dictionary of Facts, and Useful Knowledge*. 4 vols. London: Murray & Highley.

Wilson, Stephen. 2000. *The Magical Universe: Everyday Ritual and Magic in Pre-Modern Europe*. London: Hambledon & London.

Wolfe, Charles T., and Ofer Gal, eds. 2010. *The Body as Object and Instrument of Knowledge: Embodied Empiricism in Early Modern Science*. Dordrecht: Springer.

Wolfe, Jessica. 2004. *Humanism, Machinery, and Renaissance Literature*. Cambridge: Cambridge University Press.

Wrapson, Lucy, et al. 2012. *In Artists' Footsteps: The Reconstruction of Pigments and Paintings: Studies in Honour of Renate Woudhuysen-Keller*. London: Archetype.

Yinglong, Lu. 1985. *Xian chuan kuo yi zhi* [Leisure Collection of Marvels]. Beijing: Zhonghua shuju.

Young, Tim. 2005. "The St. Fagans Experimental Bloomery: The First Six Years." *HMS News* 59 (Spring): 1–2.

Zenetti, Christiano, ed. 2016. *Janello Torriani: A Renaissance Genius*. Cremona: Museo del Violino.

Zilsel, Edgar. 2000. *The Social Origins of Modern Science*, ed. Diederick Raven, Wolfgang Krohn, and Robert S. Cohen. Dordrecht: Kluwer Academic.

Zimmermann, Karin. 2014. "The Electors Palatine at Heidelberg in the 16th Century: Collectors and Compilers of Recipes." Paper presented in the conference "Reading How-To: The Uses and Users of Artisanal Recipes," Max Planck Research Group "Art and Knowing in Pre-Modern Europe," Max Planck Institute for the History of Science, September 19–20.

Zindel, Christophe. 2010. *Güldene Kunst-Pforte: Quellen zur Kunsttechnologie*. Bern: Hochschule der Künste Bern.

Index

Page numbers followed by *f* refer to figures.